hristopher Kennedy 2025
iddleton-Wells Publishers
ndon

thus Enigma: Technology, Science, and Policy for a Fragile Earth
979-8-316-66831-1

ed on acid-free paper

hts reserved. No part of this publication may be reproduced or
itted in any form or by any means, electronic or mechanical, including
opying, recording, or any information storage and retrieval system,
t permission from the publishers.

For future generations

CHRISTOPHER KENNEDY

Malthus Enigma

Technology, Science, and Policy for a Fragile Earth

MIDDLETON-WELLS PUBLISHERS

London

Contents

Preface   vii

1 The Gift of Apollo   1

2 Malthusians and Cornucopians   23

3 Feeding the World While Saving the Planet   41

4 One Big Greenhouse   75

5 The Science of Sustainability   107

6 The Global Problématique   131

7 Ecological Economics   157

8 Industrial Ecology   187

9 Conclusions   215

Notes   231

Bibliography   263

Index   291

Preface

Inspiration to write this book began in 2011, when I was seconded to the Environment Directorate of the Organisation for Economic Co-operation and Development (OECD) in Paris. Prior to joining, I had been engaged in fundamental ideas underlying sustainable development through interaction with other industrial ecologists. Most of my work had focused on sustainability challenges in the context of the world's rapidly growing cities. Building upon research on urban metabolism, I had worked on methods for inventorying greenhouse gas emissions for cities that were adopted by the World Bank and the United Nations Environment Programme, which later evolved into a global protocol. Through this work, and other work in infrastructure economics, I had come to know climate policy expert Jan Corfee-Morlot at the OECD and headed to Paris to join her team.

At the OECD, I provided assistance to colleagues studying policies on aspects of cities and green growth, but I was primarily engaged in broader aspects of how to encourage large-scale investment in low-carbon, climate-resilient infrastructure. It was an intriguing and enlightening year as a civil servant in an influential international organization: I travelled to London to attend the launch of a green bonds initiative, presented at a meeting to over one hundred mayors in South Korea, and attended a United Nations Framework Convention on Climate Change meeting in Bonn. I also contributed to the OECD work for a Roundtable of Mayors and Ministers, the Clean Energy Ministerial, and the Rio +20 Conference. Perhaps the most revealing meetings were those back at OECD headquarters around the beautiful Château de la Muette. I principally supported the Working Party on Climate, Investment and Development, contributing analysis and – learning from Jan – chairing a meeting on mobilizing investment in low-carbon infrastructure. But more than this, I spent a year deeply immersed in the details of the world's struggles to overcome formidable environmental challenges – examining them from economic, financial, political, and other perspectives. I attended meetings on green growth and sat in the wings as the ministers of environment for the thirty-four OECD governments came to Paris for

Malthus Enigma

their triennial meeting.

Prominent in the OECD's work that year was the release of its alarming *Environmental Outlook to 2050*. This critical study, produced for the ministers, warned that progress in overcoming environmental degradation would be overwhelmed by the sheer scale of growth in population and the economy. Drawing upon best available science and computer systems models, the *Environmental Outlook* assessed the expected environmental impacts under continuing trends in the urgent areas of climate change, biodiversity, water resources, and health impacts of air pollution. By 2050, global population is expected to be over nine billion – with a quadrupling of economic activity, accompanied by huge increases in the demand for energy and natural resources. Unless major actions are taken, temperature increases of 3°C to 6°C will be locked in, with impacts including the melting of glaciers, ice caps, and permafrost, as well as human suffering from increases in the intensity and frequency of extreme weather events. Loss of biodiversity is projected to continue to 2050, with the diversity of terrestrial species expected to decrease by ten percent, including mature forests shrinking by thirteen percent.[1] By 2050, moreover, climate change is expected to be the fastest-growing cause of biodiversity loss. The study predicted that 2.3 billion more people than in 2012 will be settled in river basins, experiencing severe water stress. Groundwater depletion will threaten water supplies in cities and agricultural regions, while nutrient pollution will further degrade aquatic biodiversity. Meanwhile, air pollution will become the most prevalent environmental cause of premature human mortality, with premature deaths reaching an expected 3.6 million per year.

Even though I was familiar with many aspects of these environmental challenges, it was the scale of the rapidly growing, highly coupled environmental damage that was so concerning. The OECD report clearly stated that urgent actions would require international cooperation. It urged that well-designed policies could draw upon the interdependencies of the challenges, finding ways to maximize synergies and co-benefits of actions. Nonetheless, the prognosis was clearly daunting.

Toward the end of my year, I attended the Gordon Research Conference on Industrial Ecology held in picturesque Les Diablerets, in Switzerland. The program that year took a big-picture view of the Earth's global resource challenges. The conference included sessions addressing questions such as: How can we provide adequate land, water, and nutrients to feed the Earth's population of 2050? Where will sufficient supplies of affordable metals and minerals be found for a more populous and wealthier world? How can industrial ecology be used to establish

Preface

environmentally benign industry in rapidly developing countries?[2] Excellently planned by Sangwon Suh from the University of California, Santa Barbara, and Helga Weisz from the Potsdam Institute for Climate Impact Research, the conference program added further stimulus for this book.

The year in Paris was a happy one. Working for the OECD was inspiring and empowering. After short stays in the 7th and 1st Arrondissements, we settled in an apartment in the 3rd. My children, William and Clarisse, attended the local French school, École Saint-Martin – and my wife, Denise, enjoyed her leisure time. There were frequent visits to patisseries, cafes, parks, and museums. It was like living in a big candy box for a year. Many friends came to visit, and we also took excursions by TGV for short vacations in smaller French cities, at beaches, and in the countryside. There was discussion of my staying longer at the OECD – for a change in career – but the potential position was in another area outside of the Environment Directorate. For a variety of reasons, I did not pursue the opportunity and returned to my job as an engineering professor at the University of Toronto. I had been a tenured professor for many years, so I had the flexibility and resources to soldier on in writing this book.

What I set out to do was write a thesis that sought to understand the fundamental roots of global environmental challenges. Such a book could be written from the perspective of a variety of academic disciplines, such as philosophy, sociology, political science, economics, or, perhaps obviously, environmental studies, but my approach is different. It is largely informed by technical and scientific perspectives of the substantial challenges through my work in industrial ecology and engineering, with a touch of economics and environmental sciences. This book is all about the development of systematic ways of thinking about global environmental challenges. There are also large doses of history. Similar to the style of my previous book,[3] I find exploration of influential people and wider events an interesting way to understand fundamental ideas. In this book, the approach is made more explicit. In its simplest interpretation, what I tell here are the stories of people and organizations that have contributed to our understanding of the Earth system and humanity's environmental struggles within it, beginning with Thomas Malthus. In researching the book, I satisfied a desire to go back and read some of the classic works on the environment – by influential authors such as Rachel Carson, Paul Ehrlich, Herman Daly, Eugene and Howard Odum, and Robert Ayres. But *Malthus Enigma* is not just about the people who have provided new systematic insights into environmental challenges. It includes inventors, engineers, and scientists, who, through their endeavours, have changed technologies that affect Malthusian environmental struggle. It also includes a handful of movers and shakers – actors in the policy world – who have influenced environmental agendas, as well as a few skeptics who have contributed to the scientific process.

Any book that attempts to examine broad global environmental challenges will necessarily have some shortcomings – and mine is no different. Reviewers of the book have noted that I do not include any discussion of capitalism, nor the role of the capitalist system in our environmental woes. Were this book written by a political scientist, a political ecologist, or a Marxist geographer, that might be a central topic. It is quite understandable why many environmental scholars are critical of the capitalist system of the global economy – and, in particular, the way that it permits various nefarious actors to continue to undermine the sanctity of the planet. I get it. Other than wrestling with market mechanisms – and questioning when they should be employed – I do not pursue a deep inquiry into capitalism.

There are many other topics where all I do is barely scratch the surface. Some ecological economists may be disappointed that I do not dive deeper into topics such as sustainable consumption or de-growth, or social issues such as inequities in wealth. These are not ignored, but my goal is to describe the wider discipline within which they are studied. Other readers might ask why I do not go further into systems theory, measurement of biodiversity, global water challenges, biogeochemical cycles, or Earth systems science more broadly, for example.

There are many people I need to thank for helping me craft *Malthus Enigma*. Foremost are my publication manager Fiona Holt and my editor Amy Haagsma, as well as the anonymous reviewers. The text of many chapters was tightened following careful reading by Ted Sheldon. Other feedback on individual chapters or sections of the book was given by Thomas Homer-Dixon, Jan Corfee-Morlot, Heather Buckley, Astrid Brousselle, Axel Kleidon, Tamara Krawchenko, Cora Hallsworth, Matt Murphy, Reid Lifset, and Dan Hoornweg. I own any remaining mistakes and all the points of contention.

[1] Measured in terms of mean species abundance.
[2] The full program is available at https://www.grc.org/industrial-ecology-conference/2012/ (accessed November 17, 2024).
[3] Kennedy, *The Evolution of Great World Cities*.

Chapter 1: The Gift of Apollo

*The Moon is an escape from our Earthly responsibilities
... it leaves a troubled conscience.*

New York Times columnist Anthony Lewis, July 20, 1969

There was an unusual silence at Pad 39 as the three astronauts arrived at the monstrous Saturn V rocket. Usually there would be NASA ground crews bustling around the pad at the Kennedy Space Center, but today was different. Early morning of December 21, 1968, just three technicians accompanied the astronauts to their thirteen-foot-diameter command module, sitting some 363 feet above the ground.[1]

Apollo 8 was NASA's riskiest mission yet. In the race with the Russians to put a man on the Moon, each of NASA's missions had gradually built upon previous expeditions, learning about space flight and improving technology as they went. Encouraged by the great success of the Apollo 7 mission and responding to news that the Russians were planning a lunar flyby, albeit unmanned, the objectives for Apollo 8 had been accelerated. The goal – to send a manned spacecraft on an orbital flight around the Moon – had many risks. On top of launching the world's heaviest and most powerful rocket, these included perfectly timed thrusts into and out of the Moon's orbit and re-entering the Earth's atmosphere at far greater speed than any previous manned space flight.

Among the three astronauts, Commander Frank Borman was particularly aware of the risks. He, like his fellow astronaut Jim Lovell, had been a commander of a previous Gemini space mission. Borman had also been part of the investigation committee that studied the cause of a tragic fire that killed the crew of Apollo 1 while they were on the ground conducting a simulation launch countdown. The third member of Apollo 8 was a rookie, Bill Anders, a devout Roman Catholic fresh out of the Air Force Institute of Technology.

Liftoff aboard the massive Saturn V rocket went smoothly. Ignition of the first-stage engines occurred at 7:51 a.m., almost perfectly on schedule. The sensation of motion was slow at first while the 3,000 tonnes of fuel and metal picked up momentum. The noise grew to a deafening level, before the rocket passed through the speed of sound and all became eerily silent. The rocket was an incredible feat of engineering. Its designer was the legendary German rocket scientist Wernher von Braun, the father of the V-2 rockets that had terrorized Britain during World War II, who had escaped both Nazis and Russians to join the Americans at the end of the war. In Saturn V, he reached another level. It was taller than the Statue of Liberty, had a takeoff weight greater than that of twenty-five fully loaded jet airliners, and produced the power of eighty-five Hoover Dams.[2] The first stage of the rocket was fuelled by an incredible 203,000 gallons of kerosene and 331,000 gallons of liquid oxygen – all of which was burned up in the first two and a half minutes.[3] The more modest second-stage engines burned for six minutes before falling away. Then, with a short burst of the third-stage engine, the ship reached an orbiting speed of 17,522 miles per hour.[4]

A critical part of the mission occurred close to three hours into the flight, with the re-firing of the third rocket. From an orbit of 114 miles above the Earth, the third-stage engine, powered by a mixture of liquid hydrogen and liquid oxygen, burned for over five minutes, thrusting Apollo 8 to a speed of 23,226 miles per hour.[5] This, the fastest speed that humans had ever travelled, was necessary to propel the one-hundred-tonne spaceship on a gravity-escaping trajectory toward the Moon.[6] Borman was required to make some unexpected small manoeuvres to avoid the ejected third-stage rocket. The spacecraft then successfully passed through the Van Allen radiation belt, which some scientists had incorrectly predicted would give a fatal dose of radiation to the astronauts. Lovell navigated the ship's trajectory using a telescope and sextant aimed at target stars and lunar landmarks, aided by calculations from Apollo 8's on-board computer. After two and a half days, travelling 239,000 miles, Apollo 8 successfully arrived in orbit about seventy miles above the highest point of the Moon's surface.

One key objective of Apollo 8 was to take pictures of the Moon's surface to help establish suitable landing sites for later missions, but the camera and video camera were important for other reasons as well. The crew made a total of six live TV broadcasts for viewers back on Earth, including an emotional reading from the *Book of Genesis* on Christmas Eve. The astronauts took many pictures of the Moon, describing its "awesome, forlorn beauty – desolate beyond belief," [7] but it was a view of Earth that proved to be so much more important. The ship made ten lunar orbits, each one requiring it to pass behind the dark side of the

The Gift of Apollo

Moon, out of sight of the Earth and out of radio contact for thirty-six minutes. On the third of these rotations, Borman and Anders caught a glimpse of the Earth rising above the lunar horizon and captured the awesome moment on camera. Later, in his autobiography, Borman described the Earthrise as follows:

> *It was the most beautiful, heart-catching sight of my life, one that sent a torrent of nostalgia, of sheer homesickness, surging through me. It was the only thing in space that had any color to it. Everything else was either black or white, but not the earth. It was mostly a soft, peaceful blue, the continents outlined in a pinkish brown. And always the white clouds, like long streaks of cotton suspended above that immense globe.*[8]

Apollo 8 returned safely to the Earth. Borman and his crew survived the perilous re-entry into the Earth's atmosphere and were hailed as heroes. They charted the course for the successful Moon landings that would follow, but the most enduring impact of the Apollo missions was not about the Moon, but a revelation in humankind's conscience about the Earth. The TV pictures of the Earth and the reading from *Genesis*, soon followed by the photographs of the Earthrise on the Moon, had profound effects on the inhabitants of our planet, awakening simmering environmental concerns about the Earth. The sounds and images beamed to millions of viewers were reinforced by the words of poet Archibald MacLeish: "To see the Earth as it truly is, small and blue and beautiful in that eternal silence where it floats, is to see ourselves as riders on the Earth together, brothers on that bright loveliness in the eternal cold."[9] The *New York Times* published MacLeish's celebrated words on Christmas Day, and they were soon reproduced in other newspapers and magazines in the coming days, accompanied by the stirring photograph of the Earthrise. The photograph was everywhere, from the cover of *Time* Magazine to a double-spread in *Life* Magazine, reputedly read by one in four Americans.

To many, the Apollo 8 mission was like a second Copernican Revolution – a Columbus-like mission that began a new Renaissance. Without doubt, it was a watershed in the development of the global environmental movement – heralded by some as a new age of ecology. *Earthrise* became "the most influential environmental photograph ever taken."[10] The list of those inspired by it included John McConnell, the founder of Earth Day; James Lovelock, creator of the Gaia Hypothesis; René Dubos, co-author of *Only One Earth*; and many others.[11] University of Georgia professor Eugene Odum, one of the fathers – or, rather,

3

brothers – of systems ecology, is said to have kept a poster of *Earthrise* on his office wall; it was also on the front of his pioneering textbook *Fundamentals of Ecology*.[12] Remarkably, even decades before the Apollo missions, the significance of such a photo of the Earth had been prophesized. In the 1950s, science fiction writer Arthur C. Clarke foresaw the emotional force of humans seeing the Earth.[13] In 1931, social reformer David Lasser argued that the sight of the Earth would break down racial divides;[14] going back even further, the likes of Jules Verne, Edgar Allan Poe, and Johannes Kepler recognized its significance in their writings. The impacts would be enduring, too. In 1975, when the United States Congress was debating the future of the space program, it was poignantly remarked that "what was most significant about the lunar voyage was not that men set foot on the Moon, but that they set eye on the Earth."[15] Even space guru Carl Sagan, in his book *Cosmos*, based on the popular 1980s TV series, surmised that "the inescapable recognition of the unity and fragility of the Earth … is the unexpected gift of Apollo."[16]

The Environmental Revolution of the 1960s

The Apollo missions occurred against the backdrop of the 1960s, which was a transformative decade in the emergence of a global environmental movement.[17] This was particularly so in the United States, where a growing activist political body sometimes mixed concerns over threats to the human environment with civil rights and anti-war protests. Early in the decade, US government marine biologist Rachel Carson published the now-legendary *Silent Spring*, which raised alarm over the misuse of synthetic chemicals, pesticides, and insecticides.[18] Carson provoked aggressive attacks from the US Department of Agriculture and several large chemical companies, but she was already a well-established, best-selling author and was aided in her fight by the serialization of her book in the *New York Times* prior to its release. A special panel of President Kennedy's scientific advisory committee vindicated Carson's work, and all twelve of the most toxic chemicals discussed in *Silent Spring* were later banned or restricted. Other issues, too, began to stir an environmental movement, particularly on concerns that impacted human well-being. Fallout from nuclear weapons testing had in many respects become the first truly global environmental issue; a partial test ban treaty was signed by the United States, the USSR, and the UK in 1963, eliminating all but underground testing. There were also several high-profile environmental disasters that stirred environmental passions.[19] In

The Gift of Apollo

March 1967, 875,000 barrels of crude oil spilled from the wreck of the *SS Torrey Canyon* supertanker after it struck a reef off the southwest tip of England. The cost of cleanup and the lack of governmental preparedness in part led to the creation, in 1969, of the Royal Commission on Environmental Pollution. Another highly public oil spill occurred off the coast of Santa Barbara, California, in early 1969. Large, but unknown, quantities of oil seeped onto nearby beaches following two blowouts at a Union Oil Company platform. Such incidents incensed the public.

Environmental concerns arguably reached a crescendo in the four years from 1968 to 1972, which correspond to the era of Apollo missions to the Moon. The first Earth Day took place in San Francisco on March 21, 1970,[20] soon followed by a second Earth Day on April 22, organized by Wisconsin Senator Gaylord Nelson and involving rallies in major cities throughout the United States. Friends of the Earth was founded in 1969. The US Environmental Protection Agency was founded in 1970 and soon produced a range of environmental legislation, including the *Clean Air Act*. Meanwhile, at the international level, the United Nations Educational, Scientific and Cultural Organization (UNESCO) held an important biosphere conference in Paris in September 1968 and launched the Man and the Biosphere Programme in November 1971. The following year saw the founding of the United Nations Environment Programme (UNEP) and the landmark United Nations Conference on the Human Environment in Stockholm.[21] That year, 1972, also witnessed the last manned mission to the Moon – Apollo 17 – during which the *Blue Marble* photograph, another iconic shot of the Earth, was taken.

A number of influential environmental writers also emerged in the 1960s. Alongside Carson, who died in 1964, then US Secretary of the Interior Stewart Udall is credited with sowing the seeds of the environmental movement. In 1963, his best-selling book *The Quiet Crisis* helped establish an American environmental philosophy and warn of the pollution and resource challenges ahead.[22] In 1966, building upon a notion developed by Buckminster Fuller and Barbara Ward, University of Michigan economist Kenneth Boulding penned an essay, "The Economics of the Coming Spaceship Earth." The essay contained the seeds of the notion of a future steady-state economy – bound by Earth's limits – and became influential to the field of ecological economics. Two university biologists, Barry Commoner and Paul Ehrlich, also emerged as spokesmen for the environment. Commoner was heavily concerned about the potential effects of nuclear fallout – and more broadly the pollution resulting from flawed technologies. *Time* Magazine hailed him as the Paul Revere of ecology and put him on the front cover

Malthus Enigma

in February 1970. Commoner's book *The Closing Circle* (1971) was one of the first to develop ideas about sustainability.[23] Interestingly, Commoner would often clash with Ehrlich, whose concerns were more focused on the resource challenges of meeting population growth. His 1968 book *The Population Bomb*, commissioned by the Sierra Club, is one of the best-selling environmental books ever.[24] Commoner and Ehrlich became environmental evangelists, travelling the country to speak to crowds of thousands – although their differences in perspective would sometimes lead to "vitriolic debate of questionable merit."[25]

In *The Population Bomb*, Ehrlich was deeply concerned with the challenge of feeding a large and growing global population on a planet that is of limited size and is poisoned by pollution. In focusing on population as a critical issue, he was clearly picking up the mantle of nineteenth-century Englishman Thomas Malthus, who was similarly concerned. Ehrlich was particularly alarmed by population growth in developing countries and noted that, during the 1960s, population increases in these countries were outgrowing their food production.[26] Moreover, populations in developing countries were rapidly growing, with doubling rates of twenty to thirty-five years, compared with values on the order of fifty to two hundred years for developed countries.[27] Ehrlich recognized that food transfers to developing countries had been occurring since 1958 but did not see them as being sustainable in the long run. Indeed, he suggested that "we will not be able to prevent large-scale famine in the next decade or so."[28] Ehrlich observed that the problem was compounded by the emissions of a variety of environmental pollutants, which he noted tend to increase in severity with the increasing population. Among the environmental challenges he raised were the clearing of forests to grow food, soil deterioration, the use of synthetic pesticides, ocean dumping, local air pollution, smog, the effects of CO_2 concentration on Earth temperatures, the death of lakes, and pollutants in the food chain. His diagnosis of the causal chain was quite straightforward: "Too many cars, too many factories, too much detergent, too much pesticide, multiplying contrails, inadequate sewage treatment plants, too little water, too much CO_2 – all can be traced easily to too many people."[29]

Epitomizing his neo-Malthusian perspective,[30] Ehrlich described some possible scenarios for the 1970s and early 1980s. Looking back with about fifty years of hindsight, it is hard not to be perplexed by some of the possible pictures that Ehrlich paints – reflecting in part the politics of the times. In his most pessimistic scenario,[31] food riots around the globe turn into anti-American protests, the whole of Latin America turns

The Gift of Apollo

communist, bubonic plague kills sixty-eight percent of already starving Egyptians, and killer smog in Los Angeles wipes out 98,000 people. Melting of the polar ice caps causes sea level rise, and there is an irreversible decline in the Atlantic and Pacific fisheries. America is so short of food that births are first restricted to one child per couple; then, later, compulsory sterilization is enforced upon all persons with an IQ less than ninety. The scenario ends in the 1980s with China and Russia ganging up on America, resulting in a nuclear war that, along with radiation fallout and climate change, destroys all of the humans and other mammals on the planet – leaving only the cockroaches. Not to be overly negative, Ehrlich does present a more optimistic scenario.[32] Yet even this "cheerful scenario," which requires the Pope to encourage Catholics to use contraception and Jews and Arabs to make peace in the Middle East, still sees half a billion people (one-fifth of the global population at the time) die of starvation.

Perhaps Ehrlich was fear mongering; perhaps his crystal ball was poorly calibrated in time; perhaps he fundamentally misunderstood humankind's ability to respond to the resource challenges; perhaps not. Certainly, the scenarios he painted were way off the mark, but they were not entirely wrong. The United States did not introduce a one child per couple policy – and I can't imagine that it ever would – but China did. The Chinese policy, introduced in 1978, is estimated to have prevented as many as 400 million births between 1979 and 2011.[33] An interesting twist is that some writers have tied the genesis of China's population control policy back to Western environmental and resource concerns of the 1960s and '70s.[34]

Another of the issues that Ehrlich mentioned, sea level rise from the melting of the ice caps, continues to be of great concern. In its *Sixth Assessment Report*, the Intergovernmental Panel on Climate Change (IPCC) noted that mean sea level rise somewhere between 28 cm and 102 cm is likely to occur by the end of the twenty-first century.[35] A study co-authored by my former colleague Jan Corfee-Morlot at the Organisation for Economic Co-operation and Development estimates that the costs of flooding in the world's 136 largest cities will be over US$50 billion per year by 2050; Guangzhou, Miami, New York, New Orleans, and Mumbai are expected to have the greatest losses.[36] The longer-term perspective is even more troubling. I remember sitting at a session held by glaciologists at the United Nations Climate Change Conference in Paris (COP 21)[37] in 2015; the scientists noted that the melting of the West Antarctic Ice Sheet will raise global sea levels by three metres.[38] Most of the scientists

Malthus Enigma

thought the ice sheet was already so unstable that its melting was likely – they were just unsure about how long it would take.

A further critical global environmental issue that was barely understood in the Apollo era is loss of biodiversity. While human population and economic activity have ballooned over the past fifty years, other species on the planet have been going extinct at a rate thought to be on the order of one hundred to one thousand times above normal.[39] The Millennium Ecosystem Assessment, released in 2005, described a dramatic transformation of the Earth's ecosystems with loss of native habitat in major biomes – and serious declines in many types of ecosystem function that humans rely on. More recent studies estimate that thirty-two species of mammals have become extinct since 1900, and 515 or more terrestrial vertebrates are on the brink of extinction.[40] Ecologists are so concerned, they warn that the Earth is on track to experience a mass extinction within the next few centuries.[41] In other words, the impacts of humans on the planet are so severe that we may cause an extinction of other species comparable to an asteroid hitting the Earth and wiping out the dinosaurs.

Ehrlich may have got most of the details wrong, but there is something of sobering substance to the essence of his concerns. Humankind faces a continual struggle – both technologically and organizationally – to feed a growing population using the limited resources of the planet, just as Malthus had worried about. Moreover, there are global environmental challenges – such as climate change and biodiversity loss – that make the task even harder. How humanity copes with this struggle is an *enigma*, which is what this book is all about. I will go beyond simple conclusions that Malthus's population concerns were wrong because he overlooked our ability to innovate. I will review some of the key inventions, such as fertilizers and steam engines, that have helped humans thrive, but in doing so have created further challenges that perpetuate the Malthusian struggle. I will describe how science – such as the field of thermodynamics – has been important, not only for developing technology, but also for understanding the world around us, so that we are able to assess our environmental burden on the fragile planet that Borman and his Apollo crew first set eye upon. Without good science – and sometimes even with it – the political ability to respond to environmental and resource challenges is hampered.[42] This policy response is, I argue, a further layer to Malthusian struggle. To explain these complex, interwoven layers of the Malthus Enigma, I am going to take a 250-year perspective – and piece together the stories of groups and individuals who have made deep contributions to humankind's eternal struggle since Malthus was born.

The Malthus Enigma

The central question of this book is: *How do we sustain a growing global population within the carrying capacity of the planet?* The purpose of the book, however, is more about trying to *understand* the question than to answer the question. This is because the question itself reveals an *enigma* – that is, something that is puzzling, mysterious, and difficult to explain or understand. I have named the enigma after Malthus, as it builds upon the population problem that he wrestled with. The phenomenon we are dealing with is more complex than a *dilemma*, which is a situation where a difficult choice between undesirable outcomes has to be made. With the Malthus Enigma, there are no clear choices, and hence no definitive answers. Nonetheless, by seeking to understand the Enigma, with some help from the field of ecology, we might find our way.

The Malthus Enigma is the ongoing challenge of sustaining a growing global population within the carrying capacity of the Earth. Many have recognized Malthus, an economist and member of the clergy, to be one of the earliest, or at least the most influential, writers to understand the challenges of sustainable human development.[43] He is primarily known for his concerns about feeding a seemingly ever-growing human population. Some have criticized Malthus for failing to recognize the role of human innovation and technological change in addressing the *population problem*. This, however, is a simplistic rejection of Malthus, which I seek to move beyond. In particular, while recognizing that Malthus's view was in a simple sense wrong, or perhaps incomplete, the process of Malthusian struggle is very much still with us. This challenge has subtly evolved from the population problem that Malthus observed and has several complex layers to it. Malthus was concerned that a growing population would outstrip its available food supply. In the 1960s, the challenge became recognized as one of sustaining humans within the bounds of Planet Earth – that is, upon "Spaceship Earth," as Boulding put it,[44] or within planetary boundaries, as now called.[45] On top of the issue of providing food and other resources from the limited area of the planet, the challenge has grown to include maintenance of the Earth's climate and biodiversity, as well as assimilation of wastes – ranging from toxic pollutants to the carbon emissions that undermine the stability of the climate system. The Enigma

Malthus Enigma

is all about sustaining a flourishing human population without destroying the ecosystems and stable climates that we rely on.

Neo-Malthusian concerns that the human population cannot be sustained have been highly controversial. There is a long history of Malthusian ideas being rejected – although this really just adds to the Enigma. The essential argument of the Cornucopians, who oppose the neo-Malthusian viewpoint, is that humanity will find a way to innovate through its resource challenges.[46] The confrontation between neo-Malthusians and Cornucopians has taken many forms, although typically it has pitched environmental scientists against economists. Ehrlich's position on population was firmly attacked by University of Illinois professor Julian Simon – the showdown resulting in a bet on the future price of resources.[47] The highly prominent *Limits to Growth* study of the early 1970s led to a prolonged feud between MIT professor and systems guru Jay Forrester and Yale economist William Nordhaus.[48] Forrester's work played an important role in the development of environmental systems models used today. Another confrontation was that between early ecological economists Nicholas Georgescu-Roegen and Herman Daly on one side, and neoclassical economists Robert Solow and Joseph Stiglitz on the other.[49] The debate was over why conventional economic growth theory ignored energy and material resources. Malthusian and Cornucopian positions have even escalated to political levels – President Reagan's rejection of resource concerns in the Carter Administration's *Global 2000* study, in favour of a free-market approach, being one example.[50] Such contention over the Malthusian position goes only to magnify the complexity of the Malthus Enigma.

In this book, I set out to look at the debate between Malthusians and Cornucopians from a higher level, to understand its complexities. I aim to look beyond the simple idea that Malthusian concerns can be dismissed due to human innovation. The Cornucopians have a fair point that humans have tended to find innovative ways of increasing their food supplies, but this is part of the Enigma. The need to innovate to sustain the human race is constantly with us. Like the literary classic *Catch-22*, however, anything we do to improve the human condition on this planet seemingly makes it harder to do so. This Malthusian struggle is part of what it means to be human. Another aspect of the Enigma is that undertaking the science to understand the global environment around us is an ongoing process – and this, too, can be contentious. It is like humanity is playing a long, complex, never-ending board game, but we are still learning the rules as we play along. With uncertainties in the science – where they still exist – and continual evolution of technology and other innovations, a third

The Gift of Apollo

complex layer of the Enigma then arises. This is the difficulty of developing policies for addressing pressing global environmental challenges and resource constraints. So technology, science, and policy are the three complex layers of the Enigma that I will examine further.

Technology plays an essential role in the Malthus Enigma, though this, too, comes with contention. Looking back over the past 250 years, changes in technology have consistently helped feed an ever-growing global population. Malthus lived during the Industrial Revolution. This was the era that saw remarkable improvements in the efficiency of steam engines – increasing access to Britain's coal reserves. Britain's use of coal was intimately tied to the onset of human-induced global climate change, but combined with advances in steelmaking and other industrial developments, it changed the world in ways that Malthus was blind to. There were also significant increases in crop yields per area of farmland in Britain during the Industrial Revolution that helped address the challenge of feeding a growing population.[51] A further massive boost to food production occurred early in the twentieth century, when the German scientists Fritz Haber and Carl Bosch devised a highly efficient process for producing the active nitrogen that is a key ingredient of synthetic fertilizers. Their efforts helped approximately double the productivity of agricultural land.[52] Then, into the 1920s, mass-produced tractors began to replace horses on American farms. This reduced the land area required on farms to produce fodder crops for feeding the horses. Global population, nonetheless, continued to grow rapidly throughout the twentieth century, putting more stress on food supplies. This was met, however, by advances in crop breeding, which produced higher-yield, disease-resistant strains of wheat. Led by American agronomist Norman Borlaug,[53] the Green Revolution in agriculture seemingly blew away Malthusian concerns about feeding a growing population.

Despite the clear historical record of humanity innovating its way out of the population trap, this does not actually solve the Malthusian problem. Rather, it points to a key first layer of the Malthus Enigma – that humankind must continue to innovate to sustain itself within the carrying capacity of the Earth. Such a statement is not without challenges, though. Ehrlich and Commoner, for example, clashed on the importance of technology in solving the global challenges of the late 1960s.[54] Ehrlich professed a need to reduce absolute population, while Commoner was more concerned with reducing the environmental impacts of technology. This begs the question: *What kind of technological innovation do we want?* Technological change has come with unintended consequences that add to the Enigma. The coal that powered the Industrial Revolution

11

also began an era of human-induced climate change. The tractors that replaced the farm horse are powered by gasoline or diesel, which, along with other fossil fuels used for myriad purposes, perpetuate today's climate crisis. Synthetic fertilizers may have boosted food production, but a high percentage of the nutrients they provide go to waste – interfering with natural waterways and causing algal blooms, decreasing water quality, and impacting biodiversity. Similarly, the Green Revolution increased crop yields, but at the expense of crop diversity, which may undermine the long-term resilience of agricultural lands. The technological trade-offs are often complex, though. With lower crop yields, more land area is required for food production – all else unchanged – and this might come at the expense of natural ecosystems and biodiversity.

The technological challenges of feeding a growing global population without destroying massive areas of the Earth's ecosystems are daunting. American ecologist David Tilman projects that global food demand could approximately double from 2005 to 2050.[55] This is not only due to increased global population, but also as a result of richer diets that accompany increased wealth. If the productivity of agriculture in developing countries remains at existing low levels, Tilman warns that about one billion more hectares of farmland would have to be created globally. This is about the land area of the entire United States. Clearing of such land would be devastating for natural ecosystems – and there would be a large increase in global greenhouse gas emissions too. This situation is particularly crucial to Africa, where the largest population growth is expected – and natural habitats may be most under threat. The technological solutions for feeding the planet are highly contested, though. Sharing of Western agricultural technologies to intensify yields may prevent widespread land clearing. Food researchers such as Catherine Badgley, Verena Seufert, and South African farmer-professor Raymond Auerbach argue that regionally appropriate agricultural approaches will be important.

Another position that may be taken is that we already have the technology we need to solve today's environmental challenges. There are merits to such a position with respect to addressing global climate change. Studies by government agencies in the United States and China, as well as the International Energy Agency, among others, indicate that much of the world's energy needs could be met with renewable energy sources, such as solar and wind.[56] This is the soft path forward that American energy guru Amory Lovins called for over forty years ago, in the wake of the 1970s oil crises.[57] Of course, technological progress has been required to bring the costs of wind and solar down – and further improvements in energy storage would also help. Nonetheless, by 2010, the energy generated by a solar panel typically exceeded the energy required to manufacture it by a factor of seven.[58] For wind turbines, the energy return was typically a factor of about nineteen over their lifetime, and these energy returns are increasing with improvements in technology.[59] With such "energy return on energy invested," there is the potential to create positive-reinforcing net-energy-generating systems – similar to the way that James Watt's steam engine did during the Industrial Revolution, by accessing coal seams. This is despite some claims that technology cannot advance sustainable development due to thermodynamic constraints.

The science of thermodynamics plays an intriguing role in the Malthus Enigma. Thermodynamics is the study of energy in relation to heat, work, and temperature. Development of this science will be covered later. At a deep level, thermodynamics is also a science of sustainability; not only do its laws tell us how humanity can access and use energy, but they also dictate how complex systems can grow, sustain themselves, or die. The laws of thermodynamics, however, have been discovered only over the past 200 years. Malthus's first essay on population was written almost fifty years before the first law – conservation of energy – was rigorously established. French scholar Nicolas Léonard Sadi Carnot developed the principles of the second law in 1824, but it was not until the mid-nineteenth century that thermodynamics coalesced as a discipline through the works of Rudolf Clausius and William Thomson (Lord Kelvin).[60] Nineteenth-century thermodynamics, moreover, was applied only to inanimate systems, such as steam engines and diesel engines. It was not until the twentieth century, through the works of Austrian physicist Erwin Schrödinger and Belgian chemist Ilya Prigogine, that the thermodynamics of living systems was developed.[61] So even now, in the twenty-first century, we are still developing this fundamental science, which gets at the heart of sustainability.

The emergence of the science of thermodynamics is one example of a second, broader characteristic of the Malthus Enigma. This is that humanity has to continually conduct science to understand the world we inhabit – to learn how Spaceship Earth works. How can we sustain ourselves within the carrying capacity of the planet if our knowledge of its carrying capacity is incomplete? In 2009, when a group of environmental scientists led by Johan Rockström pieced together our understanding of the planetary boundaries within which humanity can sustainably operate, the picture was quite uncertain.[62] Scientists today have a well-developed understanding of the limits of the atmosphere to receive human-produced emissions of greenhouse gases without causing instability to the climate. This has come, of course, through decades of work by thousands of scientists – and with a fair amount of contention along the way. For other environmental stresses, however, the science is arguably less advanced. Before the Millennium Ecosystem Assessment, published in 2005, data on the state of the world's biodiversity was fairly slim – and even today, scientists have named only approximately two million species out of an estimated nine million or more on the planet.[63] Still, tracking of species at risk and those going extinct, through the International Union for Conservation of Nature's *Red List of Threatened Species*, supports legitimate concerns over loss of biodiversity globally. Continuing to conduct the science to further our knowledge of the world's ecosystems and the species within them is clearly necessary. So is furthering our understanding of other environmental systems, from the impacts of chemical pollutants, to changes in the nitrogen cycle, to stresses on the world's river systems and groundwater aquifers. Undertaking the science to understand environmental systems on a global scale is not straightforward and deserves to be recognized as part of the Malthus Enigma.

The history of humankind's struggle to understand the greenhouse effect – and our role in it – provides a further example of the second layer of the Malthus Enigma. The first scientist to experimentally observe the greenhouse effect was probably the American Eunice Foote in the 1850s, although her work unfortunately gained little attention at the time.[64] By the end of the nineteenth century, a handful of scientists, including the influential Swedish scientist Svante Arrhenius, had recognized that society's emissions of CO_2 could cause an increase in the greenhouse effect.[65] Due to errors in distinguishing the infrared spectrum of CO_2 from that of water vapour, however, Arrhenius's theory fell out of fashion. Not until the late 1930s, against considerable opposition, did British engineer Guy Stewart Callendar revive the CO_2-based theory of warming by piecing together primitive meteorological data sets.[66] Through the work of further scientists, notably including accurate measurements of atmospheric CO_2 concentrations by Charles Keeling, by the early 1970s, humankind's role in causing the greenhouse effect was clear to many.[67] In the mid-twentieth century, however, the Earth experienced a period of global cooling, now understood to be associated with sulphate emissions, but which caused uncertainty. Development of computer models of the global climate system proved to have an essential role in the progress of climate science.[68] Beginning in the late 1940s as toy models for John von Neumann's early electronic computers, by the end of the century, global climate models evolved into complex scientific hypotheses about the workings of the planet. Some policy-makers were reluctant to accept the results of the models, as they did not conform to simplistic ideas about reductionist science. Persuaded by ever-convincing evidence that the models were predicting climate change as it was clearly being experienced, the global community now overwhelmingly accepts the science of climate change.

Development of policy to address climate change, as well as other global environmental stresses such as excessive destruction of biodiversity, brings us to the third layer of the Malthus Enigma. The third layer of the Enigma is crafting appropriate policies to respond to environmental challenges. Even when scientific understanding of the challenge is well established and technological solutions exist, there are many complexities to policy-making: the logic of the political process balancing the interests of constituent groups; the potential for unintended consequences or rebound effects from policies; and philosophical differences in the use of different mechanisms, such as when or when not to use markets. Progress on environmental policy has sometimes occurred due to perverse self-interests of groups or individuals. When the world's first international agreement on an environmental issue – preservation of wildlife in Africa – was reached in 1900, many of the national representatives were game hunters.[69] The motive behind the Convention for the Preservation of Animals, Birds and Fish in Africa was to maintain hunting rights for the wealthy elites. Another quirky occurrence was that the leader of the United Nations Conference on the Human Environment, held in Stockholm in 1972, was a Canadian oil executive. Maurice Strong, who was also Secretary-General for the Rio and Rio+20 Earth Summits, made his fortune as owner of a natural gas company. To his credit, Strong was passionate about sustainable development. He recognized, as have many others, that concerns over the environment and resources cannot be addressed in the face of poverty and social inequities.

Beyond actors and constituents, the mechanisms of policy also have challenges. One example of this is the *rebound effect*, by which efforts to save energy through efficiency measures can lead to some or all of the energy savings being eliminated by increased consumption. This phenomenon is sometimes known as Jevons' Paradox, after the English economist William Stanley Jevons.[70] In the nineteenth century, Jevons observed that continued progress in the energy efficiency of steam engines led only to greater consumption of energy – in particular coal. In hindsight, Jevons' observation was quite understandable, as the most fundamental use of steam engines was for dewatering of mining sites –

and more efficient engines could access coal more easily. Rebound effects have since been observed in other contexts, however, leading to contention in the effectiveness of energy-efficiency policies. Underlying unintended consequences, such as rebound effects, are market mechanisms – which leads to another policy dilemma that is part of the Enigma. Markets are a human creation, and those for goods and services have been in existence for centuries. Since the 1990s, we have also taken the novel step of creating markets for "bads" – that is, markets for air pollution and carbon emissions. The use of such markets is still somewhat of a policy experiment. Knowing when, if ever, to use market mechanisms for environmental policy is something we are still learning. There are even some misplaced ideas about creating markets for biodiversity[71] – but this is a topic where ecologists have important lessons to share.

The Age of Ecology

Rather than structure this book around the three layers of the Malthus Enigma, I have organized it around seven broad questions:
1. How have Malthusians and Cornucopians disagreed?
2. How do we sustain humanity without destroying global biodiversity?
3. How did we discover the greenhouse effect, and how will we adjust our energy systems to accommodate it?
4. How did we discover the laws of thermodynamics, and of what relevance are they for sustaining humanity?
5. How do we conceptualize the interrelationships between global problems?
6. How do we bring ecological principles into economics?
7. How do we develop technological systems to be in harmony with Nature?

The logic to the order of the questions – as well as the nature of the three layers of the Enigma – will emerge through the book. Briefly, though, the three layers become apparent from wrestling with the contention that is manifest in the first question – and the other questions lead us to understand the Enigma through the lens of ecology.

The Malthus Enigma will be explored through a series of stories interspersed with some technical explanations and summaries, structured around the seven questions. In telling the stories of influential scientists, inventors, and policy influencers, I will necessarily jump around a bit in history depending on the topic. Essentially, the three layers of the Enigma are developed from the history-based nature of this inquiry. Innovating, conducting science, and crafting policies are things that we have long been doing – we just need to do them with a better sense of the whole, and with an ecology-based mindset.

For the reader looking for a more straightforward take-away from this book, there is a simple, single one – that is, *ecology*. The Malthus Enigma – sustaining humanity within the Earth's carrying capacity – is ongoing. It is part of the human condition and perhaps cannot ever be overcome. Emerging out of the twentieth century, nonetheless, the field of ecology potentially provides guidance on how humanity might come to better understand the Enigma – and prosper within its context. The last two chapters of this book provide introductions to the fields of ecological economics and industrial ecology for new readers to these fields, drawing upon a thread of *systems thinking* that runs through all the chapters.

In his textbook *Fundamentals of Ecology*, Eugene Odum described ecology as the first *holistic science*.[72] Ecologists apply systems approaches to understand the structure and interrelationships of the components of natural systems. This is important for determining how ecosystems function – and clearly helps support the challenge of protecting biodiversity. The methods of ecology, which include analysis of the flows of energy and matter in natural systems, can, moreover, also be applied to human society. Indeed, it was Eugene Odum's younger brother Howard who was one of the earliest to realize this. In his Apollo-era book *Environment, Power and Society*, Howard Odum wrote: "A study of humanity and nature is thus a study of systems of energy, materials, money and information."[73] Howard Odum developed some early techniques for analyzing the flows of energy and materials in human systems, which made him an early motivator of the modern fields of ecological economics and industrial ecology. By the end of the twentieth century, these two groups of interdisciplinary scientists – ecological economists and industrial ecologists – were applying principles of ecology to human economies and industrial systems, respectively. Thus, the Age of Ecology offers potential guidance on the development of technology, science, and policy in the face of the Malthus Enigma.

The roots of ecology are deep, going back to the likes of Swedish naturalist Carl Linnaeus, who worked on taxonomy; German scientist Alexander von Humboldt, who studied relationships between climate and plant species; and Charles Darwin, famous for his text *On the Origin of Species*. Ecology coalesced as a discipline over the twentieth century, with the development of concepts such as the biosphere, ecological succession, and mathematical models of species populations. Sir Arthur Tansley, who founded the British Ecological Society and was the first editor of the *Journal of Ecology*, is credited with developing the term *ecosystem*.[74] Tansley considered the ecosystem to be a complex system capturing the interactions between communities of organisms and the non-living (abiotic) physical environment in which they live. Another highly influential ecologist was Aldo Leopold, who wrestled deeply with topics such as the interrelationships between population dynamics and consumption, carrying capacity, and environmental degradation. Marrying ideas from ecology and ethics, Leopold inspired subsequent generations of environmentalists through his collection of essays, *A Sand County Almanac*.[75]

Ecology was just one field in which systems theory was being developed and applied in the mid-twentieth century. Systems theorists study the relationship between a whole and its constituent parts. The area of general systems theory was developed in the 1950s and '60s by theorists such as Boulding and Ludwig von Bertalanffy.[76] Application of systems methods came with the development of computing technology. A notable example of this was the work of Forrester, who, after inventing a higher-speed computer early in his career, went on to build systems models with applications for industry and cities.[77] In the late 1960s, Forrester's techniques were picked up by the Club of Rome, which encouraged their use for the *Limits to Growth* study.[78] With expert promotion by savvy industrialist Aurelio Peccei, results from the computer systems modelling of humankind's predicament made front-page news around the world in 1972. Peccei and his collaborators at the Club of Rome also had a role in establishing the International Institute for Applied Systems Analysis at the Laxenburg Palace near Vienna, which continues to further systems models of the global economy today.[79]

The early global models of Forrester and his associates were quite contentious, in part because they lacked a grounding in thermodynamics. They ignored conservation of energy – and conservation of mass too. When ecologists such as Eugene and Howard Odum began to develop mathematical models of ecosystems, however, these key physical laws were central to their work. This was also the case in the establishment of global climate models – led by climatologists such as Carl-Gustaf Rossby and Jule Charney starting in the late 1940s.[80] The Odums at first applied their physics-based models to natural ecosystems, such as a Florida spring and a Pacific coral reef. Later, however, Howard Odum recognized that physics-based systems models – employing laws of thermodynamics – could also be developed for human economies.

Alongside Howard Odum, another thought leader who influenced the field of ecological economics was Romanian economist Nicholas Georgescu-Roegen. Late in his career, Georgescu-Roegen came to the realization that economies are subject to the laws of physics. Focusing on the second law of thermodynamics in particular, he wrote a mighty treatise on the subject, entitled *The Entropy Law and the Economic Process*.[81] Moreover, Herman Daly, Georgescu-Roegen's doctoral student, emerged as one of the early leaders of ecological economics when it coalesced as a discipline in the 1980s. Daly wrote an influential book, *Steady-State Economics*, in which he encouraged a mixture of market-based approaches to control capital and population, combined with measures for redistributing wealth, in addressing neo-Malthusian concerns.[82] Daly became one of the two co-founding editors of the journal *Ecological Economics*. The other was Robert Costanza – a former student of Howard Odum – who led research on the value of natural ecosystems, among other topics.[83]

A significant aspect of the field of ecological economics is that it spans between physical sciences and social sciences. Ecological economists recognize that the economy is bounded by physical constraints, as expressed by the laws of thermodynamics. Physical laws alone, however, are insufficient to understand economic processes – and human value systems also come into play. Hence, ecological economics provides a paradigm for policy-making that tackles the huge environmental challenges of biodiversity loss and climate change while addressing the social challenges of meeting basic human needs, closing large inequalities in wealth, and averting overconsumption by the wealthy.

The importance of the field of ecological economics has come to a fore in recent decades with the emergence of market-based approaches for managing air pollution and carbon emissions. Stemming from influences beyond ecological economics, markets for air pollution were first started in California in the 1990s.[84] They were riddled with many challenges at first, including issues of social justice, over-allocation of permits, fraud, and manipulation by companies.[85] In the late 1990s, nonetheless, when the Kyoto Protocol on greenhouse gas emissions was under negotiation, the United States pushed heavily for carbon-trading mechanisms to be included. Various experiments with carbon markets have been conducted since, including some failures, such as the Chicago Climate Exchange, and others more persistent, such as the European Union's Emissions Trading System, now in its fourth experimental phase. While perhaps not universally liked by ecological economists, the

Malthus Enigma

philosophy of carbon markets is consistent with the market-based approach to environmental challenges that Daly, among others, subscribed to. A more difficult issue, however, has been proposals to develop markets for biodiversity. The proposed markets have many flaws, in particular the fallacy of representing diversity through a single dollar metric. In spite of their work on evaluation of natural ecosystems, most ecological economists would likely reject the notion of markets for biodiversity, based on the strong sustainability principle that natural capital and financial capital are not substitutable.[86] This is one example of how development of the language and ideas of ecological economics has become essential for protecting the Earth's natural ecosystems.

While ecological economists have found ways of understanding the economy within the context of the Earth's wider supporting ecosystems, a parallel effort has been envisioning industrial systems that both mimic and sustainably coexist with natural ecosystems. The basic founding idea of industrial ecology is that we can learn to build industrial systems without waste.[87] If the residual masses produced from one type of industry provide the feedstock for another, then there is no waste – just like in natural systems. Inspired by the notion of industrial metabolism from the physicist Robert Ayres and encouraged by former NASA administrator Robert Frosch, industrial ecologists at first set out to design industrial systems in this fashion.[88] With growing understanding that households – as the consumers of goods and services – were equally part of the industrial system, the field of study began to widen. Industrial ecology evolved to be an applied science, studying the energy and material flows of industrial society and the associated environmental impacts. Having started with a focus on technology, industrial ecologists began to contribute to environmental policy – establishing methods of material accounting for national environmental reporting and providing technical expertise to the United Nations' International Resource Panel. The ideas of the field came to prominence with the crafting of policies on circular economies by the European Union, China, and other governments. Such policies caused contention among industrial ecologists, however. After several decades of work, they had begun to understand the many challenges of developing truly circular economies.[89] Such contention is normal, however; it has long been part of the Malthus Enigma and was very much apparent in the life of Malthus himself, which we turn to shortly.

The fields of ecological economics and industrial ecology – along with systems thinking more broadly – have a lot to contribute to the development of policies for addressing critical global environmental challenges. The objectives of this book, however, do not include prescribing specific policy recommendations. The topic of this book is an enigma, which, by definition, is a mysterious and complex phenomenon that is difficult to understand. My objective is to explain the Malthus Enigma – not to solve it. I am not going to prescribe any particular technological solutions or policies for saving the planet. Nor will I make any strong predictions about how humanity will sustain itself in the future. My aim is to describe the philosophy, ingredients, and approaches to policy-making under the Malthus Enigma, rather than to give specific policies.

Overall, from the chapters that follow, we will see how the perspective of ecology can help us navigate the layers of the Malthus Enigma. The need to innovate, to undertake science to understand the Earth, and to craft appropriate policy to sustain humanity will be demonstrated. Adding principles of ecology to the storyline – including systems thinking, and the laws of thermodynamics – on top of human values, will point to ways that humans can potentially flourish under the Enigma. We will see how rekindling the spirit of the Age of Ecology, inspired by the Apollo 8 mission to the Moon, can comprehensively guide us in tackling the great environmental challenges of our time.

Chapter 2: Malthusians and Cornucopians

Malthus has been buried many times, and Malthusian scarcity with him. But ... anyone who has been buried so often cannot be entirely dead.

Herman Daly, echoing Garrett Hardin [1]

Who was Malthus? The author of an important, alarming, and controversial essay on the problem of population growth – a pivotal work in understanding the sustainable development of human civilization – was surprisingly a rather obscure and mild-mannered English clergyman. Thomas Robert Malthus was born on February 13, 1766, near Dorking in Surrey, as the sixth child of an eccentric, but wealthy, country gentleman, Daniel Malthus, and his wife Henrietta.[2] Though raised in an Anglican family, he unusually attended an academy run by dissenters, before undertaking a program of study at Jesus College, Cambridge, focusing primarily on mathematics. Despite having a speech defect, Malthus was ordained upon leaving Cambridge, taking a position as a curate at Okewood in Surrey, near his family home. For the next ten years, he lived with his parents and his two unmarried sisters, enjoying the subdued life of a country curate and continuing his studies as a non-resident fellow at Cambridge. In 1798, he published his first *Essay on the Principle of Population*, which, after further research during travels in Scandinavia, France, and Switzerland, was followed in 1803 by a second, expanded edition. That year, he became rector in the picturesque village of Walesby in Lincolnshire; the following year, 1804, he married Harriet Eckersall, and three children soon followed. Malthus's late marriage, at age 38, and modest number of children for the times were remarkably consistent with one of his solutions for the population problem. In 1805, Malthus became professor of history and political economy at the East India Company College in Hertfordshire, and he was recognized as one of the leading political economists in England until his death in 1834.

Malthus Enigma

Malthus's concerns over a growing population – which are central to the Enigma – have been controversial, pitching Malthusians against opposing techno-optimistic Cornucopians.[3] As this chapter will show, in recent decades, the differences in perspectives of these opposing groups have manifested in different ways. Malthusians and Cornucopians have disagreed on many things, from simple expectations about the future prices of resources to notions of the carrying capacity of the planet. These disagreements capture the twentieth-century evolution of the Malthus Enigma to include global environmental concerns. The differences in views on whether the human population is doomed to suffer or technological innovation can save the day have been quite stark. By studying the nature of the disagreements between Malthusians and Cornucopians, however, we can begin to develop an understanding of the three layers of the Malthus Enigma. Also, while rational assessment of the disagreements is difficult, we will end the chapter by discussing "systems thinking," which further helps with framing the differing perspectives. We should begin, however, with the population problem as Malthus saw it.

To understand, perhaps, why Malthus was so concerned with the question of feeding a growing population, it is useful to note the context of his time. Malthus wrote his first *Essay* during the turbulent years of the French Revolution, which blazed from 1789 to 1799. Although partially related to the bankruptcy of the French state due to its participation in the American Revolutionary War, the major cause was hunger and malnutrition among some of the French population when food prices rose after several years of poor harvests. Periods of grain scarcity, moreover, occurred in England in 1794 and again in 1800.[4] Subsequent editions of the *Essay* – there were six in total – were written during the Napoleonic Wars, during which food prices remained very high, and Britain became a net importer of food. Despite Britain's economic power during these times, a large and growing segment of its population was supported by the Poor Laws. Access to food was a major issue in Malthus's time.

At the heart of Malthus's work was a concern that prospects for improvements to society in the long term depended upon an ability to expand food production to satisfy population growth. The conventional view was that the population of a nation was a measure of its power, and Malthus further considered that "there is not a truer criterion of the happiness and innocence of a people than the rapidity of their increase."[5] By his reasoning, however, the population of a community, in the absence of constraints, would tend to increase at a geometric rate – that is, at an increasingly accelerating rate – while the ability to produce food could grow only at an arithmetic rate – that is, at a constant rate. Thus, Malthus argued that difficulties in provision of subsistence would eventually exert a check on population, which, moreover, always entailed some form of misery or vice for society. He identified two types of checks: positive ones of war, pestilence, and famine, and preventive ones of abortion, infanticide, and prostitution. He also theorized that population growth rates might follow cycles, with rates dropping during hard economic times, before, at the bottom of the cycle, cheap labour stimulating further food production and hence rising population. Overall, his prognosis of the human condition was extremely bleak.

Malthus may have identified a fundamental struggle of humanity to improve its lot, but hindsight shows that he underestimated human ingenuity and the ability to cope with the stress of further population growth. Even in his lifetime, the Industrial Revolution was well underway, contributing to increases in agricultural productivity. Wheat yields in Britain averaged about twenty bushels per acre in 1800, up from around fifteen bushels a century earlier, and continued to rise, reaching thirty bushels per acre by 1850.[6] Similar increases were observed for barley and oats. Agricultural output per worker increased about thirty percent over Malthus's lifetime.[7] In 1846, twelve years after Malthus's death, Britain also began importing large quantities of grain from bountiful North America, thereby alleviating immediate stresses on the food supply – and doing so with decreasing food prices. Malthus's *Essay* also came before a whole host of other revolutionary technologies, from the advent of fertilizers to modern crop breeding and contraception. Indeed, from a basic science perspective, Malthus was writing even before understanding of conservation of energy had been fully discovered. Malthus seemed to fundamentally underestimate the ability of humankind to rise to the population challenge through scientific and technological progress.

While Malthus has posthumously received criticism for underestimating human ingenuity, this was nothing compared with the reactions he received in his day toward the policies he subscribed to to address the population trap.[8] This was particularly the case with Malthus's position that the poor should be denied the right to long-term relief under the English Poor Laws. Various forms of Poor Laws had existed in England since Elizabethan times, requiring parishes to provide support to children, the elderly, and both able-bodied and non-able-bodied persons in need. From 1750 to 1820, there was a rapid increase in relief expenditures. By 1803, over one million people – about eleven percent of the population – were receiving poor relief, at a cost of around two percent of gross domestic product.[9] Malthus, however, was strongly against such policies, stating: "Those that came to nature's mighty feast without the means of paying for their meal had no right to sit at the table."[10]

Malthus's stance against the Poor Laws quite understandably drew heavy criticism, especially from the likes of Karl Marx and Friedrich Engels, who saw Malthus as a spokesperson for hard right conservative country gentlemen. This is perhaps unfair to Malthus, though, as his position was not arrived at lightly. He seemed to have a deep conviction that providing relief to the poor had the undesirable effect of encouraging the poor to reproduce at levels above their natural rate, thereby exacerbating the population problem for wider society. As he saw it, the Poor Laws alleviated misfortune while "spreading the evil over a much larger surface."[11] Moreover, except when it came to matters of food and population, Malthus's political philosophy was much more that of a *laissez-faire* liberal than a Tory gentleman. Malthus largely subscribed to the capitalist philosophy laid down by Adam Smith, only his work on population left him concerned that at the heart of political and economic liberalism was a mechanism that undermined the human perfectibility or advancement that it promised.

That Malthus was at least to a point of liberal sensibilities is evident from the fact that he published his work in the Whig journal *The Edinburgh Review*. This changed, however, around 1815, with Malthus's position on the establishment of the English Corn Laws. Under the Corn Laws, the price of grain in England was maintained at an artificially high value due to steep import duties imposed on more competitive foreign imports. This essentially meant that the landed gentry continued to make substantial returns from their agricultural holdings, but to some extent at a cost to the country's industrialists. The costs of manufacturing were held high, as the price paid for the key input – labour – had to be high enough that workers could afford to buy bread. The repeal of the Corn Laws in 1846 perhaps marked, from a political economy perspective, the transition from an agricultural society to an industrial one. Despite his liberal leanings toward free trade, in 1814, Malthus came out with a pamphlet strongly in support of the Corn Laws. Some saw this as an indication that he was showing his true colours – he was, after all, from a family of landed gentry. Indeed, following his stance against free trade in grain, he was no longer welcome by *The Edinburgh Review* and turned instead to publishing his work on political economy in the Tory periodical the *Quarterly Review*.[12] Again, however, Malthus's support of the Corn

Laws was also consistent with his belief that high corn prices would restrain population growth. He saw this bad medicine as necessary.

The difficulty of forming policy to address rising population growth in the context of food scarcity demonstrates what I call the third layer of the Malthus Enigma. This policy layer follows on top of the technological and scientific layers and is complicated by these other layers. Setting policies or laws for feeding the poor or trading practices that impact the price of food is by no means straightforward. Policy-makers are seemingly caught between a rock and a hard place. Similar dilemmas continue to persist with modern-day environmental challenges. Deforestation, for example, has terrible impacts on biodiversity loss, but many Indigenous people rely on wood biomass for cooking, and agricultural land is required to grow crops. Responding to such tough dilemmas is part of our continuing Malthusian struggle and requires ecology-based systems thinking. We will return to policy aspects of the Malthus Enigma in later chapters, but in this chapter, we will review some of the wrestling of ideology between Malthusians and the dismissive Cornucopians.[13] We ask: *How have Malthusians and Cornucopians disagreed?*

An early challenger to Malthus's conception of the population problem was American political economist Erasmus Peshine Smith (1814–1882). Educated at Columbia University and Harvard Law School, Smith progressed from his Rochester-based law practice to a variety of jurisdictional positions in the United States – and also served as an advisor to the Japanese government on international law. In 1853, Smith published his *Manual of Political Economy*, which, in some respects, was a precursor to the modern field of ecological economics. Smith attempted to build an American system of political economy based primarily on physical laws. The first three chapters of his *Manual* are quite remarkable for a book on economics. The first outlines laws of the endless circulation of matter and force, with particular focus given to food production. The second chapter is concerned with the formation of soils, and the third describes the role of natural energy sources, such as wood and water flows, and the expansive power of steam, in performing useful work in civilized societies.

In the first of these chapters, Smith recognizes that Malthus's theory is widely accepted, but he argues that it ignores the phenomenon of circulation of nutrients, which had been established by organic chemists. He notes: "Malthus's theory of the relations between population and subsistence is obviously founded upon the false notion, that man's consumption of food is its destruction – that having once served the purpose of supporting animal life, its capacity to contribute to that object is absolutely spent and exhausted."[14] Smith's main point – based on a somewhat primitive understanding of nutrient cycles – is that the food-producing power of the soil can be increased to meet needs through the return of nutrients to the Earth. He concludes: "Nature nowhere teaches a system which results in continuous and permanent exhaustion, though the Economists of the Malthus school have done so." Many economists would later follow Smith in critiquing Malthus, though typically not on ecological grounds.

Just a decade later, however, another American scholar, George Perkins Marsh, wrote an important treatise, *Man and Nature*, which would prove to be pivotal in adding broader environmental dimensions to Malthus's concerns about agricultural constraints. Marsh had many talents: initially a lawyer from a small town in Vermont, he helped found the Smithsonian Institution in Washington, DC; became the US ambassador to Italy under President Lincoln; and even had a hand in the architectural design of the Washington Monument.[15] *Man and Nature*, published in 1864, was foremost a book on conservation. Marsh was primarily concerned with the impacts of forest clearing, the damming of streams, the destruction of sand dunes, and the burning of wood. He devoted chapters to topics such as plant and animal species, the woods, and the waters. Environmental historian Paul Warde considered *Man and Nature* to be "the canonical text of American conservationism up until Rachel Carson's *Silent Spring*."[16] What made it stand out was Marsh's ability to produce a high-level synthesis – to see beyond individual actions and provide a total sense of humanity's aggregate impacts on the Earth.[17] He understood that, for better or for worse, humans had become a force above Nature – but were degrading the Earth upon which they depended.[18] Marsh's insights were so profound that they would fuel neo-Malthusian concerns in later centuries. Almost one hundred years after publication of *Man and Nature*, the hefty 1,200-page proceedings of a historic symposium – Man's Role in Changing the Face of the Earth,[19] held at Princeton in 1955 – were dedicated to Marsh.

Between the time of Malthus and the environmental revolution of the 1960s, Malthusian concerns still persisted. This was apparent, for example, in the views of Henry Fairfield Osborn and William Vogt, both of whom published strongly neo-Malthusian books shortly after World War II.[20] Osborn and Vogt were both trained in biology. Osborn was the director of the Bronx Zoo and the New York Zoological Society, while Vogt worked as a field naturalist and was a well-established ornithologist. In 1948, Osborn's *Our Plundered Planet* and Vogt's *Road to Survival* became national bestsellers in the United States. This prompted *Time* Magazine to proclaim that the ghost of Malthus had returned.[21] With the spectra of World War II still fresh, both authors connected the war with challenges of overpopulation and excessive exploitation of natural resources. No surprise, then, that they also expressed deep concerns that humanity was overwhelming the Earth's carrying capacity – and drew upon the works of Aldo Leopold and other ecologists on topics such as soil health and wildlife ecology. Their sentiments had some close similarities with those of Paul Ehrlich, Barry Commoner, and others that led the environmental movement two decades later, including "a distrust of progress and human technology, a sense of apocalyptic urgency, and a focus on overconsumption, sustainability and limits to growth."[22]

The Simon Versus Ehrlich Wager

When Ehrlich resurrected Malthusian concerns about the population problem in the late 1960s, he was joined by others, too, and the issues were broadened to include concerns over resource scarcity and environmental pollution. A group of concerned industrialists, academics, and diplomats formed the mysterious Club of Rome, which, among its other exploits, supported a study of the limits to growth conducted by systems analysts from the Massachusetts Institute of Technology. Their particular systems approach, which will be discussed further in Chapter 6, entailed building computer models to understand relationships between population, resource consumption, and pollution. Also during this period came the writings of scholars such as Kenneth Boulding, Howard Odum, Nicholas Georgescu-Roegen, and Herman Daly, which would later influence the field of ecological economics. In spite of almost 200 years of seemingly limitless growth, a good number of like-minded people were turning again to the challenges with which Malthus was so concerned.

One man, however, who was not in the least worried by the Malthusian concerns was Julian Simon, a professor of business administration at the University of Illinois Urbana-Champaign.[23] From an early age, Simon had a fetish for collecting facts and data. He got his bachelor's degree in experimental psychology from Harvard University, going on to obtain an MBA and a PhD in business economics from the University of Chicago. For several years, he ran his own mail-order business and even produced a successful book on the subject: *How to Start and Operate a Mail-Order Business*. After becoming a professor, Simon initially worked on solutions to problems of airline overbooking. When he watched Ehrlich pushing his doomsday message on the Johnny

Carson show in the late 1960s, he was irritated; Simon had already come to the conclusion that this Malthusian stuff was nonsense. By the 1970s, his focus had turned to challenging the hysteria, as he saw it, of the environmental revolution, and he began to collect data on the issues.

Simon thought through the population problem as Malthus had seen it. If population were to grow geometrically and food production could grow only arithmetically, then he agreed that humanity had a serious problem. He argued, however, that since food was produced from populations of plants and animals, then in theory they, too, could grow with a mathematical form similar to that of human population growth.[24] Moreover, when Simon looked up data assembled in the Food and Agriculture Organization Statistical Yearbook, he found that, from 1948 to 1976, global food production per capita had increased at a rate of about one percent per year.[25] Simon gave some further arguments in a paper he published in *Science* in 1980. He called out a United Nations official for exaggerating deaths from hunger in West Africa when there was little data to back up the claim. He summarized: "This is an example of a common phenomenon: Bad news about population growth, natural resources, and the environment that is based on flimsy evidence or no evidence at all is published widely in the face of contradictory evidence."[26] He singled out Ehrlich for proliferating such false mythology before attempting to debunk a range of environmental issues, from urban sprawl to fish catches in Lake Erie. On the question of whether the planet was running out of natural resources, he noted that "the only meaningful measure of scarcity in peacetime is the cost of the good in question."[27] Furthermore, he continued that "raw materials have been getting increasingly available and less scarce relative to the most important and most fundamental element of life, human work time."

Simon and Ehrlich were fast becoming staunch adversaries, and this eventually resulted in a showdown that was played out in the March 1981 edition of *Social Science Quarterly*. Ehrlich published a paper, a keynote from a conference in Houston in April 1980, in which he discussed challenges of ecologists and social scientists working together to address the population–environment crisis. In doing so, he picked on Simon and his "naïve treatment" of market mechanisms for eliminating long-run concerns over dwindling resources.[28] Simon rebutted in his usual style with a bunch of time-series data addressing a subset of the issues raised, along with economic dogma on the working of markets. He then proceeded to call out Ehrlich on some of his somewhat ridiculous past predictions before proposing a wager of up to $10,000 on the future prices of raw materials that were free of government control.[29] Goading Ehrlich, he wrote, "How about it, doomsayers and catastrophists? First come, first served."[30] Ehrlich responded with a strong defence on environmental issues including air pollution, water quality, and land use – and then an acceptance to take up the wager, along with colleagues John Holdren (University of California, Berkeley) and John Harte (Lawrence Livermore National Laboratory). "I and my colleagues ... jointly accept Simon's astonishing offer before other greedy people jump in."[31]

Under the terms of the wager, Ehrlich, Holdren, and Harte selected five metals – chromium, copper, nickel, tin, and tungsten – which they believed would undergo substantial price rises. They each bet a total of $1,000, with $200 on each metal. The metal prices on September 29, 1980, were set as an index, with inflation-adjusted prices on September 29, 1990, used to determine the outcome of the bet. If metal prices rose over the decade, Simon would pay Ehrlich and colleagues the difference in the index price; if prices fell, Ehrlich et al. would pay the difference to Simon. Ten years later, Ehrlich's poor track record of making predictions had been extended. The world had not entered an age of scarcity (or at least not the developed world), and the price of metals had fallen. In October 1990, Ehrlich mailed Simon a cheque for $576.07. Much to Simon's chagrin, however, in that very same year, Ehrlich was awarded a MacArthur Foundation "genius" award for promoting public understanding of environmental problems.[32]

Global 2000: The Report to the President

The Simon–Ehrlich wager was just one battle in a wider ideological war that played out among respective proponents from about 1980 onward. The struggle was between the existing system of large, high-spending central governments and rising *laissez-faire* free-market liberalism. Big government with an emphasis on central planning, supported by high taxes, had been in place since the 1930s and the beginning of World War II. The return to a smaller, hands-off style of government, as per earlier decades of the twentieth century, was championed by the Reagan and Thatcher governments in the US and UK, respectively, and was soon followed by other Western countries. In economic terms, it was about replacing the Keynesian planned economy with a more free-market approach, as espoused by Friedrich Hayek and Milton Friedman. The transformation also meant a fundamentally different approach to tackling environment and resource challenges.

In 1977, President Jimmy Carter considered environmental worries to be of sufficient concern to warrant a massive study. He directed the

Malthus Enigma

State Department and the Council on Environmental Quality, along with a host of other agencies, "to make a one-year study of the probable changes in the world's population, natural resources, and environment through the end of the century."[33] The study was to be the foundation for long-term planning. Employing the latest techniques in computer modelling and supported by the full weight of the US government, the production of *The Global 2000 Report to the President* was a serious undertaking. The study cost around US$1 million,[34] and contributions were made from at least a dozen government departments. The three-volume report was published in 1980, selling over 1.5 million copies before a second edition was released in 1988. It included projections to the year 2000 on global population growth and a full suite of environmental issues.

The overall findings of *Global 2000* were described by the opening paragraphs of the executive summary:

If present trends continue, the world in 2000 will be more crowded, more polluted, less stable ecologically, and more vulnerable to disruption than the world we live in now. Serious stresses involving population, resources, and environment are clearly visible ahead. Despite greater material output, the world's people will be poorer in many ways than they are today.

For hundreds of millions of the desperately poor, the outlook for food and other necessities of life will be no better. For many it will be worse. Barring revolutionary advances in technology, life for most people on earth will be more precarious in 2000 than it is now – unless the nations of the world act decisively to alter current trends.[35]

The alarming conclusions of the study were echoed by *Newsweek* Magazine:

It reads like something out of "The Empire Strikes Back." The time: the year 2000. The place: Earth, a desolate planet slowly dying of its own accumulating follies. Half the forests are gone; sand dunes spread where fertile farm lands once lay. Nearly 2 million species of plants, birds, insects and animals have vanished. Yet man is propagating so fast that his cities have grown as large as his nations of a century before.[36]

Up against *Global 2000*, however, was a rival study. Funded with a modest $30,000[37] from the right-wing Cato Institute, with some chapters freely contributed by academics, *The Resourceful Earth: A Response to*

Global 2000 was spearheaded by Julian Simon and Herman Kahn. The senior of the partners, Kahn, was a military strategist and systems thinker who wrote extensively on nuclear warfare and future socio-technical systems, but he died in 1983, leaving Simon to finish the study. In a remarkable piece of writing, akin to expert defence lawyers, Simon and Kahn went through each of the specific conclusions of *Global 2000* – and turned every one of them on its head. Some of their arguments were clever, if not entirely convincing: the Earth was becoming "less crowded" because people could spread out more due to increases in income.[38] In other cases, they slammed the work of the Carter Administration for shoddy data – on biodiversity loss and urbanization of farmland. Simon and Kahn drew upon chapters written by academic experts and brought their own data to bear. Still, many of their findings would be hard for modern-day Malthusians to stomach, such as "The fish catch, after a pause, has resumed its long upward trend" and "The climate does not show signs of unusual and threatening changes."[39] They clearly got these wrong!

On matters where *The Resourceful Earth* and *Global 2000* did make predictions, however, the free marketers were generally more successful. A post-2000 evaluation conducted by researchers Jonathan Chenoweth and Eran Feitelson was published in 2005. The actual world population in 2000 ended up between the high and low values predicted in *Global 2000*, although the study surprisingly underestimated US population growth. *Global 2000* predicted global food production quite well, with only a three percent error, but it was far off in grossly overestimating costs of food. *Global 2000* also greatly over-predicted energy use and energy costs for 2000 – and deforestation rates. Fewer predictions were made in *The Resourceful Earth*, and these were mainly just extrapolations of previous trends. While generally more accurate, *The Resourceful Earth* was less successful than *Global 2000* on fisheries predictions, and it did not really take on issues of atmospheric CO_2 concentrations or the depth of the ozone layer. That said, while *Global 2000* recognized that increasing CO_2 concentration in the atmosphere could be problematic, it considered that the consequence could be either global warming or global cooling, with equal likelihood.

Of course, the confrontation between *Global 2000* and *The Resourceful Earth* was really more about political ideology than Earth and environmental science. There were some accusations that the results from the main chapters of *Global 2000* were being distorted in the executive summary.[40] The government authors of *Global 2000* were also arguably somewhat naive about the problems of using computer models to make future predictions. Unlike the authors of the *Limits to Growth* work, who made it very clear that they could not make predictions, the *Global 2000* team was essentially required to do so. Simon and Kahn, meanwhile, added free-market ideology to their recommendations, indicating that the production and pricing of natural resources should be left in the hands of the private sector and, moreover, that governments should get out of the business of scientific assessments of the *Global 2000* variety altogether:

> *We believe that the government should not take steps to make the public more aware of issues concerning resources, environment and population. We consider that the public has been badly served by having been scared by a very large volume of unfounded and/or exaggerated warnings about these matters.*[41]

Underlying their approach was a belief that due to human knowledge, ingenuity, and ability, the resources of the planet were essentially unlimited. They even went so far as to scandalously conclude:

> *Because of increases in knowledge, the earth's carrying capacity has been increasing throughout the decades and centuries and millennia to such an extent that the term carrying capacity has by now no useful meaning.*[42]

Julian Simon ended up on the winning side of this political battle, too, indeed even before his wager with Ehrlich had made it to print. When Ronald Reagan defeated Jimmy Carter and became president in 1981, he thoroughly rejected the *Global 2000* report.

The animosity and divisiveness of the *Global 2000* and *The Resourceful Earth* perspectives – on top of the Simon–Ehrlich wager – strongly demonstrated the need for better science to address environmental challenges. Understanding the carrying capacity and resource constraints of the Earth – indeed understanding the Earth as an environmental system – is an important aspect of contemporary Malthusian struggle. This is especially the case because the policy process substantially relies, or at least *should* rely, upon good science. The weakness of the science in the 1970s and '80s, largely due to a lack of data, was an underlying motive for new scientific disciplines such as ecological economics and industrial ecology to emerge.

Planetary Boundaries

Almost thirty years after Simon and Khan's challenge to *Global 2000*, a more enlightened scientific appraisal of the Earth's carrying capacity was published in the journal *Nature*.[43] A group of environmental scientists led by Johan Rockström of the Stockholm Resilience Centre conducted a study raising heightened concerns about the impacts of humanity on the functioning of the planet. The team, including many well-known scientists, concluded: "Anthropogenic pressures on the Earth System have reached a scale where abrupt global environmental change can no longer be excluded."[44] In particular, they presented nine non-negotiable planetary boundaries that humanity needed to respect to avoid potentially catastrophic environmental change at continental or planetary scales. They called this the *safe operating space for humanity*. Three of the boundaries pertain to systemic processes on the planetary scale: climate change, ozone depletion, and ocean acidification. Actually, the third boundary here also relates to climate change: higher CO_2 concentrations in the atmosphere causes CO_2 dissolution into sea water, thereby increasing the acidity of the ocean surface. The other six boundaries relate to processes that occur on local or regional scales but provide resilience to the Earth as a system, and in aggregate manifest themselves as global measures of concern. The six processes are changes to global nitrogen and phosphorus cycles, atmospheric aerosol loadings, fresh water use, land-use change, biodiversity loss, and chemical pollution. For seven of the nine, Rockström and colleagues were able to quantify, albeit with some notable uncertainty, long-term boundary values within which humanity is "safe." In three cases – climate change, biodiversity loss, and interference in the nitrogen cycle – the safe long-term thresholds were being exceeded. Forty years after humans had first left the boundaries of the Earth on Apollo 8, we were realizing that we had crossed some different types of global boundaries.

The planetary boundary of perhaps greatest concern – and greatest scientific understanding – is climate change. Safe limits to climate change have been vigorously studied and debated through the many conferences of the United Nations Framework Convention on Climate Change. Through a combination of scientific projections, value judgments on the impacts of climate change, and political realities of already committed emissions, a maximum tolerable rise in global mean temperature of 2°C above pre-industrial levels has been arrived at. Many scientists today, however, are pushing for global society to limit the rise to 1.5°C. By 2012, the average global temperature rise was approaching 1°C above

Malthus Enigma

pre-industrial levels.[45] The fear is that a mean rise of greater than 2°C would possibly cause abrupt and irreversible changes to the Earth system, including the main circulation patterns and sea level rise. Drawing upon the best available science, Rockström and colleagues adopted an atmospheric CO_2 concentration of 350 parts per million[46] and radiative forcing of one Watt per square metre above pre-industrial levels as boundary values to keep temperature rise within 2°C. Radiative forcing is essentially the difference between the energy from incoming sunlight that is absorbed by the Earth and the amount radiated back to space. Higher concentrations of CO_2 and other greenhouse gases block the radiation out from the Earth, causing it to heat up. In 2014, the Intergovernmental Panel on Climate Change reported that the total human contribution to radiative forcing from 1750 to 2011 was already around 2.4 Watts per square metre.[47] By May 2013, moreover, daily averaged atmospheric CO_2 concentrations at the famous Mauna Loa Observatory in Hawaii exceeded 400 parts per million for the first time.[48] Thus, climate change scientists argue that CO_2 and other atmospheric greenhouse gas emissions have to be reduced.

The work on planetary boundaries provoked a huge discussion among scientists and policy-makers, in part due to the uncertainty in some of the boundaries. Many scientists commented or offered critique on the planetary boundaries, which encouraged some of the original authors to provide an updated version in 2015.[49]

One of the most interesting responses to the original work was a proposal for a single planetary boundary by Steven Running, a professor of ecology at the University of Montana. Running argued that net primary production by plants on the Earth's land surfaces was a simple, useful measure, as it captures many aspects of the Rockström boundaries and is well supported by existing data sets.[50] Primary production is the transformation of sunlight into chemical energy by biota – living organisms – on the Earth's surface, primarily through photosynthesis.[51] Solar radiation, water, and CO_2 from the atmosphere combine to make the plant carbohydrate that sustains global food webs. There are two related ways of expressing primary production. Gross primary production is the total uptake of carbon by all biota. It is usually expressed in units of petagrams of carbon per year. Some of the chemical energy stored in the carbohydrates is simultaneously used up in plant respiration – that is, in meeting the energy needs of plants. If the carbon lost through respiration is subtracted from gross primary production, the resulting quantity is called net primary production.

Running noted that, over the past thirty years or more, net primary production on the Earth has been relatively constant, with approximately fifty-four petagrams of carbon produced each year. This value could possibly change by a small amount due to human activities, but broadly speaking, life on Earth is constrained by a production of fifty-four petagrams of carbon per year. Research has found that about thirty to forty percent of net primary production is already co-opted by humans – providing food, fibre, and fuel.[52] This might suggest that humans are currently living well within the net primary production boundary of the planet. The challenge, however, is that approximately fifty-three percent of net primary production is considered to be unharvestable – that is, it is in plant root systems, in protected national parks, or not accessible to transportation. Moreover, we may not want to exploit it anyway, because of the biodiversity that it holds. Potentially, this means that humans can increase their exploited net primary production by only another ten percent – that is, by five petagrams of carbon. This really is not a large breathing space.

We will return to the topics of climate change, planetary boundaries, and primary production in later chapters. First, however, we can use the conflicting perspectives of Malthusians and Cornucopians to start developing the layers of the Malthus Enigma.

How Have Malthusians and Cornucopians Disagreed?

From the disagreements between Simon and Ehrlich, the conflicting conclusions of *Global 2000* and *The Resourceful Earth*, and other contrasting views discussed so far, we have a few snapshots of how Malthusians and Cornucopians have disagreed. Further examples, such as the heated reaction to the 1970s *Limits to Growth* study, will be added later, so let us turn to identifying and building the three layers of the Malthus Enigma that follow from such contention.

The first challenge, which adds a further twist to the Enigma, is that Cornucopians do not even recognize that there is a problem of sustaining humanity within the carrying capacity of the planet. The attitude of free marketers such as Simon and Khan was that society will just innovate its way out of any perceived resource or environmental constraints. Just leave it to the free market. Contrasting this is the neo-Malthusians' mistrust of technological innovation – expressed by many of the environmentalists of the 1960s and 1970s, for example. The first layer of the Malthus Enigma involves resolving these contradictory positions. How are we going to do that?

Malthus Enigma

Having been trained in engineering – as well as social sciences – I have a book on my shelf by Henry Petroski called *To Engineer Is Human: The Role of Failure in Successful Design*.[53] The title of this book captures the essence of the first layer of the Enigma. Humans are innovative. There are several million engineers on this planet, not to mention other innovative professions such as scientists, architects, artists, designers, and others in business, for example. So Cornucopians are onto something in recognizing the role for human innovation. The subtitle of Petroski's book, however, would be of concern to the neo-Malthusians. If I recall correctly – it was over twenty years ago that I read it – Petroski's book was focused mainly on the engineering of bridges and other structures, but the history of engineering has involved a lot of learning from failure in other areas too. Given the worrisome state of the world's climate and ecosystems – not to mention polluted cities, overstressed aquifers, etc. – Malthusians could legitimately argue there has been a lot of failed innovation dating back over 200 years to the Industrial Revolution. The role of innovation is central, then, to the Malthus Enigma – and we will wrestle with it further in the next chapter, when we ask: *How do we sustain humanity without destroying global biodiversity?*

There is a second dimension to the disagreements, which is more concerned with science. Some of this is healthy questioning of the quality of scientific data, which is a legitimate part of the scientific process. Beyond questions about data, however, are radically different scientific conclusions with regard to the carrying capacity of the planet. The Cornucopian perspective that carrying capacity is increasing to a point where the term has no useful meaning is very much at odds with the position of Rockström and colleagues with respect to planetary boundaries, some of which we have already overstepped. Recognition of the high level of uncertainty in many of the planetary boundaries, however – and propositions for alternative planetary boundaries, such as net primary production – is an indication that scientific progress plays a key part in the Malthus Enigma. We will pick that thread up again in subsequent chapters.

When there is disagreement on science, it is not surprising that policy perspectives might differ. There is something deeper, however, in the disagreements between Malthusians and Cornucopians, which is the way that conceptual models are formed and the nature of the economy is framed. Simon disagreed with Malthus's assumption that plant and animal food supplies could grow only arithmetically. What is perhaps more interesting here, though, is the critique of Malthus's theory by Erasmus Smith. Both he and Malthus were political economists. In criticizing Malthus for ignoring the potential for nutrients to be recycled, Smith was essentially introducing some ecology into his economic frame. This is somewhat ironic, because subsequently, it was far more common that neo-Malthusians – such as Osborn and Vogt – would invoke ecology in the face of mainstream economists. This raises a question that is central to the third level of the Enigma – policy-making – which is just how do we incorporate ecology into economics? We will build toward answering this in a later chapter, incorporating some essential groundwork on thermodynamics and systems thinking on the way.

Systems Thinking

The scientists working on planetary boundaries; those who worked on the *Global 2000* study and *The Resourceful Earth*; and others such as Malthus, Simon, and Ehrlich all have something in common. In making their assessments on the state of the Earth – or human conditions on it – they all used a systems perspective. By this I mean they developed some form of conceptual or mathematical model that explains relationships between components of the Earth. Indeed, anyone aiming to assess or make plans to address global environmental challenges has to develop some form of systematic model of how the Earth functions. The many layers and components of the biosphere, including humans, flora, and fauna, as well as non-living (abiotic) components, intersect in complex ways. Any conception of how these components interact is inherently taking a systems perspective, whether it is expressed in a mathematical form, a conceptual scientific model, or even a policy framework.

Systems models can differ significantly in their degree of complexity. Malthus's model of scarcity was quite straightforward: human population grew geometrically while the food source grew arithmetically; hence, the relationship between them was stressed. Beyond his work on *The Population Bomb* and his bet with Simon, Ehrlich is well known for developing a simple systems model known as the IPAT equation, which relates human impacts to population, affluence, and technology.[54] The Earth systems models used in the *Global 2000* study were a bit more complex. The study relied on a set of common projections about growth in population and the economy that were fed into various models developed by different US federal government agencies.[55] The planetary boundaries work of Rockström and colleagues did not involve such complex mathematical models; it was more about development of a conceptual scientific model – based on a weaving together of influences from Earth sciences, ecological economics, and an area in ecology

Malthus Enigma

focused on the resilience of living systems.[56] In tentatively defining the safe operating space for humanity, Rockström and colleagues drew upon results from a whole slew of other mathematical models, including long-established global climate models. So systems models can take many forms.

There has been wide-ranging variation in the degree of acceptance of conclusions derived from Earth systems models. Malthus's simple model has been attacked – and also defended – for over 200 years. The competition between *Global 2000* and *The Resourceful Earth* showed that framing of the Earth system can be strongly influenced by political ideology. One of the most controversial Earth systems modelling studies was the *Limits to Growth* work of the early 1970s, which will be reviewed more fully later. Other Earth systems models, however, are now considered scientifically robust. Though they certainly have been challenged in the past, the models used for assessing global climate change have moved into the realm of holistic science. Having a rigorous physical basis – based on thermodynamics – global climate models have grown to be well-corroborated scientific hypotheses about the workings of the Earth. Part of the reason for ending this book with the fields of ecological economics and industrial ecology is that they, too, draw heavily upon the laws of physics. This will lead me to propose that more rigorous systems models of human–environment interactions could potentially develop from these fields.

As we delve further into the Malthus Enigma and the Earth system, the next chapter will focus on the planetary boundary that, according to Rockström and colleagues, is being most overstretched – the loss of biodiversity. Decades of work in ecology have established the importance of species diversity in maintaining healthy ecosystems. Through a range of mechanisms, biodiversity of species makes ecosystems more resilient to disturbance and therefore less prone to collapse. This is important because the very existence of ecosystems provides underlying resilience to other planetary systems. Moreover, humans inherently rely on the Earth's ecosystems to provide for our basic needs, such as food, timber, and medicines, which are called *ecosystem services*. Maintaining the biodiversity of the Earth's ecosystems while continuing to feed a growing global population is a massive challenge for humanity. This combination of classic Malthusian concern with modern-day global environmental challenges is what we turn to next.

Chapter 3: Feeding the World While Saving the Planet

Hunger, he warned, was humankind's oldest companion; every society lived within its shadow.

Elvin Stakman, inspiring Norman Borlaug [1]

The encounter with the elephants occurred about a month after David Livingstone first set eyes on Victoria Falls, known to locals as *Mosi-oa-Tunya* ("The smoke that thunders"). The Protestant missionary and nineteenth-century explorer – an actor from the era of colonial exploitation – was on the first of his expeditions to uncover the interior of south and central Africa. Livingstone had bravely paddled to a lush green island at the top of the falls, straddled between two great curtains of water. There, he had planted some peach and apricot stones, and then, in an unusual show of vanity, had carved his initials and the year, 1855, into a tree.[2] Naming the falls for his queen, he then headed north from the Zambezi River and to the east, accompanied by his band of 114 locals, with oxen in tow. Over the next few weeks, the party generally followed a ridge to the northeast of the falls; they crossed a few smaller rivers, such as the Kalomo and Mozuma, and skirted to the north of some white mountains.[3] They negotiated their way through an area occupied by many villages where the locals had never set eyes on a white man before; then, on December 14, they entered a beautiful valley "abundant in large game."[4]

The young elephant and his mother were flailing around in a mud pit at the end of the valley – flapping their ears, tossing their trunks, and wagging their tails seemingly in enjoyment.[5] Then the sound of the pipes began, made by young men blowing through tubes, accompanied by others simply blowing through their hands. The elephants' ears expanded, and they quickly moved out of their bath. It was too late, however, as the crowd of men was already upon them. The young male animal – barely two years old – made a run for the end of the valley, but doubled back to his mother as further men appeared across his path. The mother tried to place herself on the danger side of her calf, as they moved to cross a small stream. The first volley of spears went out as the men closed to twenty yards. Ten or twenty or more spears punched through the female's leathery hide, causing considerable bleeding. To Livingstone's dismay, as he looked on from afar, several spears hit the baby elephant as it cowered in the water, and it soon passed away. The mother shrieked in rage, and she turned to charge at the men. Three or four times she rushed at them, but she was wounded, and her charges were slow and short. Except for one man wearing a bright cloth on his shoulder, the men easily avoided the distraught animal. They cheered in celebration as more spears pierced the magnificent animal, and she finally staggered to a halt, sank to a kneeling position, and died.

Livingstone expressed some regret at the killings, especially of the younger elephant. He had asked that it be spared but had been too far away to influence the proceedings. As an onlooker, he was nauseated by the experience, but he had not felt any sickness the previous day when he himself had shot a large adult elephant. Indeed, he had been "right glad to see the joy manifested at such an abundant supply of meat."[6] The reality was that to feed the men upon whom he relied for his explorations, he needed at times to hunt. Moreover, the prize of the elephant tusks would help finance his missionary enterprise. This was God's work.

Livingstone's mildest expression of remorse at the killing of the baby elephant generally cut against an imperialistic nineteenth-century culture that exalted the virtues of the game hunter. Many men gained notoriety for their hunting exploits. One example was Captain William Cornwallis Harris (1807–1848),[7] who, having arrived in South Africa with the British East India Company, set out to further recorded knowledge of geography and natural history while fulfilling his passion for the chase. He and his colleagues shot game in large numbers: elephants, hartebeest, hippos, impalas, quaggas, rhinos, waterbucks, and wildebeest, and Harris was particularly excited by the killing of giraffes. Accounts of his activities were recorded with venerable institutions such as the Bombay Geographical Society, the Geographical Society of London, and the Zoological Society of London. Like many nineteenth-century gentlemen, Harris was an amateur artist; he published several lithographs from his adventures, as well as a book, *The Wild Sports of Southern Africa*.

Possibly the most famous of the hunters of southern Africa was Roualeyn Gordon-Cumming (1820–1866),[8] who undertook five expeditions between 1843 and 1848, shooting over one hundred elephants and a vast array of other animals, often with primitive guns. He once boasted to have killed thirty hippos within a few days. Many of Gordon-Cumming's trophies were exhibited at the Crystal Palace in London during the Great Exhibition of 1851, and he toured Britain lecturing under the nickname "The Lion Hunter," promoting his book, *The Lion Hunter of South Africa*. In an era before middle-class sensibilities became unsympathetic to hunting, the exploits of men like Gordon-Cumming and Harris were held in high regard. This was in part because of the perceived association with the acquisition of scientific and geographic knowledge. In retrospect, historian John MacKenzie noted that "hunting represented the most perfect expression of global dominance in the late nineteenth century."[9] It required "all the most virile attributes of the imperial male – courage, endurance, individualism, sportsmanship, resourcefulness, a mastery of environmental signs, and a knowledge of natural history."

The result, of course, was the awful, unbridled butchery of thousands of animals, including the extinction of several species. Focus on the virtues of the nineteenth-century imperial male misses the economic dimension. Victorian hunters made great profits on the sale of ivory from their slaughtering of elephants. American and European demand for ivory was driven by growth in the manufacturing of cutlery (with ivory handles), pianos, billiard balls, and various ornaments. Prices fluctuated, as they do in any resource market, but growing demand for a scarce resource made the sale of ivory profitable. In peak years around 1880, an estimated 12,000 elephants per year were being killed.[10] As the century progressed, hunters had to go deeper into Africa in search of elephants. First southern Africa was cleared, then central Africa. Only in east Africa did elephants survive in greater numbers, because of the later arrival of white hunters and the migration of elephant herds to the remotest areas. The rapid retreat of elephants upward through Africa was soon followed by the extermination of other game – such as rhinos, hippos, and antelopes. In MacKenzie's view, "[t]he game was simply worked out, like a mineral seam."[11]

With its game population being the first to be wiped out, it was in southern Africa that the first calls for conservation arose. The Cape region of South Africa has a particularly special ecology; it is by far the smallest of the six unique floral kingdoms that cover the Earth's land area. In other words, the majority of the plant species in the Cape region cannot be found anywhere else on the planet.[12] The Cape has such an extraordinarily high diversity of plant species – in particular its shrubland known as *fynbos* – that it became a UNESCO World Heritage Site in 2004. The region also used to have an abundance of large animals, but it was one of the earliest areas to be settled by whites. In a period of just

Malthus Enigma

150 years, the Boers and British essentially cleared the entire colony of game – and this was using relatively primitive muskets, before the advent of more accurate breech-loading rifles. The Dutch East India Company had introduced game legislation for the Cape as early as 1657, and the British also created legislation in 1822.[13] These had both been quite ineffective, though; elephants, for example, were soon made exempt from British hunting restrictions that applied to royal game. It was not until 1886 that the more robust *Act for the Better Preservation of Game* was introduced, but by then, there was little game left in the Cape Colony.

Concerns over the destruction of African wildlife led eventually to an international convention on the matter, held in London in 1900. The initial instigator of the meeting was a German, Hermann von Wissmann, but the British also strongly promoted the issue. The Convention for the Preservation of Wild Animals, Birds and Fish in Africa was hosted by Sir Clement Hill at the Foreign Office on May 19, 1900.[14] All of the European powers with African colonial interests attended,[15] including Britain, Germany, Spain, France, Italy, and Portugal, as well as the Congo Free State.[16] The strongest promoters were the Germans, led by von Wissmann and Baron von Lindenfels, and the British, represented by Hill, the Earl of Hopetoun, and Ray Lankester, director of the Natural History Museum in London. Emerging from the convention was an agreement that the colonial nations should establish game regulations. The convention included some details: the killing of young female animals was to be illegal; there should be closed seasons and licences for hunting; and the use of nets, pitfalls, and poison was to be prohibited.[17] There were also regulations on types of guns and ammunition, minimum tusk sizes, animal classification schemes, and dimensions for game reserves. Unfortunately, not all of the nations attending went on to ratify the agreement, and enforcement was, frankly, difficult for those that did. Nevertheless, the convention was significant, and precedent setting, in that it was the first international agreement on an environmental issue.

For many modern readers, reflections back on the era of colonization by Europeans will cause some distaste. Many countries are thankfully now independent of their former European colonial-era rulers. Challenges remain, nonetheless, for the many Indigenous Peoples on colonized lands, which prompted the United Nations to adopt the Declaration on the Rights of Indigenous Peoples in 2007.[18] An academic literature base connecting environmental damage with colonization has also developed.[19] Moreover, an intriguing sign of the changing tides of time is that Britain's environmentalist King Charles III now frequently engages with, listens to, and champions Indigenous perspectives on

Feeding the World

global environmental issues. In one of his many statements on the matter, he asserted: "It is high time we paid more attention to the wisdom of Indigenous communities and First Nations people all around the world."[20] In another statement, he reiterated: "We simply must learn practical lessons from traditional knowledge, through deep connections to land and water, about how we should treat our planet."[21] Charles has a deep sense of the profound sadness of global biodiversity loss.

Beyond their illegitimate authority, another troubling aspect of the 1900 convention in London was the motives of the attendees. With so much African game destroyed, it was politically easy to engage British and German aristocracy on the topic of conservation, because so many of them were hunters. Sir Clement Hill, for example, had spent much of his time hunting when on a visit to east Africa with the Foreign Office. The conservation efforts that began with the 1900 convention essentially began to restrict the hunting of African game except to the wealthy elites. Following on from the convention, on December 11, 1903, was the founding of an institution: the Society for the Preservation of the Wild Fauna of the Empire (now known as the Fauna and Flora Preservation Society).[22] Although strongly supported by the Zoological Society of London, the members of the society soon became known as the *penitent butchers*, as so many of them were former – and, in some cases, continuing – game hunters.

One notable example was President Theodore Roosevelt. He was one of thirty honorary members that joined the society in its first year, along with seventy regular members. Roosevelt was well known as an outdoorsman.[23] With a Wyoming heritage, he personified the frontier spirit of the rancher, soldier, cowboy, and big-game hunter. As president, he championed the environmental agenda, declaring that "the conservation of our natural resources and their proper use constitute the fundamental problem which underlies almost every other problem of our national life."[24] Yet just after relinquishing the American presidency, travelling in the service of the Smithsonian Institution and the American Museum of Natural History in New York, Roosevelt went on a hunting expedition to British East Africa. The 512 head of large animals his party shot was far more than were required as museum specimens. Moreover, Frederick Jackson, the deputy commissioner of the East Africa Protectorate, observed that the "ex-President was overweight and suffering from poor eyesight."[25] He was such a bad shot that he caused many wounded animals to escape.

There were many other complexities in stemming the destruction of African wildlife. Foremost of these was the need for humans to feed

45

Malthus Enigma

themselves. In the 1920s and '30s, for example, there were reported cases of excessively large herds of elephants in Zambia causing damage to crops.[26] A more difficult issue arose in Southern Rhodesia in 1922, when it was hit by a famine. The removal of hunting rights for locals – and the decline in wildlife – meant that game meat could no longer be used as a hedge against starvation.[27]

Horrific as the butchery of wildlife has been, there is a much bigger, macro-scale problem of species loss that goes well beyond animal hunting. As will be discussed further in this chapter, massive increases in human population accompanied by changes in diets and lifestyles have had much larger negative impacts on other species than hunting. With population growth and technological change has come widespread clearance of natural habitats for agriculture and forestry; roads and railways have been cut through wilderness, weakening communities of animals and historic migration patterns through fragmentation; and human invention has produced toxins that accumulate in food chains and plastics that more directly impact birds, fish, and other animals. And on top of all these, changes in climate due to human activities have disturbed the habitats of numerous species around the globe. The impacts of an exploding human population on the Earth system have been huge.[28] The twentieth century witnessed humans destroying plant and animal species at a level that far surpassed the impacts of the nineteenth-century hunters.

The Convention for the Preservation of Wild Animals, Birds and Fish in Africa never really entered into force, and later, in 1933, it was superseded by the Convention Relative to the Preservation of Fauna and Flora in their Natural State. After its questionable start, however, the Society for the Preservation of the Wild Fauna of the Empire did at least help with the establishment of an international organization that keeps watch over the biodiversity of the planet. In 1948, the society was one of the founding members of the International Union for Conservation of Nature (IUCN).[29] Formed in Fontainebleau, France, the IUCN claims to be the world's oldest global environmental organization. Among its most notable activities is the publication of its *Red List*. Since its inception in 1964, the IUCN's *Red List* has become recognized as the world's most comprehensive source for information on the status of the world's biodiversity. Tracking animals, fungi, and plants, the book provides national governments, international agencies, and others information on population, diversity, threats, and the state of species' habitats around the world. Unfortunately, most of what the *Red List* has to report is bad news. One of the consequences of humanity's Malthusian struggle is an increasing threat to the existence of other species on the planet.[30]

Feeding the World

In this chapter, we pick up the topic of innovation again, which is central to the first layer of the Malthus Enigma. As we discussed in the last chapter, the role of technological innovation in sustaining a growing population is a topic that neo-Malthusians and Cornucopians have fundamentally disagreed on – and Malthus himself has been criticized for failing to recognize technological progress in his thesis on the population problem. We will approach the topic of innovation through examining the same problem that Malthus wrestled with – that of feeding people within resource constraints. Only to address the issue in a more current context and on a planetary scale, we ask: *How do we sustain humanity without destroying global biodiversity?*

This chapter seeks to understand how it is that a growing global population has managed to be fed – and whether it can continue to do so without massive loss of other species on the planet. According to data from the United Nations, the global population was about 8.2 billion in 2024 and will likely reach ten billion later this century.[31] Such population numbers make a mockery of Malthus's concerns over the population problem. He had not foreseen the potential for technology to increase food production. In particular, he had not counted on the development of chemical fertilizers due to the work of twisted German scientist Fritz Haber, nor the incredible persistence of American agronomist Norman Borlaug in growing higher-yielding, disease-resistant crops. Although other scientists participated too, the lives and works of these two are so extraordinary that we will explore them in more depth.

Sustaining a massive human population on a finite planet has had an enormous toll on natural ecosystems. This was apparent from the Millennium Ecosystem Assessment study, released in 2005, and some scientists are now predicting a possible mass extinction of flora and fauna within a few hundred years. To understand how we might continue to feed the world's population without severe loss of biodiversity, we will turn to the work of modern-day ecologist David Tilman, who argues for the intensification of agriculture in developing countries. Through exploring this challenge, we will eventually end up where this chapter began – in the continent of Africa. With far higher fertility rates than any other continent, it is in Africa where Malthusian struggle will be greatest this century – and the battle to preserve biodiversity the hardest.

As we explore the connections between technology, population growth, and biodiversity impacts in this chapter, several ways of framing the Earth system will be drawn upon. The Millennium Ecosystem Assessment study has a particularly critical conceptual scientific model linking biodiversity, ecosystem services, and human well-being to drivers

Malthus Enigma

of change.[32] To this, we will add economic models that connect biodiversity loss to global trade – and Tilman's Earth systems model estimating the potential future impacts of a growing population on land use. But before we get to these, we start with a simple systems understanding of how the world population is changing.

Global Population Growth

In 1800, when Malthus was rewriting his *Essay*, global population was just shy of one billion people. By 1900, when the world's first environmental convention on preserving endangered species was hosted in London, the population had risen to 1.65 billion. This was a substantial increase compared with previous centuries but was nothing compared with what happened in the century following. Over the twentieth century, global population exploded, reaching six billion by 2000. A further billion people were added just in the first eleven years of the twenty-first century. Intuitively, it would perhaps seem that global population is growing in some exponential fashion, but this is not the case – the growth rate is actually declining. In 2019, medium projections from the United Nations suggested that population would reach 9.7 billion by 2050 and sluggishly move to 10.9 billion by 2100.[33] Global population is continuing to rise, but the rate of increase is actually slowing down. To understand the dynamics of this population growth in the Earth system, we need to unpack it.

Underlying population growth are two key processes: the birth rate and the death rate. The former is often expressed as a birth rate per thousand people or, alternatively, homing in on those who actually give birth, as a fertility rate – births per female. The birth rate is highly dependent on economic circumstances, family planning practices, and access to education. Rural families are more likely to have more children, who add to the agricultural labour force. Women with higher levels of education or living in an urban environment are likely to have relatively fewer children. The death rate, or mortality rate, can also be expressed as deaths per thousand people. The mortality rate is dependent on access to food, clean water, sanitation, and medical services. It is also subject to the prevalence of disease and war – just as Malthus recognized – but it is a bit subtler than that. Ultimately, everyone dies; it is just a matter of when. Through advances in medical science, the engineering of clean water and sanitation services, advances in agricultural production, and the ability to transport and store food, human life expectancy (i.e., average human lifespan) has increased in many countries. An era in which human

Feeding the World

life expectancy is increasing will experience a decline in death rates, but if human life expectancy then levels off, at eighty to ninety years, for example, then the death rate will eventually catch up again. The massive explosion in population in the twentieth century was all about changes in the death rate – although there were differences between so-called developed and developing nations.

The expected slowing down of population growth in the twenty-first century is due primarily to a decline in the birth rate. Intriguingly, the genesis of this decline was the same transformative decade – the 1960s – in which humans first saw Earth from the Moon and stirred our environmental consciousness. The swinging sixties were, of course, a decade of social transformation – including liberation of women and invention and uptake of birth control pills. Invention of "the pill" by Gregory Pincus, John Rock, and Carl Djerassi was another human innovation that Malthus was unable to foresee. Up until the end of the 1960s, the global average fertility rate had been fairly constant at just under five children per woman. Then, starting in the last few years of the 1960s, with family planning and contraception, fertility rates began to decline rapidly. The rate dropped to four children per woman by the late 1970s, under three per woman by the mid-'90s, and down to an estimated 2.52 per woman as of 2010.[34] Again, there is significant variation between countries. Fertility rates in the forty-one least developed countries – thirty-two of which are in Africa – peaked at 6.71 children per woman around 1970 and have since fallen to about 4.41.[35] In the most developed countries, fertility declined to around 1.6 children per woman in 2000. It has since risen slightly but remains below the level of 2.1 children per woman that is considered necessary to sustain a population over the long run.[36]

The importance of fertility rates in the United Nations' population projections cannot be understated. The medium projection assumes that fertility rates continue to decline, levelling off at about 1.9 children per woman by 2100.[37] This projection sees an increase in global population of two billion from 2019 to 2050 – which is almost like adding the combined populations of China and India today – and then a further 1.2 billion by 2100. Interestingly, half of the world's population increase by 2100 is expected to be shared by just six countries – India, Nigeria, the US (largely due to immigration), the Democratic Republic of the Congo, Tanzania, and Uganda.[38] The dominance of African countries here is quite noticeable. As of 2009, access to family planning in the least developed countries was still very low – only twenty-five percent of women of reproductive age had access to modern contraceptive

Malthus Enigma

techniques. If this were to remain the case and fertility levels over this century were to remain constant at the 2005–2010 level, the United Nations projects that the population of less developed regions would explode to around twenty-six billion by 2100.[39] Fertility rates, however, are continuing to decline in developing countries – and the United Nations expects that the global population will likely be between 9.4 billion and 12.7 billion in 2100.[40]

Population projections are, however, highly uncertain, and very much dependent on the underlying assumptions and methods employed.[41] A more recent study published in *The Lancet* is much more optimistic about the global population stabilizing or even declining by the end of the century. Using different statistical techniques from those of the United Nations, health metrics professor Stein Vollset and his team estimated that global population could peak at 9.7 billion around 2064 and then decline to 8.8 billion in 2100.[42] Moreover, the authors note that if the United Nations Sustainable Development Goals for education and meeting contraceptive needs are realized, then global population could be as low as 6.3 billion by the end of the century. Projections of fertility rates are again key to these calculations. In their base scenario, Vollset's team estimated that fertility rates would reach as low as 1.66 children per woman by 2100. The United Nations used a higher value, noting that even in countries with low fertility rates today, surveys of child-bearing preferences show women still desire to have close to 2.0 children on average.[43]

The analysis by Vollset and colleagues provokes an intriguing question: If global population stabilizes or even declines in the future, would this essentially solve the Malthus Enigma? The answer is maybe, but it depends on how the future population lives – how wealthy they are, how much they consume, what technologies they use, and how dependent they are on non-renewable resources. One problem we noted in the last chapter is that even at current population levels, we are already exceeding planetary boundaries for greenhouse gas emissions, biodiversity loss, and the nitrogen cycle. Furthermore, there is a high likelihood that future generations will be wealthier – and without significant changes in technology and lifestyles, this will further increase our environmental burden on the planet.

Back in 2013, I teamed up with two of the World Bank's waste management experts to assess when, and at what level, human civilization will reach its peak production of solid waste.[44] The total production of solid waste is another broad measure of human environmental impact and has some clear ties to consumption. The more stuff we consume, the more

Feeding the World

waste we ultimately tend to generate. Our analysis, published in *Nature*, recognized that the generation of waste tends to increase with affluence, but waste disposal per person in some wealthy countries is gradually declining due to increased recycling and technological change. In a baseline projection, we found that the generation of solid waste will likely rise throughout the twenty-first century. Even if the population stabilizes over the century and good progress is made toward sustainability goals, we still expect the annual production of solid waste to be about three times higher in 2100 than in 2000.[45]

Although not straightforward, nor one-to-one, the tendency for environmental impacts to increase with wealth adds to the Malthus Enigma. Solid waste production, intertwined with consumption in economies, exemplifies the link between wealth and environmental impacts. Efforts are being made to decouple the impacts through greater recycling and the creation of circular economies. A wide range of environmental impacts have, nonetheless, been linked with growing affluence.[46] As we will see later in this chapter, moreover, increasing wealth – along with a growing population – also has a profound impact on human food consumption and the associated threat to global biodiversity.

Fritz Haber and Nitrogen Fixation

With the technology available in Malthus's lifetime, it would have been very difficult – if not impossible – to support the current global population of over eight billion people.[47] How, then, do we manage to feed so many people today? More than anything else, it was failure to recognize potential increases in the productivity of agriculture – for feeding people – where Malthus was most astray. We now know that increases in agricultural yields can be achieved by several means, such as the use of machinery, crop selection, and genetic engineering, but most significant of all is the application of fertilizers.

There are several ingredients in a typical fertilizer, including nitrogen, phosphorus, and potassium. Phosphorus and potassium can be mined, for example, from phosphate rock and potash. So, too, can nitrogen, from relatively rare deposits. Of course, nitrogen also occurs in abundance in the air – only this is in the form of inert nitrogen gas, and to make fertilizer, it is necessary to have a more active form of nitrogen. The process of obtaining active nitrogen from nitrogen gas is known as nitrogen fixation – and it is absolutely key to the manufacture of chemical fertilizers. So when, in 1908, Fritz Haber invented a new, much more

Malthus Enigma

efficient method of nitrogen fixation, it had a profound impact on agricultural productivity. From 1908 to 2008, the number of humans fed per hectare of arable land increased from 1.9 to 4.3, with something like thirty to fifty percent of this increase attributable to the application of nitrogen-based mineral fertilizers.[48] Put another way, by 2008, chemical fertilizers produced via the Haber-Bosch process of nitrogen fixation were responsible for feeding an estimated forty-eight percent of humanity. A few billion people have been fed as a result of Haber's invention, although perversely at the same time, millions have also died in armed conflicts – killed as a result of explosives manufactured from the same active nitrogen.

By all accounts, Haber had a complex life – full of industry, genius, sadness, irony, and contradiction. He was a strongly patriotic German Jew, who served his country in both glorious and terrible ways, but ended up having to flee his homeland when Hitler came to power. Haber was born on December 9, 1868, in Breslau, Prussia (now within Poland).[49] The birth was tragic in that his mother, Paula, died three weeks later due to complications. His father, Siegfried, became remote, leaving Fritz to be largely brought up by an aunt and uncle. He became interested in chemistry at an early age and often got in trouble for creating strange smells and small explosions in his bedroom. After attending a liberal arts school, Haber pursued studies in chemistry during a golden era of German science – his instructors included household names such as August Wilhelm von Hofmann, Hermann von Helmholtz, and Robert Bunsen. His undergraduate studies were interrupted by a year of military service with the 6th Field Artillery Regiment, during which he was promoted to the rank of a non-commissioned officer. Following doctoral studies in Berlin, Haber spent several years gaining practical experience at a variety of leading schools, including Budapest, Zurich, and Jena. This practical knowledge, combined with his strong mathematical abilities, would later prove important. At about age twenty-four, Haber took the unusual step of converting to Christianity. The motives for his conversion are unknown, although one friend thought it might help achieve a permanent university position. Sure enough, in 1894, Haber settled at the Karlsruhe Institute of Technology, where, over the next seventeen years, he did his most brilliant work.

It was at Karlsruhe that Haber discovered a new and significantly cheaper means for fixing nitrogen from the air.[50] At the turn of the century, there were industrial processes that could produce active nitrogen – a necessary ingredient for fertilizer. Haber, for example, was familiar with oxidation in electric arc furnaces, which he observed at a

Feeding the World

facility near Niagara Falls during a trip to America. Such processes were very expensive, however, as they were heavily energy intensive. Even at places with low energy costs, they could barely compete with natural sources of active nitrogen, such as Chile saltpetre. Nitrates were also produced as a by-product of coal distillation in gas and coking plants, but this could meet only about half of Germany's demands. It was well recognized that a low-cost means of producing active nitrogen – such as ammonium – would be highly valuable, and several eminent German scientists had tried. Wilhelm Ostwald had applied for a patent on the synthesis of ammonium from his famous laboratory at Leipzig, but his method was flawed. Thermodynamicist Walther Nernst had recognized that ammonium might be produced more effectively at higher pressures, but he had given up trying due to a lack of belief by industry. Haber, though, was in a unique position of understanding Nernst's thermodynamics while having strong ties to industry – in particular Badische Anilin- und Soda-Fabrik (BASF) in Karlsruhe. Haber had also established a highly able staff with excellent engineering skills. In 1908, he filed a patent on a procedure to produce ammonium at high pressures (100–200 atmospheres) and relatively modest temperatures (600°C). But the most remarkable breakthrough occurred when he added the rare element osmium as a catalyst, substantially increasing the rate of reaction. BASF gave immediate and generous support to scale up the process for industrial-level production, and it was their researcher, Carl Bosch, who discovered that iron worked well as a much cheaper and equally effective catalyst. Bosch's work was delayed somewhat by World War I, but by 1917, a full-scale plant was opened at Oppau in Germany, producing ammonium using the Haber-Bosch process.

Within three years of his breakthrough in nitrogen fixation, Haber moved to the Kaiser Wilhelm Institute in Berlin and began the darkest period of his life.[51] He was made director of the Institute of Physical Chemistry and Electrochemistry, which formally opened in October 1912, but less than two years later, World War I broke out. Haber's knowledge of chemistry and organizational skills were invaluable to the German army – and his entire operation essentially became consumed within the German war effort. At first, he had the task of directing German industry to produce nitrates, which were necessary for manufacturing ammunition; Germany had only six months' supply, and the British were blockading its naval trade. Meanwhile, his institute was researching chemicals for manufacturing new explosives. Haber then became head of Germany's chemical warfare initiative – elevated to the rank of captain – and in charge of the research, development,

Malthus Enigma

manufacture, and *deployment* of poisonous and irritant chemical weapons. Although poisons had previously been used in warfare, Haber was responsible for taking chemical warfare to a level of full-scale battlefield deployment. On April 22, 1915, he directed the first use of chlorine gas blown by the wind over enemy trenches, poisoning an estimated 7,000 Allied soldiers.[52] Ironically, nine days later, Haber's wife Clara committed suicide; apparently depressed over the breakdown of her marriage, she shot herself using Haber's army revolver. Barely pausing in his loss, Haber continued his dedication to his work on chemical weapons, for which he had no compulsions, going on to manufacture the notorious mustard gas, as well as defensive equipment against the Allies' retaliatory chemical weapons. At the end of the war, Haber was put on the Allies' list of war criminals, but he was later exonerated, and then awarded the Nobel Prize in Chemistry.[53] Nonetheless, he continued to work on chemical weapons for several years after the war, involved in secret work between Germany and the Soviet Union. But then, in 1933, Haber resigned his position in Germany in the face of new Nazi laws against Jewish scientists. Having left Germany, he died in Basel, Switzerland, in January 1934.

Haber's ingenious way of fixing nitrogen has had a massive impact on humanity.[54] By 2008, close to half the world's population was fed due to the existence of Haber-Bosch nitrogen. Approximately one hundred million tonnes of nitrogen fertilizers are produced by the Haber-Bosch process every year. Unfortunately, however, our use of nitrogen fertilizer is quite inefficient, with only seventeen million tonnes actually consumed by humans in crop, dairy, and meat products.[55] In other words, eighty-three percent of the nitrogen produced goes to waste. About forty percent of this waste nitrogen is denitrified back into unreactive, atmospheric nitrogen. The rest cascades through atmospheric, terrestrial, and aquatic systems – causing a variety of environmental stresses, from declining water quality to loss of biodiversity due to excessive fertilization. There are also environmental impacts related to the production of the ammonium itself – not least of which are those associated with the energy used in the production. The great success of the Haber-Bosch process has been due to its lower use of energy relative to other means of nitrogen fixation. Nonetheless, it is still, in absolute terms, a highly energy-intensive process. Approximately one percent of global energy supply is used in the manufacture of Haber-Bosch nitrogen.[56] In fact, about ninety percent of the cost of manufacturing fertilizer is the cost of the energy supply – which is typically natural gas.

Feeding the World

The Man Who Fed the World

On top of the increases in crop yields achieved by chemical fertilizers, further boosts to agricultural productivity have been achieved by the breeding of new crops. Many agricultural scientists have made contributions to the practice, but the exploits of American agronomist Norman Borlaug are the most remarkable. Borlaug was born into rural poverty; spent a lifetime's work developing high-yield, disease-resistant strains of wheat – and became known as the father of the Green Revolution. He is one of only five people to have received a Nobel Peace Prize, the US Presidential Medal of Freedom, and the Congressional Gold Medal.

Borlaug was born on March 25, 1914, in the remote Norwegian-speaking community of Saude in the state of Iowa.[57] The "little house on the prairie" at the centre of the Borlaug family farm was a humble abode with no running water or electricity. In those days, yields of twenty-five bushels per acre were common for Iowa farms, though as a young farm boy, Borlaug witnessed the destruction of his grandfather's wheat crop by wild fungus. It was hard for his family to make ends meet, and although Borlaug never starved as a child, he was undernourished.

The 1920s were a decade of possibly unprecedented technological change, which lifted millions of impoverished rural American farmers up to middle-class incomes – at least until the great financial crash of 1929. One of the new technologies was radio – which first began broadcasting in 1921 – substantially improving communication. It was the combination of the development of tractors and chemical fertilizers, however, that had the most immediate impacts on farmers. As Henry Ford began rolling out tractors at close to 100,000 per year by 1926,[58] they "sent the farm horse into history."[59] This development alone more than doubled the production of rural farms, as previously sixty acres of a hundred-acre farm were required just for feeding the farm horses. The arrival of chemical fertilizers both doubled crop yields and eliminated the need to keep farm animals for providing organic fertilizers. Farmers such as the Borlaugs made comfortable revenues from the sale of more abundant crops – and found themselves with more time to spare with the labour of keeping farm animals reduced. By the end of the decade, however, it all blew up. The Wall Street collapse brought banks crashing down with it, including the local bank holding the Borlaug family savings.

Borlaug was fortunate to get a place at the University of Minnesota, first studying forestry and then crop science. He was a smart kid, but it was his athletic abilities – as a talented wrestler and football player – that

Malthus Enigma

helped open the door to the farm lad. Being a financially disadvantaged student in the 1930s was tough; he worked in a coffee shop to eat and took several months out of school whenever he was lucky enough to gain some income. He even took an assignment as a fire watcher with the Idaho Forest Service, alone for three months in a remote ranger station. A few weeks before finishing his forestry degree, Borlaug attended a life-changing lecture by Elvin Stakman, "These Shifty Little Enemies That Destroy Our Food Crops." Stakman was an expert on crop rusts, like the fungus that destroyed the Borlaug family wheat. "Rusts are relentless, voracious destroyers of man's food and we must fight them by all means open to science"[60] were inspiring words to Borlaug. Learning that "[b]iological science, crop diseases, soil infertility, human population and world hunger, are all interwoven," Borlaug persuaded Stakman to take him on for graduate studies. He completed master's and PhD degrees at the University of Minnesota, with Stakman, before working as a microbiologist with DuPont during World War II.

In the early 1940s, The Rockefeller Foundation was concerned about the feeble state of Mexico's food supply and resolved to open an agricultural research station near Mexico City. Borlaug was persuaded to move south and was soon heading up the wheat program. He realized that to lift the Mexican people out of near starvation, he had to beat stem rust. Having built one research station from scratch, including training of staff, he rejuvenated a second crop research facility in the Yaqui Valley of the Sonoran Desert to add an extra winter growing season to his efforts. Over a period of fifteen intense years, Borlaug worked tirelessly, cross-breeding strains of wheat, planting thousands of plants, and twisting the arms of skeptical Mexican farmers to develop high-yield, rust-resistant wheat. He worked in harrowing conditions, travelling around Mexico in an era before well-developed roads and bridges, and was almost killed in a treacherous river crossing. But he performed wonders in his work, developing a mix – the "fab five" – of rust-resistant plants. By 1956, he had increased Mexico's wheat yield to 1,370 pounds per acre – an increase of eighty-three percent.[61] The turnaround was particularly spectacular in the Yaqui Valley and the states around Borlaug's northern research centre – wheat production increased by a factor of fourteen, and the wealth of the farmers lifted the standard of living of the whole community.[62]

With the Rockefeller agricultural mission in Mexico accomplished, Borlaug's position was technically terminated, but he soon found himself on a United Nations task force to survey wheat and barley research in South Asia, North Africa, and the Middle East. He discovered conditions similar to those in Mexico twenty years earlier – farmers with miserable crops, supporting impoverished people near starvation, overseen by ineffectual scientists and clueless governments. He knew at that point he had to act, or millions would starve. His first step was to advise the Food and Agriculture Organization of the United Nations to use his Yaqui Valley research station as a training facility for young agronomists from developing countries. Meanwhile, back in the Yaqui Valley, progress had been made in breeding a new form of dwarf wheat. After six years and 8,000 cross-fertilizations, his team had developed a rust-resistant wheat with high yields and stronger stems, which, moreover, had a good taste for bread making.[63] Although it was not a legally approved wheat in Mexico, it was with this dwarf breed that Borlaug went on to feed the world. Trainees from many countries – in the first year, from Afghanistan, Egypt, Libya, Iran, Iraq, Pakistan, Syria, and Turkey[64] – came to the Yaqui Valley to learn about crop breeding and modern planting techniques; they became Borlaug's apostles. India, however, even in a critical condition with half a billion people close to starvation, notably refused any help from the Mexicans.

India and neighbouring Pakistan were in a precarious position. The productivity of their agriculture was pathetic; the Sonoran Desert was achieving a crop yield seven times higher per acre.[65] In the early 1960s, it was only the generosity of the US government, shipping millions of tonnes of grain per year, that kept India from mass starvation. When the monsoons came late in 1965, both India and Pakistan declared an official famine. Borlaug made seven trips to the region in as many years – battling misinformed, misbehaving scientists; coaching presidents; twisting the arms of deputy presidents and agriculture ministers; and, when necessary, shouting them down. He campaigned for modern agriculture – for fertilizers, Mexican seed, proper planting techniques, and market prices for farmers. He seemingly worked against incredible odds. When the famine started in 1965, both India and Pakistan agreed to buy Mexican dwarf seed. Borlaug somehow managed to guide the unauthorized Mexican seed illegally across the US border, through Los Angeles riots, and on board a ship, even with a letter of credit frozen due to a spelling error. The seed was subsequently delivered and distributed throughout the two countries just as they began shelling each other in a five-week war. The next year, the problem was worse, but Borlaug managed to deliver even more Mexican seed. By 1967, with modern agricultural production at last beginning to take hold, India's crop production increased from eleven million tonnes to sixteen million tonnes in one year,[66] and Pakistan essentially became self-sufficient in food production.

Malthus Enigma

The famines in India in the mid-1960s had made international headline news. This was furthered in part by Lyndon B. Johnson putting India's wheat relief on short tether due to Indira Gandhi's anti-US Vietnam War rhetoric. It was the situation in India that had motivated Paul Ehrlich's writing in *The Population Bomb*:

> *The battle to feed humanity is over. In the 1970s, the world will undergo famine – hundreds of millions of people are going to starve to death in spite of any crash program embarked upon now.* [67]

But Ehrlich had not reckoned with the skill, tenacity, and endurance of Norman Borlaug. Aided by his apostles, variants of Borlaug's cross-pollinated dwarf wheat were rapidly spreading throughout the world. By 1968, there were 6.7 million acres of the Mexican high-yield wheat in India, 1.8 million in Pakistan, and 1.3 million in Mexico, with further acreages in Turkey, Afghanistan, and Nepal, and it was just beginning to take root in Australia, Argentina, Brazil, Chile, the US, Canada, Morocco, Tunisia, South Africa, and elsewhere.[68]

Stepping back to our broader inquiry on the role of innovation in the Malthus Enigma, it might be tempting at this point to conclude that the Cornucopians are right. While Ehrlich was beavering away writing his *Population Bomb*, Borlaug was spreading his high-yield, rust-resistant varieties of wheat throughout the world, adding to the increases in crop productivity that had been achieved using Haber's fertilizers. But, of course, things are not that simple – we are wrestling with an enigma. The increased population supported by the innovations of Borlaug and Haber just adds further environmental stresses to the planet. Ehrlich's concerns over the environmental impacts resulting from a growing global population still remain, even if he was far off track in pronouncing that hundreds of millions of people were going to starve to death in the 1970s. As we delve further now into the issue of threats to global biodiversity, these environmental impacts are extremely worrying.

The Millennium Ecosystem Assessment

Widespread use of fertilizers in developed countries and the Green Revolution have substantially increased agricultural production per area of land. Yet while increases in agricultural productivity have helped support the massive increases in population since 1900, they have been insufficient, and large swaths of natural lands have also been converted to farmland. From 1700 to 1990, the amount of global land used by

Feeding the World

human agriculture increased from approximately eight to nine million square kilometres to forty-six to forty-nine million square kilometres,[69] with most of the increase occurring since 1900.[70] Conversion of natural habitat for human agricultural purposes clearly has a negative impact on the Earth's flora and fauna, but it also has subtler long-term consequences for human well-being on the planet. Land-use change, along with other factors such as species intrusion, overexploitation of natural resources, and excessive nutrient loadings, has caused serious reductions in biodiversity. Such losses are ultimately bad for the human species, too, because we rely on healthy ecosystems to provide for our basic needs – food, water, medicines, fibre, and fuel – and to regulate our environment and satisfy cultural needs.[71] In short, humans require biodiversity to provide so-called *ecosystem services*.

Some simple early insights into human impacts on biodiversity were observed in Aldo Leopold's 1949 classic, *A Sand County Almanac*. Leopold (1887–1948) was a professional naturalist and a professor at the University of Wisconsin, known to be one of America's leading experts on wildlife management. In his *Almanac*, Leopold recounts experiences from visiting his eighty-acre farm in an area known as the Sand Counties in rural Wisconsin. It is a rich and vivid book, being a landmark text in the field of environmental ethics, while teaching fundamental ideas of ecology, some ahead of their time. The entries in the *Almanac* are almost poetic in places: the following of skunk trails in January's snow, the history of the old farm told by the rings of the Good Oak, the return of geese in March and high waters in April, trout fishing in June and grouse hunting in October, and even November's contemplation by Leopold on his preference for pines over birches. It is in the July entry, however, where Leopold reveals an experiment on species diversity. He presents data on the number of wild plant species coming into flower, by month, around the campus and suburbs near his university and at his backwater farm. In April, twenty-six wild plants blossom at the farm, but just fourteen at the campus; in May, there are fifty-nine at the farm and only twenty-nine on campus; and so it continues. The data collected over a decade shows the diversity of species to be almost double at Leopold's remote farm and woodlot, far away from people and cars. In the peak month of June, seventy species of wildflower came into bloom at the farm, compared with just forty-three at the university.[72] Leopold goes on to explain that in the campus and suburbs, the shrinkage of flora is due to clear farming, grazing of woodlots, and a proliferation of roadways.

Leopold's counting of species is quite straightforward, but there is a bit more to the quantification of biodiversity than that. Biodiversity is a complex phenomenon, which in some respects defies precise measurement. Biodiversity refers to the variability of living organisms, including diversity within species, between species, and of ecosystems themselves. It pertains to all types of organisms within terrestrial, marine, and other aquatic ecosystems.[73] The multi-dimensionality of biodiversity makes it hard to quantify well, but several simplistic measures have been developed. One useful measure, for example, is species richness – which is the number of species in a given area, similar to Leopold's approach. On a global scale, concern over biodiversity is often expressed in terms of species extinction rates. This measure has some challenges, not least of which is that there might be on the order of nine million species on Earth,[74] but scientists have described only about one to two million of them.[75] There again, this problem can be partially overcome by focusing on families of species that are well understood, like mammals, birds, and trees. Back in the 1990s, Julian Simon was fiercely critical of efforts to quantify biodiversity loss. He chastised Harvard biologist Edward O. Wilson for suggesting that global species extinction was about 100,000 species per year and then modifying the number to 27,000 species per year, when Simon claimed that data showed the rate to be just one species per year.[76] Simon noted that Wilson's claims were not based on observation; rather, they were mathematical calculations based on Wilson's theoretical systems model of island biogeography.

Any doubts about the severity of biodiversity loss were largely put to bed in 2005, with the release of the Millennium Ecosystem Assessment. The assessment of the Earth's ecosystems was a four-year undertaking coordinated by the United Nations Environment Programme, with contributions from over 1,360 scientists representing ninety-five countries. It was chaired by Robert T. Watson, chief scientist at the World Bank; Dr. Zakri Abdul Hamid of the United Nations University; and Hamdallah Zedan from the Convention on Biological Diversity. The assessment produced over thirty technical reports and six synthesis reports – one of which was entirely devoted to biodiversity. Of course, the Millennium Ecosystem Assessment scientists still did not go out and quantify every single species on the planet, but they did conduct eighteen regional assessments in various global ecosystems and drew upon information from a further fifteen associated study areas – and of course other available scientific literature.[77]

The Millennium Ecosystem Assessment held no punches in describing the impacts of humans on the planet. The biodiversity synthesis report noted:

Virtually all of Earth's ecosystems have now been dramatically transformed through human actions. More land was converted to cropland in the 30 years after 1950 than in the 150 years between 1700 and 1850. Between 1960 and 2000, reservoir storage capacity quadrupled, and as a result the amount of water stored behind large dams is estimated to be three to six times the amount of water flowing through rivers at any one time. Some 35% of mangroves have been lost in the last two decades in countries where adequate data are available (encompassing about half of the total mangrove area). Already 20% of known coral reefs have been destroyed and another 20% degraded in the last several decades. Although the most rapid changes in ecosystems are now taking place in developing countries, industrial countries historically experienced comparable changes.[78]

The report went on to spell out the extent of land conversion for human purposes that has occurred in the major types of biomes assessed. Between 1950 and 1990, four major biomes experienced between fourteen percent and sixteen percent native habitat loss. These were tropical dry broadleaf forests and three types of grasslands or savannas – temperate, tropical, and flooded. Other biomes were being transformed at slower rates, but in some cases this was because they had already been radically altered. Temperate broadleaf forests such as those found in North America, along with Mediterranean forests and temperate grasslands, had already experienced over fifty percent land conversion prior to 1980. These are the biomes that contain many of the world's cities, surrounded by extensive agriculture. Habitat conversion continues in nearly all biomes. But over the past two decades, habitat loss has been particularly rapid in several regions. The Amazon Basin and Southeast Asia are being deforested for cropland, and degradation of dry lands has been occurring in many parts of Asia, while Bangladesh, the Indus Valley, parts of the Middle East and Central Asia, and the Great Lakes Region of Eastern Africa continue to be transformed.

The impact of land-use conversion, accompanied by other factors, has led to a huge increase in the rate of species extinction. The Millennium Ecosystem Assessment concluded that human actions have caused the species extinction rate to increase by as much as 1,000 times the background rate over the past few hundred years. There is substantial uncertainty in the overall picture, but careful study of birds, mammals, and amphibians over the past one hundred years puts the extinction rate for these groups at one hundred times background. Population sizes and ranges have declined in many taxonomic groups, including all

Malthus Enigma

amphibians, African mammals, birds in agricultural areas, British butterflies, Caribbean and Indo-Pacific coral reef species, and most commonly harvested fish species.[79] The International Union for Conservation of Nature (IUCN) estimates that thirty-two percent or more of amphibians, twenty-three percent of mammals, and twelve percent of bird species are under threat of extinction. And it's not just animals under threat; twenty-five percent of conifer species and fifty-two percent of cycads (palm-like evergreen plants) are also on the IUCN's *Red List*.[80]

There is also some evidence of decline in genetic diversity within species – particularly with domesticated species. Despite the hard labour and noble intent of Norman Borlaug and others, the result of the Green Revolution has been the specialization of farmers and plant breeders in the most productive animal and plant species. With such specialization occurring across globalized food markets, the Millennium Ecosystem Assessment found a substantial reduction in the genetic diversity of agricultural plants and animals. The establishment of gene banks has fortunately helped maintain some genetic diversity among crops, but almost one-third of the 6,500 domestic animal breeds face extinction because of small population sizes.[81]

Most people can probably understand that the fragility of agricultural systems due to lack of diversity is worrying for humans. We want a food system that is resilient to disease, droughts, and other extreme events – especially given climate change. But the diversity of species in ecosystems provides many other services to humans that are perhaps not as obvious to them. The Millennium Ecosystem Assessment identified four broad categories of ecosystem services upon which humans rely.[82] First is the provisioning of food, water, fibre (such as timber, cotton, and hemp), fuel, and natural ingredients in medicines and cosmetics. Second are regulating mechanisms such as control of climate, cleansing of air pollutants, control of stormwater and soil erosion, control of pests and diseases, and pollination. Cultural services are the third type of ecosystem service; these include spiritual and religious significance, aesthetic value, recreational value, and similar services. The fourth category includes longer-term services of soil formation, photosynthesis, nutrient cycling, and water cycling. In examining trends in ecosystem services over the past fifty years, the Millennium Ecosystem Assessment established that enhancements had been made in only four services – crops; livestock; aquaculture; and, in recent years, carbon sequestration – while twenty other categories of ecosystem services are being degraded.[83]

Biodiversity loss is truly a global issue; it is not just a local issue that occurs at specific locations where species habitats are being destroyed or fragmented. To appreciate this, it is useful to consider the Earth systems work on international trade conducted by Australian industrial ecologist Manfred Lenzen. Lenzen was originally a physicist by training but joined the interdisciplinary science of industrial ecology, which studies the environmental impacts of energy and material flows using systems models. More will be said about industrial ecology as we progress through the book, culminating in Chapter 8; for now, let us focus on the important contribution of Lenzen and his colleagues in linking biodiversity threats in developing countries to global trade. Lenzen and his team used an economic systems model that tracks 15,000 commodities produced in 187 countries to link the consumption of commodities to their environmental impacts.[84] Through their analysis, the researchers related national consumption of commodities back to 25,000 threatened species on the IUCN's *Red List*. An example of this is the threat to critically endangered species in Papua New Guinea, such as the northern glider, the black-spotted cuscus, and the eastern long-beaked echidna, due to agricultural expansion for global exports of coffee, cocoa, palm oil, and coconuts.[85] Lenzen and colleagues found that the consumption of commodities such as coffee, tea, sugar, textiles, fish, and a variety of manufactured goods in developed countries causes larger threats to biodiversity in developing nations abroad than in the country of consumption. In fact, Papua New Guinea was found to be one of the most badly impacted countries, with sixty species under threat just from the export of timber to the Japanese residential construction sector.[86] Overall, excluding invasive species, global trade was found to be responsible for thirty percent of global species under threat. Historically – before the game hunters – humans had relatively low impacts on species habitat – meeting local needs for food, fuel, and shelter. The impacts are now far beyond the game hunters. International trade in a globalized economy causes widespread habitat destruction at locations far removed from consumers.

The extent and pace of biodiversity loss is such that some biologists are asking whether a new period of mass extinction has arrived. Extinction of species in geological time scales is natural. Indeed, over ninety-nine percent of all species that have evolved on Earth in the past 3.5 billion years have now gone.[87] A mass extinction, however, is very rare. It is defined by paleontologists as the disappearance of more than seventy-five percent of species in a geologically short period of time – say 500 years.[88] Scientists think that five mass extinctions have occurred in the past 540 million years. One of these came about when an asteroid hit the Yucatán Peninsula, spewing dust into the atmosphere, cooling the

Malthus Enigma

Earth, and wiping out the dinosaurs. Other mass extinctions have involved changes in glacial activities, volcanic eruptions, and uplifting of mountain ranges. All five have involved some form of significant climate change. Reasons for the suspected current period of mass extinction also include climate change, but this time human induced. And along with climate change, biodiversity loss is also caused by the destruction and fragmentation of ecosystems, the introduction of non-native species, the spread of pathogens, and the killing of species directly, like the big-game hunters.

One scientific paper published in the prestigious journal *Nature* concluded that within the next few centuries, the Earth will witness its sixth mass extinction.[89] The data supporting the conclusion is, by the authors' own admission, patchy. First, species extinction rates have to be measured, then they have to be extrapolated from a time period of a few centuries to a few million years or more. One challenge is that less than 2.7 percent of the world's 1.9 million named species have been assessed by the IUCN for extinction status. Data from studies of these species have to be used to infer global trends. Only in a few groups, such as amphibians, birds, and mammals, is there close to adequate data. For these groups, extinction rates over recent historical times are less than one percent, but twenty percent to forty-three percent of the species are threatened. If the threatened species become extinct, as feared, then mass extinction could be reached within a few centuries. Were Julian Simon alive today, he would bring attention to the inadequacy of the data – and perhaps rightfully so. Nonetheless, drawing upon the balance of current evidence, the authors find that "there are clear indications that losing species now in the 'critically endangered' category would propel the world to a state of mass extinction that has been previously seen only five times in about 540 million years."[90]

How Can We Feed the World and Save the Planet?

Pulling the various pieces together here, it is clear that humankind is in quite a predicament. Global population is over eight billion as of 2024 and is expected to reach ten billion later this century. Feeding so many people is made possible only through the application of fertilizers – thanks to Fritz Haber – and the labours of Norman Borlaug and others from the Green Revolution. But the impact of humans multiplying as a species, converting natural habitat to farmland, cutting corridors for roads and other infrastructure, burning fossil fuels, etc., is the possible mass extinction of many other species on the planet. This is the modern form

Feeding the World

of the Malthus Enigma. In finding the technological solutions to feed ourselves, we multiply and destroy the services of the ecosystems we rely upon. How can we find a way to feed ourselves without destroying the rest of the planet?

One scientist who has wrestled with this question is David Tilman – an ecologist from the University of Minnesota. Tilman has undertaken calculations of the future environmental impacts on the Earth system of producing food for a growing population. He estimates that by 2050, global food demand will double compared with 2005.[91] To be more precise, he estimates a one hundred percent increase in the demand for crop calories.[92] This may seem like an astonishing projection. For the period of 2005 to 2050, the United Nations projects a fifty percent increase in population.[93] Why, then, would the increase in food production be double that? To answer this, it is necessary to explore the metric of crop calories in more detail and its relation to the human diet.

Crop calories are a useful measure of food demand, as they include all the fruit and vegetable crops consumed by humans, as well as those fed to animals that produce meat and dairy products for humans. Converting the meat and dairy quantities back into crop calories is important because of the inefficiency of transforming feed crops into animal-based foods. To produce one kilogram of human-edible beef protein requires cattle to eat twenty kilograms of crop protein – that is, a ratio of 1 to 20.[94] The ratio is lower for white meat – 1 to 5.7 for pork and 1 to 4.7 for poultry – while aquaculture averages about 1 to 4.6. One kilogram of milk requires 3.9 kilograms of crop production, and the ratio is 1 to 2.6 for eggs. Producing animal-based foods is so calorie intensive that in richer nations, more than half of all crop production is used to feed livestock, rather than going directly to human consumption.

The challenge going forward is that humans are getting richer as well as increasing in numbers. Citizens in richer nations tend to consume more calories of food per day and have higher intakes of both animal products and so-called "empty calories," such as sugars, fats, oils, and alcohols. Tilman has established some strong statistical relationships between GDP per person and meat demand per person,[95] but typical average values can be used to make the point.[96] In wealthy countries, citizens on average have about 3,500 calories per day of foodstuffs entering household kitchens, of which about twenty-five percent is wasted. Moreover, about twenty percent of calories in rich countries come from meat and dairy products, with thirty-eight percent from empty calories. Because the meat and dairy content is so substantial, 8,000 calories per day of agricultural crops are required to feed people in wealthy countries. Contrast this with

Malthus Enigma

poorer countries, where people demand about 2,000 calories per day of foodstuff, with little wastage – and have diets averaging only three percent dairy and twelve percent empty calories. So the reason that Tilman predicted a one hundred percent increase in food production by 2050 is the change in diet that occurs as the world gets richer.

What will be the environmental impacts – biodiversity loss, water stress, carbon emissions, and nutrient pollution – of a doubling of global food production? The answer, according to Tilman, depends on how the food is grown. From 1960 to 1995, there was a doubling of the global production of cereal grains. This was due primarily to the Green Revolution and the increasing use of fertilizers that generally occurred everywhere except for large parts of Africa. With a doubling of food production was a doubling of water used for irrigation. Plus, with the increase in the intensity of production, the use of phosphorus fertilizers went up threefold, and nitrogen sevenfold.[97] The huge rise in the application of fertilizers massively increased nutrient pollution, as about half the fertilizer is wasted and fails to reach the crops. The overuse of fertilizers may erode the long-term resilience of ecosystems, but there is a trade-off here: the use of fertilizers in more intensive farming means that less land area has to be converted to farmland for growing crops. Even so, the doubling of food production during the era of the Green Revolution still involved the global conversion of 450 million hectares of natural land into cropland and pastureland – an area equivalent to half the United States.[98] The amount of natural habitat cleared, however, could have been much worse. Were it not for the increase in yield per hectare from fertilizers and the Green Revolution, even more land area would have been required to meet the same level of food demand. So the trade-offs in environmental impacts are quite complex. Intensification of agriculture causes increased nitrogen pollution that may decrease long-term resilience of ecosystems, but it avoids further conversion of habitat into farmland – the major cause of biodiversity loss.

If global agricultural practices continue under business as usual, Tilman predicts this will entail the clearing of approximately one billion more hectares of land, a 185 percent increase in global greenhouse gas emissions from agriculture, and an increase of 175 percent in the use of nitrogen fertilizers. These calculations are for a scenario where Africa continues to develop using existing low-efficiency practices, while intensive agriculture is used in other continents. Greenhouse gas emissions will be discussed further in the next chapter. They are predominantly associated with energy use. Nevertheless, agriculture, forestry, and land-use change account for about twenty-five percent of the world's approximately fifty gigatons of greenhouse gas emissions.[99] Tilman estimates that agricultural greenhouse gas emissions would increase by a further six gigatons under the current trajectory – due primarily to methane releases and the clearing of land.[100]

Tilman also considers an alternative path forward under which the technology and management practices of developed countries are used to intensify yields from croplands in poorer countries. This would still result in increased biodiversity loss, as about 350 million hectares of natural habitat would still have to be cleared for farmland. Nonetheless, this is a reduction of 650 million hectares – an area twice the size of India – compared with the current trajectory. Global greenhouse gas emissions from agriculture would also be lower, although there would be fifty percent higher use of fertilizers, potentially leading to increased nutrient pollution. So based on Tilman's calculations, the intensification of agriculture in developing countries can help slow increases in biodiversity loss and greenhouse gas emissions, but there is still a potential trade-off in nitrogen pollution.[101]

Organic Agriculture

In recent years, there has been renewed interest in the potential role of organic agriculture in contributing to global food supply. This has perhaps been driven, in part, by increasing consumer demands for organic foods – grown without the use of synthetic fertilizers, pesticides, or genetically modified organisms. A notable study, "Organic Agriculture and the Global Food Supply," was led by University of Michigan professor Catherine Badgley in 2007.[102] Badgley and colleagues summarized the key debate over the role of organic agriculture. The conventional perspective is that we can use technologies such as synthetic fertilizers, pesticides, and mechanical tillage to produce high-yielding crops. Others, however, consider such farming practices to be a *Faustian bargain* – undermining the long-term sustainability of agriculture – and argue instead for greater use of organic agriculture. Badgley and her team presented data from close to 300 studies showing that yields from organic farming in developed countries were, on average, just ten percent below those from non-organic farming, which is fairly close. Moreover, the data suggested that yields from organic farming in developing countries were on average eighty percent higher than those from non-organic farming. Having demonstrated that organic farms can produce sufficient yields, the authors also attempted to push back on objections that there are insufficient quantities of natural fertilizers for organic farming to

Malthus Enigma

substantially contribute to global food supplies. They presented calculations for temperate and tropical regions suggesting that leguminous clover crops grown during winter fallowing periods can be physically ploughed back into the soil, providing enough nitrogen to replace synthetic fertilizers. Extrapolating their results through global modelling studies, the authors concluded that organic agriculture could substantially help feed the current and future global population without requiring more farmland.

The paper by Badgley and colleagues provoked a critical response that led to further studies. In a countering paper, "Organic Agriculture Cannot Feed the World," Australian agricultural scientist David Connor questioned the interpretation of the data – especially for developing countries.[103] He argued that in most of the underlying studies, the organic farms were receiving organic nutrients, such as animal manure, from external sources. Thus, in Connor's view, the authors had incorrectly defined agricultural systems boundaries, leading to gross overestimation of the relative yield of organic agriculture. A further data set overcoming these challenges was subsequently assembled by Verena Seufert, an environmental geographer then at McGill University in Montreal. Seufert and colleagues found that organic farming yields were on average twenty-five percent lower than conventional farming yields – and forty-three percent lower in developing countries.[104] Seufert noted, however, that when farming best practices are applied, there are cases where organic farming can compete well with conventional techniques. With respect to growing nitrogen-rich legumes during winter growing seasons, Connor also argued that much productive land already carries multiple crops in a year – and where only one crop is grown per year, this is due to limitations in temperature or water. So there is less opportunity to increase supplies of nitrogen from plants than Badgley and colleagues suggested. Projecting up to the global scale, Connor reiterated the calculations of environmental scholar Vaclav Smil – that the Earth's carrying capacity with a population fed by organic agriculture would be only about 3.3 billion people.[105]

Africa: The Last Continent

The challenge of feeding the world without destroying our ecosystems and disrupting the climate is evidence enough that Malthusian struggle goes on.[106] Technological advancement, as exemplified by the efforts of Fritz Haber and Norman Borlaug, is an important – indeed, essential – part of the Enigma. But progress is not straightforward; there are subtle trade-offs, circular routes, and unforeseen consequences on the journey. Intensification of agriculture globally seems now to be a necessary way forward to hold back the destruction of natural ecosystems – to maintain the diverse species and stored carbon that they hold. But there are trade-offs – intensification of agriculture may slow down biodiversity loss through destruction of natural lands but produce further challenges of increased nitrogen pollution and homogenization of plant species – both of which require ongoing struggle to resolve.

More than anywhere else, the place where twenty-first-century Malthusian struggle will play out in abundance is Africa. To wrap this chapter up, it is necessary to reflect back on the continent where Livingstone went exploring, and for which the penitent butchers made the first – somewhat feeble – attempts at protection of species. Africa has been called the lost continent because of its slow development – and yet, perversely, its poverty might in the short term reduce the quantity of land needed by agriculture for richer diets. Africa is the last continent when it comes to technological development. How development takes place in Africa is hugely important to us all, though, because Africa is growing so fast.

While fertility rates in the developed world have flattened out below two children per woman, in Africa, the rate is about 2.5 times higher.[107] Combining this with declining death rates, the United Nations predicts that Africa's population will reach 2.5 billion people by 2050 (medium projection).[108] The population of Asia will still be larger – at over five billion – but Africa is expected to grow by greater numbers. Again using medium United Nations projections, Africa is expected to grow by approximately 1,270 million people from 2017 to 2050, while Asia is predicted to grow by approximately 750 million.[109] It will be in Africa that the world's greatest struggles to prevent the sixth mass extinction of species will be won or lost. If Africa is held back from the intensification of agriculture, then where in the world will Tilman's extra one billion hectares of agricultural land come from? Much of it would have to be found in Africa itself – to feed over a billion more people there. But at least one-third of Africa's three billion hectares of land mass is desert, about 350 million hectares is already used for agriculture – and the rest is natural habitat that is rich in biodiversity.

Development of sustainable agricultural practices in Africa is also complicated. The challenges are well described by Raymond Auerbach – a farmer turned professor at Nelson Mandela University in South Africa. Auerbach is one of several champions who have galvanized national organic agricultural movements throughout Africa. Drawing upon almost

Malthus Enigma

fifty years of experience as a commercial organic farmer, researcher, and teacher of farming practices, Auerbach understands the difficulties of providing food security while drawing upon Indigenous technical knowledge in the context of political instability, poor infrastructure, and climate change. At the heart of African agriculture are tensions between "the technology-centred process of commercial food production, and the people-centred process of household food production."[110] Many small-scale African food growers provide for their own needs – and could produce more – but do not wish to be seen as businesspeople. Wrapped around this are issues relating to land reform, education, health and nutrition, and food security under a changing climate.

Also important is the modest, but long-standing, literature studying the productivity of Indigenous agricultural systems in Africa and elsewhere. Daniel Muthee from Kenyatta University and his colleagues discuss how Indigenous knowledge systems may enhance agricultural productivity in Kenya; they point to the multi-faceted benefits of agroforestry and Indigenous forms of fallow agriculture that provide for both crop production and animal grazing, including drought-resistant crops and livestock breeds.[111] Abbas Sambo from the Federal University Dutse in Nigeria reports on the strengths of native African crops such as acha (fonio), benniseed (sesame), ginger, indigenous millet, and guinea corn that are tolerant to droughts and potentially able to adapt to climate change.[112] Another study engaged farmers in Kenya, Uganda, and Tanzania to understand the strengths and weaknesses of Indigenous soil- and water-conserving practices.[113] Drawing upon experience in West Africa, Paul Richards' book *Indigenous Agricultural Revolution* demonstrated how many of the world's most successful food crop production systems were rooted in Indigenous knowledge.[114] Other studies in the 1990s categorized Indigenous natural resource management systems around the globe[115] and argued for more innovative interactions between Western science and Indigenous knowledge to improve the effectiveness of agriculture.[116] Meanwhile, researchers Harold Brookfield and Christine Padoch penned a fabulous essay on the dynamism and diversity of Indigenous farming practices, noting how agro-diversity "may be the source of humanity's best solutions for the future."[117]

The Need to Innovate

The main conclusion from this chapter is that humans need to continually innovate to avoid Malthus's population trap. We saw how the invention

Feeding the World

of modern fertilizers, through the Haber-Bosch process, and Borlaug's development of more productive and resilient wheat strains enabled us to feed an increasingly growing population. We can add other inventions to these too – such as Ford's tractors. There is a catch, however: the more we innovate to feed an increasing population, the harder the challenge becomes to continue to do so. If Tilman's analysis is correct, what we are up against now is the possible destruction of massive areas of natural ecosystems – especially in Africa. Even if Africa applies the best Western technologies available today, the environmental destruction to ecosystems will be huge. The need to innovate to feed ourselves within the carrying capacity of the Earth is so fundamental that we should recognize it as the first level of our ongoing Malthusian Enigma.

My argument about innovation may seem slightly Cornucopian, but it is different in important ways. I reached this conclusion after recognizing the historical record of humanity innovating its way out of the population trap, but continually leaving itself up against harsh environmental and resource constraints. Cornucopians, such as Julian Simon, were completely dismissive of the neo-Malthusian perspective – whereas I am not. The Cornucopian argument is that humans are innovative, so we do not need to worry about environmental constraints. I would suggest that we *do* have to be continually worried about environmental constraints – because humans seemingly have no choice other than to keep innovating their way through the population trap. The necessity to continually innovate solves nothing; it just keeps us in the game. As we saw from assessing the global biodiversity challenges in this chapter, the neo-Malthusian concerns are as worrying as ever.

Many arguments can and will be made as to why innovation – and especially new technology – is not required to solve our environmental problems. It does sound a bit techno-centric. All sorts of questions can be raised about what types of technology are needed – and criticisms of "end of pipe solutions" or the limits of eco-efficiency measures can rightfully be made. But, innovation can also include changes in lifestyles or business practices – and embracing Indigenous knowledge. Possible alternatives to producing food using intensive modern industrial processes include avoiding or even reversing trends toward unhealthy diets and cutting back on crop or food waste. Practices that use fertilizers and irrigation water more effectively are also important considerations. A radically different approach to the intensification suggested by Tilman's analysis would be a complete return to organic farming, such as that encouraged by the slow food movement, focused on using local, sustainable foods.[118] The difficulty, however, is that organic farming has

Malthus Enigma

lower yields,[119] and, as Borlaug pointed out, vast areas of land would be required to produce cattle manure.[120] Perhaps, however, recent increased research on agroecology – the science of sustainable agriculture[121] – will help produce suitably high yields from culturally sensitive and socially just approaches. As farmer and ecological economist Gary Kleppel argues, there are better alternatives to "the current, dangerous, inhumane, and dehumanizing American industrial food production system."[122] The challenge of feeding the world while saving the Earth is daunting, but with a population of 8.2 billion and rising, at least there are many smart, innovative people on this planet.

One possible argument that could be made against the necessity to innovate to protect biodiversity is our limited ability to assess human impacts on natural ecosystems. The relationships between factors that cause biodiversity loss and the subsequent impacts on ecosystem services that humans rely on are incredibly complex. We know from the Millennium Ecosystem Assessment, the first *Global Assessment Report*,[123] and other studies that the impacts are bad, but quantification on a global scale is still difficult. There are a variety of mathematical systems models for estimating human impacts on biodiversity, but scientists are cautious about the conclusions they draw from them.[124] Tilman's calculations could yet be proven wrong, but it is most likely that technological innovation is what might change them. At its core, Tilman's system model boils down to Ehrlich and Holdren's IPAT equation, discussed in the last chapter. Environmental impacts are determined from a combination of population, affluence, and technology. There are some complex interactions between these terms – with changes in affluence and environmental impacts themselves affecting population growth. Nonetheless, with population increasing, and affluence rising too, the key way to reduce environmental impacts is through changing technology.

Given the seriousness of our current global environment challenges, we probably need more innovation today than ever before. We face a growing list of stresses on the planet, as evidenced by the planetary boundaries work – subject to some uncertainties. Meanwhile, global population continues to grow, even if the future trajectory for population growth is also uncertain. From the threats to global biodiversity alone, as highlighted by the work of Tilman, a new question in our inquiry emerges: *How do we develop technological systems to be in harmony with Nature?* The answer will involve using principles of ecology both to understand Nature and to understand technological systems. We will build toward answering this question more fully through several more chapters.

Discussion of innovation will continue into the next chapter, which is concerned with global climate change. A possible criticism of my position on the necessity to innovate is an argument that we already have the technology today to address our current environmental challenges. This argument is sometimes made with respect to reducing greenhouse gas emissions to address climate change. Two decades into the twenty-first century, this is a fair argument to make with respect to climate change – though perhaps not with respect to biodiversity loss. Of course, over the past few decades, there has been substantial technological innovation in the development of renewable energy technologies, air-source heat pumps, batteries, and electric vehicles, among other technologies and approaches for tackling climate change. So the argument that we must continue to innovate still holds.

In the next chapter, we will move to the second layer of the Malthus Enigma, which is all about conducting the science to understand the planet we inhabit. Innovation in how humans have harnessed energy from the Industrial Revolution to the current day will be wrapped into a larger story that includes how humans discovered our role in climate change. The questions we ask next are: *How did we discover the greenhouse effect, and how will we adjust our energy systems to accommodate it?*

Chapter 4: One Big Greenhouse

Since the start of the industrial revolution, mankind has been burning fossil fuel (coal, oil, etc.) and adding its carbon to the atmosphere as carbon dioxide. In 50 years or so this process ... may have a violent effect on the earth's climate.

Time Magazine, May 28, 1956[1]

In the late spring or early summer of 1856, American scientist Eunice Foote conducted a series of experiments to ascertain how the composition of a gas influences its warming response when exposed to the sun's rays.[2] The apparatus for her experiments was quite straightforward. In a first experiment, she took two glass cylinders, each thirty inches long and four inches in diameter, and used a simple air pump to extract air from one cylinder and push it into the other. The glass cylinders were then placed side by side in the sun,[3] and temperature measurements of the gas inside the cylinders were taken every two to three minutes by identical thermometers that were fixed inside the tubes. The second experiment was a repeat of the first, only Foote used dry air in one cylinder and moist air in the second. Then, in a third type of experiment, she compared the response of several different gases, including hydrogen, oxygen, gaseous carbonic acid, and common air.

This investigation was likely a response to the question of why valleys tend to be warmer than mountaintops, which had been debated in the March 1856 issue of *Scientific American*.[4] Although Foote did not take any measurements of pressure, the first of her experiments showed that the heating of air is diminished as it becomes rarified – or less dense. Her second experiment showed that heating by the sun's rays was greater in the moist air than in the dry air. Results of her third experiment were particularly notable. Foote found that the temperature of the carbonic acid gas – better known today as carbon dioxide – rose to 125°F, whereas the temperature of the common air cylinder rose to just 106°F.[5] The

Malthus Enigma

significance of this finding for the Earth's atmosphere was clear to Foote. She concluded:

> *An atmosphere of that [carbonic acid] gas would give to our earth a high temperature; and if as some suppose, at one period of its history the air had mixed with it a larger proportion than at present, an increased temperature from its own action as well as from increased weight must have necessarily resulted.*[6]

In this chapter, we turn to what I call the second layer of the Malthus Enigma – that is, undertaking the science to understand how the Earth's systems function. A notable example of such scientific discovery about the Earth is humanity learning our own role in contributing to the greenhouse effect – a phenomenon shown by the experiments of Eunice Foote, followed a few years later by those of well-known scientist John Tyndall. I am going to weave the discovery of human-induced climate change into a wider story about the development of human energy systems. This includes other cases of scientific inquiry, such as learning about constraints on how humans access energy, and discovering why efforts to conserve energy can sometimes backfire. In describing the development of human energy systems, from fossil fuels to today's renewable power generation, we will also again touch upon the first level of the Enigma – the need to innovate. But the second layer of the Enigma – undertaking scientific exploration – involves plenty of struggle. This is apparent from the scientific endeavours of Eunice Foote, whom we return to first.

Foote's family life revolved around science and social activism.[7] She was born in 1819, to parents Thirza and Isaac Newton, Jr. Her father was evidently a distant relative, centuries apart, of the great English physicist. Eunice was raised in Bloomfield, New York, and was fortunate to study at the Troy Female Seminary. Unlike many girls' schools, the Troy seminary had a broad curriculum – including the sciences, in which Foote flourished. She married a judge, Elisha Foote, who shared her interests in science and invention. Foote and her husband lived for a brief time at Seneca Falls in upstate New York, where they became acquainted with leading suffragist and abolitionist Elizabeth Cady Stanton. Eunice and Elisha both became active in the women's rights movement. They notably attended the Seneca Falls Convention of 1848, which was a landmark event in the development of the women's rights movement in the United States. They were signatories to the Declaration of Sentiments – one of the most important influences in the communication of women's rights in the country. The Footes had two children, Mary and Augusta, and later

One Big Greenhouse

moved to Saratoga Springs, New York. Mary followed her parents' passion for social activism – and married US Senator John B. Henderson of Missouri, who co-authored the Thirteenth Amendment to the United States Constitution, abolishing slavery.

As a woman scientist in an era of male bias, Eunice Foote's social activism was clearly warranted. There is an unfortunate twist in the story of her scientific endeavours, which demonstrates the discrimination that woman scientists have often faced. Discovery of the greenhouse effect has long been credited to Irish scientist John Tyndall. The reality is that Eunice Foote got little recognition for her research by contemporary physicists – and her experiments on identifying the greenhouse effect were essentially forgotten until 2011. Geologist Raymond Sorenson is acknowledged with rediscovering Foote's work.[8] How could such important experiments be forgotten?

On August 23, 1856, Eunice Foote and her husband attended the American Association for the Advancement of Science (AAAS) Annual Meeting in Albany, New York. Rather than present her experimental work on gases at such a prestigious meeting, Eunice sat in the audience while her paper was presented on her behalf by Joseph Henry, the secretary of the Smithsonian Institution.[9] The reason Foote did not present her work is unclear. She was not a member of the AAAS – very few women were in the nineteenth century, and it was rare for them to present.[10] Her husband presented his research in the same session and became a member of the AAAS that year, but she did not. Henry is noted as being a family friend, but unfortunately, he did not grasp the significance of Eunice Foote's paper at the time. According to the *New-York Daily Tribune*, Henry evidently stated that "although the experiments were interesting and valuable, there were [many difficulties] encompassing [any] attempt to interpret their significance."[11]

Regrettably, Eunice Foote's paper from the August 1856 meeting was not published in the proceedings of the AAAS Annual Meeting. This is not necessarily an indication of discrimination. Not all papers from the conference were published. Yet, strangely, neither her paper nor her husband's were mentioned in the long list of papers presented but not published.[12]

Fortunately, Eunice's paper from the 1856 meeting was published in the *American Journal of Science and Arts*.[13] Her experiments were also described in an article entitled "Scientific Ladies.--Experiments with Condensed Gases," published in *Scientific American*, which recognized her contribution toward understanding the climate of a past geological era.[14] The *Scientific American* article also noted that "the experiments of

Mrs. Foote afford abundant evidence of the ability of women to investigate any subject with originality and precision."[15]

Eunice also went on to write a second paper, detailing experiments on the static electricity of air, for the AAAS Annual Meeting in Montreal in 1857.[16] The paper was again read by Joseph Henry, but this time it was published in the conference proceedings – the first-ever paper by a woman.[17] Clearly, Eunice was not only a pioneering climate scientist, but also a pioneering woman scientist. Her two papers published in 1856 and 1857 are thought to be the only physics research papers published by an American woman prior to 1889.[18]

The question that naturally arises is: Why did John Tyndall not recognize Eunice Foote's experiments in reporting on his own research? Tyndall published a full paper on the heating properties of gases in 1861.[19] He went further than Foote, using infrared radiation in his experiments and explaining the mechanism by which the greenhouse effect occurs, but he did not mention her work. There is some contention as to whether Tyndall knew of Foote's research, but there is no evidence that he read her paper.[20] Poor citation by others can be partly blamed. Foote's experiments were briefly summarized in two European publications, but neither included her important conclusion on atmospheric warming – and one incorrectly attributed the work to Elisha Foote. The science writer David Wells republished an account of Eunice's 1856 paper in the *Annual of Scientific Discovery*.[21] He also continued to state the results of her experiments in other publications, but without proper citation.[22] Perhaps, however, the saga was best summed up by Maura Shapiro, writing in *Physics Today* in 2021, who noted: "Foote's fall into obscurity is part of a larger narrative of women's disenfranchisement in the scientific establishment."[23]

Unbeknownst to Foote, anthropogenic emissions of that gaseous carbonic acid started to rise during the Industrial Revolution and rapidly increased in subsequent centuries, leading to the change in global climate we are experiencing today. The altering of global climate through the burning of fossil fuels, among other drivers, has become a major element of contemporary Malthusian struggle – along with feeding humanity without causing mass extinction of other species. Of course, global climate change was not something that Malthus himself worried about; indeed, he had limited understanding of the implications of the Industrial Revolution despite living at its height. That said, when Malthusian sentiments rose to the fore again during the environmental revolution of the 1960s, climate change remained relatively poorly understood by the mainstream. This is despite an article in *Time* Magazine in 1956

One Big Greenhouse

describing how the Earth's atmosphere acts like "one big greenhouse."[24] The reality is that developing scientific understanding of the Earth's interwoven systems also takes time – and is itself a layer of the Malthus Enigma. Moreover, when there is scientific uncertainty – and even when there isn't – this leads to a further policy level to the struggle.

Recognizing the phenomenon of anthropogenically induced climate change was itself a struggle. Beyond the experiments of Eunice Foote and John Tyndall and the theory of Joseph Fourier, the notion that anthropogenic emissions could cause climate change was understood by Swedish scientist Svante Arrhenius in the nineteenth century – but his scientific theory fell out of fashion. The perseverance of British engineer Guy Stewart Callendar and American geochemist Charles David Keeling, among others from the mid-twentieth century, resurrected it. Then, from the 1970s onward, a political battle took place, before further science eventually established beyond reasonable doubt humankind's impacts on the climate. Thus, a new twist to Malthusian struggle emerged: the need to leave the Industrial Revolution behind and move to another form of human development – one based almost entirely on renewable energy sources.

In the last chapter, we saw that the necessity to innovate is an inherent part of the Malthus Enigma. Society has continually had to find more productive means of providing food to support a growing population. To the innovations in fertilizers, tractors, and crop breeding discussed in the last chapter, in this chapter we will touch upon innovations in other technologies. These range from the steam engines and blast furnaces of the Industrial Revolution to modern renewable energy technologies that are necessary to address climate change. The need for humanity to innovate goes beyond technologies for supplying food to technologies for supplying energy, too. Even before climate change became a critical issue, we were motivated by concerns about running out of energy supplies – another aspect of the first layer of the Malthus Enigma.

In this chapter, I demonstrate that there is a second layer to the Malthus Enigma, beyond the need to innovate. This second layer is all about scientific discovery – to learn about the state of the Earth. How can humans sustain themselves within the carrying capacity of the planet if we have poor or imperfect understanding of the nature of the interrelated systems of the Earth? Recall from Chapter 2 just how much difficulty we had even in recent decades understanding, let alone quantifying, the Earth's carrying capacity. To illustrate the scientific struggle to learn about the world around us, in this chapter, I recount how humans discovered our role in the planet's greenhouse effect. Of course, there are other aspects of the Earth system that humans have struggled to understand – and continue to wrestle with. The quest to understand how ecosystems function – discussed in the last chapter – is also part of the second level of Malthusian struggle. We will see further examples in later chapters, too – such as discovering the laws of thermodynamics and how they apply to the Earth system.

To learn how the Earth's greenhouse effect occurs – and our role in contributing to it – again requires conceptualization of the Earth system. In this chapter, I will cover the scientists who discovered the nature of the greenhouse effect – leaving fuller discussion of global climate models to Chapter 6. Another type of systems framing covered in this chapter, however, is for the development of human energy supply systems – including how we have to expend energy in the process of accessing energy supplies, as well as rebound effects that can occur when we try to be more efficient in our use of energy. Being able to frame and analyze the energy flows that cascade through societies is necessary for transforming to a sustainable energy system. Toward the aim of understanding global climate change, its inherent entanglement with energy supply, and potential possible solutions to the challenge, it is necessary to first start with understanding the Industrial Revolution.

The Industrial Revolution

The Industrial Revolution beginning in Great Britain in the eighteenth century left an unprecedented mark on world history. A series of interrelated technological inventions, combined with an entrepreneurial free-market economy, produced incredible transformations in manufacturing and transportation processes – with widespread cultural ramifications – that eventually spread to virtually all parts of the world. Arguably the most fundamental aspect of the Industrial Revolution was that it entailed a transformation of the British – and later global – energy system from one based on renewable, solar-driven energy sources (which include biomass, wind, and water power) to one based on fossil fuels, in particular coal.[25] This change of energy supply was more than just a result of the Industrial Revolution – it was the essence of the revolution. The Industrial Revolution produced new technologies, changes in economic productivity, and increases in wealth that at times seemingly blew away Malthusian concerns over the population problem. Yet, at the same time, the switch from a sustainable solar energy supply to one based on limited fossil fuels created new resource challenges and environmental stresses – including climate change – that we continue to wrestle with to this day.

Various scholars point to different decades as being the start of the Industrial Revolution. Of course, most inventions rely upon those that have preceded them, so defining a beginning is somewhat arbitrary, depending upon which inventions are considered pivotal. Of all the great and important inventions of this era, however, it is those of James Watt in dramatically improving the performance of the steam engine that most stand out. In 1763–1764, as a young engineer in Glasgow, Watt was asked to repair a model of an old and very inefficient Newcomen steam engine at the local university.[26] He had the good fortune of learning some of the thermodynamic principles upon which the engine worked from Professor Joseph Black at the University of Edinburgh. In 1765, Watt came up with an idea for drastically improving the efficiency of the steam engine – adding a separate condenser that enabled the engine to maintain its heat, rather than generating steam from cold water on every stroke. He developed a patent with Dr. John Roebuck, the proprietor of a local ironworks, but after financial difficulties, Watt moved to Birmingham, England, and went into partnership with the entrepreneurial manufacturer Matthew Boulton. Over a period of twenty-six years, the formidable partnership of engineer and financier developed, tested, and manufactured increasingly better steam engines. The improvements included double-acting engines that produced twice the power from the same stroke, rotative engines that turned shafts, and centrifugal governors that allowed control over the speed of the engine. By the end of his career, Watt had developed an engine capable of delivering ten horsepower,[27] increasing the efficiency of steam engines by a factor of five to ten.[28] These improvements to the efficiency of steam engines were fundamental to the Industrial Revolution.

Indeed, if there was a single day among the decades of the Industrial Revolution that best captured the essence of the time, perhaps it was March 8, 1776. It was on this day that one of the first two full-size steam engines constructed using Watt's principles – with a condensing boiler – was put to work at the Bloomfield coal mine in Staffordshire, England. Boulton and Watt made an event out of the occasion, attracting a small crowd of respectable onlookers. *Aris's Birmingham Gazette* produced a long article on the successful event, this being the first time that word of Watt's steam engine appeared in print.[29] In subsequent days, the news was spread by other newspapers in England, including the following excerpt in the *Bath Chronicle*:

Friday last a steam engine, constructed upon Mr. Watt's new principles, was set to work at Bloomfield colliery, near Dudley, in the presence of its proprietors, and a number of scientific gentlemen whose curiosity was excited to see the first movements of so singular and so powerful a machine; and whose expectations were fully gratified by its performance. From the first moment of its setting to work, it made about 14 or 15 strokes per minute, and emptied the engine pit (which is about 90 feet deep, and flood 57 feet high in water) in less than half an hour. ------ This engine is applied to the working of a pump 14 inches and a half diameter, which is capable of doing to the depth of 300 feet, or even 360 if wanted; with one fourth of the fuel that a common engine would require to produce the same quantity of power. The cylinder is 50 inches in diameter, and the length of the stroke is 7 feet. ------- These engines are not worked by the pressure of the atmosphere: Their principles are very different from all others. She was executed this day.[30]

It was particularly pertinent that Boulton and Watt's inaugural steam engine was installed at a coal mine. Coal mining was already established in Britain. Generally, it was easier and cheaper to use wood as a heating source for buildings, but in larger cities – particularly London – shipping of coal was more cost-competitive than pulling in large volumes of wood by horse and cart. London had been supplied by sea coal, much of it from Newcastle, since at least the thirteenth century. Many canals were also constructed in the eighteenth century to provide cost-effective waterborne transportation for heavy materials such as coal. Unless they were near a river suitable for a water wheel or were located with high elevation with potential to build a drainage shaft, most pre-1800 coal mines were relatively close to the surface. Mine shafts rarely went deeper than 200 metres – and then with great difficulty – and many pits were typically no deeper than eight to twelve metres.[31] The fundamental problem with digging deeper shafts was that one inevitably encountered groundwater. To drain the coal deposit, an old, inefficient Newcomen steam engine could be used to some extent. This, of course, entailed consuming large quantities of coal to drive the engine. Even if this could be done economically – for example, because there was cheap waste coal at a mine – Newcomen engines were still so inefficient that drawdown of the water table was limited.

This is one of the key reasons Watt's engine was so pivotal in the transformation of energy systems of that era. By installing a steam engine at Bloomfield Colliery that was four times more efficient than a Newcomen version, less energy was consumed in accessing the coal seam. In other words, there was a higher *return of energy on the energy invested*. Moreover, Watt's engine was capable of lowering the water table further, allowing access to coal seams unreachable with a Newcomen engine. By the end of the eighteenth century, Boulton and Watt were producing steam engines that were up to ten times more efficient than old Newcomen engines. So more and more coal became accessible and cost-effective to extract.

The Newcomen engines continued to be made up until 1800. This was essentially a matter of economics: Boulton and Watt charged a substantial royalty for their steam engines. When Watt's patent expired in 1800, however, the price of his steam engine dropped significantly, and it became the standard design. By this time, about ten percent of British coal production was being used to power steam engines.[32]

Watt's steam engine was at the centre of a virtuous cycle around which other layers of technology in the Industrial Revolution took off. Steam engines powering groundwater pumps at coal-mining sites produced growing quantities of cheap coal.[33] This supported a major transformation in iron production and then later a transportation revolution. The transformation in iron production was a switch in the source fuel from charcoal – derived from wood – to coke – derived from coal. The technological breakthroughs for producing iron using coke were made over the eighteenth century, but there were economic and geographic constraints. Iron smelting was traditionally conducted by heating with charcoal from wood. Mineral coal had been impossible to use as a heating source because it contained chemicals such as sulphur and phosphorus that contaminated the resulting pig iron. In 1710, Abraham Darby developed a new process for smelting iron using coke. He established a coke ironworks at Coalbrookdale in Shropshire, but it served only a niche market for cast iron goods, such as pots and frying pans, as well as cylinders, pistons, and pipes for steam engines. It took about fifty years for coke iron to become competitive with charcoal iron, as the cost of wood rose while coal became cheaper. In 1783–1784, Henry Cort invented the puddling process by which the last stage of steel production – the refining – could also be done using coke rather than charcoal. This meant that the location of iron production became fully independent from wood supplies;[34] hence, blast furnaces could expand in size with improving economies of scale – and iron could be produced in larger quantities at lower costs. The advances in the steam engine made by Watt were important to the transition from charcoal- to coke-based iron production. Not only did Watt's steam engine lower the price of coal, but most of the new large blast furnaces were also powered by steam engines. As industrial historian Rolf Peter Sieferle noted, "There should

83

Malthus Enigma

be no doubt that the steam engine accelerated the establishment of smelting with coke."[35]

The combination of more powerful steam engines and large-scale production of coke iron and coal – all at decreasing costs – led within the first decades of the nineteenth century to a transportation revolution. The first steam locomotive running on a full-scale railway track was built by Richard Trevithick, in 1804, at the Penydarren Ironworks, near Merthyr Tydfil in South Wales. Three years later, the American Robert Fulton built the first steamboat. The age of steam railways really took off in 1814, when George Stephenson built a steam locomotive at the Killingworth Colliery near Newcastle upon Tyne in England. Stephenson improved upon the design of Trevithick and other early pioneers, and in continuing to do so, he became the most successful builder of steam locomotives in Great Britain. In 1825, he built the locomotive for the first public steam service for the Stockton and Darlington Railway in northeast England; four years later, he produced the legendary Stephenson's Rocket locomotive, which was the most advanced of its day. All of these developments stemmed from Watt's improvements to the steam engine.

The rapid growth in railways throughout Britain after 1815 was indirectly aided to a large extent by economic ramifications of the Corn Laws.[36] The high tariffs that were applied to imported grain allowed British landowners to charge higher prices for domestic grains. Hence, the cost of keeping horses was high – thereby allowing trains to compete financially as an alternative means of transportation. Indeed, it is intriguing that Malthusian concerns about overpopulation were grounds for expanding the railways. In 1833 – the year before a reform of the Poor Laws – a government committee argued that railway expansion would decrease the need for horses, thereby freeing up land to feed people. Although perhaps convincing at the time – upkeep of animals does require lots of land – the expansion of the rail network was actually accompanied by increased demand for other transport modes; hence, horse traffic continued to grow until the invention of the automobile.

As the Industrial Revolution proceeded with layers upon layers of technology and ever-rising living standards, Malthusian concerns were seemingly confined to history. Yet as this chapter explains, Malthusian struggle was always there in the background. As new forms of energy supply emerged, they always had to provide sufficient energy return on investment. As technologies became more energy efficient, total energy use seemingly kept increasing, raising concerns over what is today called the rebound effect. Every so often a cry would go out: *What happens when we run out of energy reserves?* It was first made by William Stanley

Jevons, in 1865, with respect to coal, and later by M. King Hubbert, in 1956, for oil. Then, gradually, over the nineteenth and twentieth centuries, as society burned up more and more fossil fuels, came the hard recognition – the "inconvenient truth" – that the atmosphere was a greenhouse, and emissions to it altered the stability of the biosphere.

Energy Return on Energy Investment

The performance of Watt's revolutionary steam engine at Bloomfield Colliery – decreasing the amount of energy required to access an energy source – demonstrated the fundamental notion to human development of *energy return on investment*, or EROI. EROI has been more formally defined as "the ratio of how much energy is gained from an energy production process compared to how much of that energy (or its equivalent from some other source) is required to extract, grow, etc., a new unit of the energy in question."[37] To give an example, if one ton of coal is needed to power a pump for dewatering at a coal mine and one hundred tons of coal are recovered from the mine, then the EROI is one hundred to one. EROI is a general measure, applicable not just to coal or fossil fuel extraction, but also to renewable sources of energy. On the face of it, EROI is an important and straightforward measure. If humans require more energy to access an energy supply than they get out of it, then unless the energy output is of a more useful quality, it would be fruitless. Surely we would just use the initial source of energy for human comfort, sustenance, mobility, communication, etc., rather than burning it up in an attempt to exploit more energy.

Studies of EROI were at their height in the late 1970s and 1980s, as a consequence of the oil crises. Early work on the concept is apparent in the writings of systems ecologist Howard Odum and economist Kenneth Boulding. Quantitative studies began in the late 1970s, with the works of Charles Hall and Cutler Cleveland at Cornell University and Robert Herendeen from the Illinois Natural History Survey. Historical US values for the EROI of coal have been reported at eighty to one, while a study of oil and gas extraction in 1930 put the EROI at greater than one hundred to one.[38] The challenge is that EROI values have generally declined over time, as the most accessible fossil fuels get used up first. By 1999, the estimated EROI for global oil production was about thirty-five to one.[39] One study by Hall and Cleveland of US petroleum drilling, published in *Science*,[40] was highlighted on the front page of *The Wall Street Journal*. Hall and Cleveland's analysis of US data showed that there was a downward trend of two percent per year in the rate at which petroleum was being added to reserves per foot of drilling effort. Based on industry performance data for 1977, they calculated that 1.5 barrels of oil were used for each foot of drilling during petroleum exploration and development.[41] Hall and Cleveland concluded that, under the trends of the time, the energy produced by the US petroleum industry would be equal to the energy used in exploration by 2004. That would have meant an EROI of less than one to one. The US has managed to maintain domestic oil supplies for longer than that, although the EROI on oil production has been steadily decreasing.[42]

A declining EROI is an indication that society has to work harder and harder to access new energy reserves. This includes looking to alternative sources of energy, such as renewables – which are very much needed to address climate change anyway. In the US, one of the alternatives encouraged in the face of declining domestic oil supply was the production of biofuels – especially corn ethanol – for mixing with gasoline. Production of biofuels, however, requires a substantial amount of work: growing and harvesting crops, transporting them, and then refining. EROI calculations show that, at best, about two units of energy are yielded for every unit of energy that goes into producing corn ethanol.[43] This is a very poor return compared with conventional oil, which has an EROI of over ten to one.[44] From a purely energetic perspective, corn ethanol is not a sustainable alternative to oil. Moreover, its production competes for land with agricultural crops required for feeding people – another twist to Malthusian struggle. Since 1970, the response of the United States to declining domestic oil supplies was simply to import more oil from elsewhere. This changed around 2010, with advances in drilling and fracking technology that enabled the US to exploit untapped reserves of shale oil and gas. Still, shale oils are reported to have an EROI of about five to one[45] – and one day, too, their production may peak and begin to decline, or simply be surpassed by carbon-free renewable sources.

Predicting when, or if, supplies of non-renewable resources such as oil will peak, decline, or essentially run out is fraught with difficulty. The American geologist M. King Hubbert made a remarkable prediction about peak US oil production in a paper presented to the American Petroleum Institute in 1956.[46] Based on historical data and assumptions about known reserves, Hubbert correctly predicted that production from conventional sources would peak around 1965 to 1970. Domestic oil production peaked in 1970, at 10.2 million barrels per day, and moreover followed a curve very much like Hubbert had predicted up until the mid-1990s. At this point, a classical development under Malthusian struggle occurred: new technologies that could extract from unconventional sources, such as shale oil, began to yield additional production. There was a surge in fracking from 2010, which led to US oil production surpassing ten million barrels in November 2017 for the first time in almost fifty years.[47]

The Coal Question

Fossil fuels play a central role in Malthusian struggle – not only because they lead to greenhouse gas emissions, but also because they are of limited supply and increasingly hard to reach. Concerns about when they will run out are older than the US concerns about oil production in the 1970s, going back to at least the nineteenth century. *The Coal Question*, by English economist William Stanley Jevons – first published in 1865 – was concerned with the sustainability of Britain's coal resources. Jevons was a leading nineteenth-century political economist who contributed to the marginal revolution and was a professor at Owens College, Manchester, and later University College London. By the middle of the nineteenth century, the Industrial Revolution was transforming Britain, fuelled to a large extent by the country's substantial coal resources. Jevons recognized that England's economic power was significantly tied to exploitation of coal: "The greatness of England much depends on the superiority of her coal, in cheapness and quality, over that of other nations."[48] The statistics show that around 1860, about eighty million tons of coal per year were mined in Great Britain, which was more than sixty percent of the entire world's total. The United States was a distant second, mining fourteen million tons, with German states mining twelve million tons, Belgium nine million, and France eight million.[49] The amount of coal mined in Britain was, moreover, rapidly increasing, at an average rate of two and three-quarter million tons per year. Hence, the question had been raised as to whether – or indeed when – Britain's coal supplies would become exhausted. The full title of Jevons' classic was *The Coal Question: An Inquiry Concerning the Progress of the Nation, and the Probable Exhaustion of our Coal-Mines*. Jevons tackled the question from multiple perspectives – wrestling with geology, economics, and technological change. Based on an estimate that eighty billion tons of coal lay above a depth of 4,000 feet, he calculated that Britain's coal supply could last for only 212 more years if the annual recorded rate of increase were to continue into the future.[50] He was concerned that exhaustion of Britain's coal mines might occur even

Malthus Enigma

faster, diminishing its manufacturing abilities and, with it, public wealth and the comforts of Britain's "high civilization."[51]

In wrestling with the "coal question," Jevons made explicit connections back to Malthus's concerns over population. He showed some remarkable calculations that revealed that Britain's population growth rate over the eighteenth and nineteenth centuries was – in percentage terms – greatest around 1821.[52] He connected the population growth rate to the progress of the steam engine and other inventions that consumed coal.[53] He also showed that the population growth was primarily in manufacturing towns rather than in agricultural regions. Jevons recounted Malthus's law that human populations increased at a geometric rate when under consistent circumstances[54] – and the difficulty of increasing agricultural productivity to meet population demands. Jevons argued, however, that the ascendancy of manufacturing in Britain – based upon coal – enabled it, at least for a while, to evade Malthus's predicament. Aided by the repeal of the Corn Laws in 1846, which lowered tariff barriers on the import of corn, Britain was able to import the food it required from abroad through exchange for manufactured goods. Jevons stated that "the momentous repeal of the corn laws throws us from corn upon coal"[55] and, further: "For the present our cheap supplies of coal, and our skill in its employment, and the freedom of our commerce with other wider lands, render us independent of the limited agricultural area of these lands and apparently take us out of the scope of Malthus' doctrine."[56]

Jevons also went beyond Malthus in recognizing that technological change occurs. He provided statistics on the increasing efficiency of steam engines, noting how the pounds of coal consumed per horsepower per hour went from thirty in 1769 to one in 1900.[57] He remarked on the extraordinary impacts of the efficiency gains – enabling many more collieries to open up – and the evolution of the steam locomotive. He also presented a hypothesis that the pace of innovation tended to accelerate over time. Inventions often build upon and multiply upon each other, such that "[t]he tendency of progress … is to quicken progress."[58]

Ultimately, though, Jevons arrived back at Malthusian concerns. He recognized that increasing technological progress and increasing wealth led to increases in the rate of coal consumption, bringing the time of exhaustion closer. He was relatively pessimistic about exploiting deeper coal deposits below 4,000 feet and saw no realistic opportunities for substituting coal with other forms of energy supply. Oil, natural gas, nuclear, and modern renewables, lay decades away. Hence, he wrote, "there is no probability that when our coal is used up any more powerful substitute will be forthcoming."[59] Thus, he arrived back at the same bleak Malthusian position: "So far, then, as our wealth and progress depend upon the superior command of coal we must not only cease to progress as before – we must begin a retrograde career."[60]

Jevons' Paradox

One of the most enduring aspects of Jevons' inquiry – supporting his Malthusian leanings – was his view that increases in the fuel efficiency of technologies lead only to increased energy consumption. This position is now known as Jevons' Paradox. His basic reasoning is an economic one:

> *Economy multiplies the value and efficiency of our chief material ... [and] renders the employment of coal more profitable, and thus the present demand for coal is increased.*[61]

Jevons gives an example of the paradox using the blast furnace:

> *[If] the quantity of coal used in a blast furnace, for instance, be diminished in comparison with the yield, the profits of the trade will increase, new capital will be attracted, the price of pig iron will fall, but the demand for it increases and eventually the greater number of furnaces will more than make up for the diminished consumption of each.*[62]

This rebound effect to energy efficiency was clearly of importance to Jevons due to his questioning of how long Britain's coal supplies would last.

Interest in Jevons' Paradox re-emerged in the late 1970s and 1980s, but with considerable debate. Among those concerned about rebound effects were Daniel Khazzoom, from the San Francisco–based Institute for Research in Energy and Economic Modeling and the Business School at the University of California, Berkeley, and Leonard Brookes, who was with the UK Atomic Energy Authority in London. Khazzoom was concerned with the economic rebound from implementing efficiency standards for household appliances,[63] while Brookes's interest was in mitigating the effects of large hikes in energy prices during the 1970s oil crises.[64] Their conclusions on the significance of rebound effects were controversial, as they flew against policies that aimed to reduce energy use through increasing energy efficiency. Yet further studies emerged that supported the hypothesis that energy efficiency can backfire, in accordance with Jevons' Paradox. Examples included studies of

Malthus Enigma

household appliances in Denmark, commercial buildings in the United States, and lighting efficiency and residential home insulation in the UK.[65] Broader studies pointed to the rebound effect that occurred when the Bessemer process made steel manufacturing more efficient and the energy use increases in the US that came with productivity gains from using more efficient electric motors. Theoretical understanding of Jevons' Paradox also developed, with various mechanisms recognized.[66] The rebound can be a direct effect – where energy demand increases as a consequence of the lower energy costs resulting from energy-efficiency improvements. Various indirect effects can also occur, ranging from increases in energy use to manufacture energy-efficient devices, to increased energy consumption from households re-spending saved income on other goods and services.

In the broader context of this book, Jevons' Paradox – where it applies – can be seen as emblematic of the Malthus Enigma. As we find innovative ways to make our use of resources more efficient, this sometimes results in a rebound effect by which we use even more resources. We make our factories, buildings, appliances, and light bulbs more efficient, but our efforts come back and bite us. This is similar to the situation faced in agricultural production discussed in the last chapter: as we increase the efficiency of farming – producing more food per acre – the human population just grows more, unconstrained by any limitations in food supply.

But there is a second dimension to the struggle that is also apparent with Jevons' Paradox – and this relates to the uncertainty in the science. Despite a good number of studies demonstrating possible rebound effects, strong theoretical foundations, and consistency with the history of technology, many energy analysts are skeptical about the size and significance of rebound effects. One review of the literature by Steven Sorrell of the Sussex Energy Group in England found that the debate "tends to be polarized, theoretical and inconclusive."[67] Arguments made in support of the paradox are based on "a mix of theoretical argument, illustrative examples and 'suggestive' evidence."[68] The arguments are so inconclusive that "the possibility of large economy-wide rebound effects has been dismissed by a number of leading energy analysts."[69] The challenge is that teasing out the rebound effect – if or when it exists – from other influences on the economy and on energy consumption is difficult.

Whether or not Jevons' Paradox matters may depend largely on context. One commentary published in *Nature* claimed that the rebound effect is overplayed, but it was drawing upon analysis of consumer response to changes in gasoline prices.[70] It may well be the case that the rebound effect is small for energy-efficiency improvements to automobiles. People do not necessarily drive twice as much when the price of gasoline is halved. But Jevons' Paradox could still apply in other contexts. Recall that Jevons primarily observed an increase in coal consumption in response to marked increases in the efficiency of the steam engine. The context was an efficiency gain in a technology that was used for drawing down water to mine coal – that is, to produce energy. There can be no doubt that a rebound effect occurred, as that was the objective! In concluding his review, Sorrell suggests that the rebound effect is possibly more significant for general purpose technologies such as steam engines and computers.

The continued scientific debate over the significance of rebound effects is just one example of humanity seeking to understand the workings of the world. This is the second layer of the Malthus Enigma. An even greater scientific struggle, which we turn to now, has been our efforts to understand humanity's role in the greenhouse effect.

Svante Arrhenius and the Greenhouse Effect

On May 28, 1956, *Time* Magazine published a short article describing how CO_2 and water vapour can trap long-wave heat radiating from the Earth's surface, causing the atmosphere to act like "one big greenhouse." This was not the first time such a theory had been proposed – Eunice Foote, John Tyndall, and Svante Arrhenius had similar ideas in the nineteenth century – but it signalled a resurrection of the greenhouse effect as a mainstream scientific theory – and an issue of considerable concern to humanity.[71] The article indicated that continued burning of fossil fuels could result in violent changes to the Earth's climate in about fifty years.[72] *Time* Magazine observed that in 1950, nine billion tons of CO_2 were poured into the atmosphere from anthropogenic sources, and that if trends continued, emissions would reach forty-seven billion tons per year by 2010. The actual emissions for 2010 were later estimated to be 30.3 billion tons, excluding other greenhouse gases.[73] Moreover, the *Time* Magazine article warned of a possible unknown "chain of secondary effects," whereby as ocean waters warmed, more CO_2 and more water vapour would enter the atmosphere, compounding the greenhouse effect. These reinforcing effects could raise temperatures "enough to melt the icecaps of Antarctica and Greenland, which would flood the Earth's coastal lands."

Malthus Enigma

This level of understanding of the existence, cause, and impacts of climate change is overwhelmingly accepted by climatologists today,[74] but the manner in which the theory arose, went out of fashion, and then came back is emblematic of the human struggle to understand the world around us. As early as 1859, John Tyndall had, through an interest in glaciology, established that both water vapour and CO_2 in the atmosphere had the ability to trap heat rays and impact climate. Svante Arrhenius took the idea further, concluding in 1908 that "any doubling of the percentage of carbon dioxide in the air would raise the temperature of the earth's surface by 4°."[75] Scientists today might haggle with Arrhenius's numbers, although he may not be too far off. The bigger question, though, is why it took fifty to one hundred years more for anthropogenically induced climate change to be accepted by mainstream scientists.

Svante Arrhenius was a heavyweight among scientists. Born near Uppsala in 1859, he studied at the Royal Swedish Academy of Sciences in Stockholm, where he produced a thesis on electrolyte conductivity.[76] One of the key ideas he later developed, based on his thesis experiments, was a theory that many readers will readily recall from high school chemistry: when a solid crystallized salt is dissolved in water, it dissociates into charged particles (for example, Na^+ and Cl^- in the case of sodium chloride, ordinary table salt). The notion of ions had been established before Arrhenius, by Michael Faraday. But the idea that ions existed in solution – even in the absence of an electric current – came from Arrhenius. Taking matters further, Arrhenius established that the chemical reactions occurring in solutions could be understood as reactions between ions. He also extended his theory to provide today's broad understanding and definitions of acids and bases – and he developed an equation for the energy required to activate a reaction between two molecules. For these developments, Arrhenius is recognized as one of the pioneers of physical chemistry. Indeed, he was awarded a Nobel Prize for his work on ionic theory in 1905 – although there's more to say on this.

Arrhenius was an extremely controversial figure. As is often the case with revolutionary scientific theories, Arrhenius's ideas were met with considerable skepticism at first. When his thesis developed in Stockholm was submitted to Uppsala University, it was snubbed by his former teachers; he received a fourth-class degree, which was only later raised to a third class.[77] Unable to find paid work at a university in Sweden, Arrhenius was fortunate that his research was valued by the German chemist Friedrich Wilhelm Ostwald. Arrhenius eventually moved to Germany and, through collaboration with Ostwald, German physicist

One Big Greenhouse

Walther Nernst, and Dutch chemist Jacobus Henricus van 't Hoff, formed the "wild army of the Ionists"[78] – the first physical chemists. Nernst and Arrhenius were close drinking friends at this stage, although they later fell out. Arrhenius returned to Sweden in 1891, rejoining the Högskola in Stockholm, where he became chair of physics in 1895.[79] Around this time, Arrhenius, who was now a scientist of notoriety, had the good fortune to become involved in setting up the Nobel Foundation. Alfred Nobel had left a sizable fortune to establish international prizes for physics, chemistry, medicine, literature, and peace. There was some confusion in interpreting Nobel's will, and Arrhenius was involved in resolving matters. In doing so, he really had a conflict of interest – not least of which that he was a candidate for one of the prizes.[80] Arrhenius became so involved that he set up the rules for nomination and the selection of prize winners – and also diverted funds toward his school. As one commentator noted: "He used the Nobel prize and foundation to build a personal fiefdom ... he would use his power to reward his friends and punish his enemies."[81] Ostwald, for example, though no doubt an exceptional scientist, was possibly awarded a Nobel Prize due to Arrhenius's influence. Arrhenius himself declined to share the first Nobel Prize in Chemistry with van't Hoff – perhaps in part because he wanted the prize for physics. He was eventually awarded the third Nobel Prize in Chemistry in 1903.

By the time of his appointment at the Högskola in Stockholm, Arrhenius had, however, given up physical chemistry, and he turned to other scientific questions. One of these was an investigation of how ice ages occur, which led him to perform the first calculations of how changes in atmospheric CO_2 concentration could alter surface temperatures through the greenhouse effect. To do the calculations, Arrhenius was able to draw upon estimates of CO_2 emissions from natural processes, as well as factory and industrial processes determined by his colleague Arvid Högbom. The data at that time suggested that anthropogenic emissions were of similar size to natural emissions.[82] Arrhenius came to the realization that emissions of CO_2 from human activity could change the future climate, and he concluded that the effects could occur within a few centuries.[83] In a world of competitive scientists, however, Arrhenius's work was challenged by his countryman Knut Ångström. In 1900, Ångström published the infrared spectrum for CO_2,[84] concluding, incorrectly, that the spectrum was identical to that of water vapour and therefore CO_2 could not account for climate change.

It took almost fifty years for Ångström's mistake to be rectified. Writing in 1956, the same year as the *Time* Magazine article, Canadian scientist Gilbert Plass observed: "Although the carbon dioxide theory of climatic change was one of the most widely held fifty years ago, in recent years it has had relatively few adherents."[85] There were several competing theories that could explain the Earth's temperature rise since 1900, such as fluctuations in solar radiation received from the sun or the influence of volcanic dust in the atmosphere. Another theory even considered changes in the average elevation on continents, although that could occur only on geological time scales. The most significant arguments against CO_2-induced climate change were still, following Ångström, based on the absorption spectrum of CO_2. It was known that water vapour could absorb infrared radiation – similar to CO_2 – and many scientists thought that water vapour would dwarf any impacts that CO_2 concentrations might have on the greenhouse effect. The main person to revive the CO_2 theory of climate change was Guy Stewart Callendar, discussed further below, who reviewed scattered measures of the absorption spectrum of CO_2. Several further developments also enabled scientists in the mid-twentieth century to show the importance of fossil fuel combustion on climate change, including improved measurement of infrared absorption and access to computers. Gilbert Plass, for example, worked on the revived CO_2 theory of climate change while on faculty at the physics department at Johns Hopkins University (including a sabbatical year at Michigan State University, where he had access to a large computer). Plass had made accurate infrared measurements showing that the frequencies at which water vapour and CO_2 absorb infrared radiation were relatively independent of each other. He furthermore observed that concentrations of water vapour in the atmosphere rapidly reduced with height above the Earth's surface, whereas CO_2 was relatively uniformly distributed. In other words, CO_2 absorption is significantly larger than water vapour absorption beyond a short distance above the Earth's surface.

Callendar's Revival

After Arrhenius's nineteenth-century theory of human-induced climate change had been rejected, the man whose work largely led to its revival had an unlikely background. Guy Stewart Callendar (1898–1964) was a talented British engineer who worked on combustion processes, specialized in infrared physics, and made some notable contributions to British military research during World War II, on top of being an avid meteorologist. He was born in Montreal, Quebec, in 1898, but soon moved to England, when his father, a highly accomplished physics

One Big Greenhouse

professor, returned to take a chair at University College London, later moving to Imperial College London.[86] Guy was brought up in Ealing, West London; attended St. Paul's School in Hammersmith; and worked at his father's laboratory at Imperial College London, before completing an engineering degree at the City and Guilds of London Institute. In his early career, he worked with his father on establishing international steam tables, which defined the standard properties of steam at various temperatures and pressures. A few years after his marriage in 1930, Callendar travelled to America to participate in an international steam table conference, and he continued to investigate properties of steam prior to the outbreak of World War II.

Callendar began to take a significant interest in climate change in the late 1930s. He had read early works on the subject, such as those by Arrhenius and Nils Ekholm,[87] and became a meticulous keeper of meteorological data, filling "dozens of research notebooks"[88] with temperature data from stations around the world. Callendar published his first paper on climate change in 1938, "The Artificial Production of Carbon Dioxide and Its Influence on Temperature." In the paper, he first presented results for the concentration of CO_2 in the Earth's atmosphere as of 1900 (274–292 parts per million) and in the late 1930s (289–310 parts per million), showing a six percent increase.[89] He estimated that human actions over the past fifty years had added about 150,000 million tons of CO_2 to the atmosphere, of which three-quarters were still remaining. The bulk of Callendar's paper then showed new calculations for the radiative effect of CO_2 on the atmosphere, bringing to bear new information on the vertical structure of the atmosphere, the impacts of pressure on absorption, and detailed measurements of the infrared absorption spectrum of water vapour. Thus, he was able to conclude that two-thirds of the world's warming over the past fifty years was due to anthropogenic emissions. Moreover, he predicted that a doubling of CO_2 concentrations in the atmosphere would produce a 2°C rise in global temperatures.

Callendar's revival of the CO_2 theory of climate change was met with considerable skepticism, but he continued to publish a huge body of research on the topic. Members of the Royal Meteorological Society were among the most questioning at first, but the opinions of many critics began to change with Callendar's further efforts. One of his papers in 1941 was particularly influential, in which he reviewed spectroscopic measurements of the absorption bands of CO_2 that had been scattered throughout a dispersed scientific literature.[90] The results showed only limited overlap between the absorption frequencies of CO_2 and water vapour. This was followed by further study, published in 1944, with Gordon Sutherland, in which the infrared spectrum of the atmospheric gases other than water vapour were measured.[91]

There was some irony in Callendar's life circumstances during his early work on climate change. He was clearly not an environmentalist: his 1938 paper discussed the benefits of burning fossil fuels to stimulate plant growth and perhaps indefinitely delay the "return of deadly glaciers."[92] Moreover, during World War II, he was employed as a researcher by the Petroleum Warfare Department and the Ministry of Supply. The most important project Callendar worked on was development of the Fog Investigation and Dispersal Operation (FIDO) system, which involved combustion of huge quantities of petroleum.[93] The problem of fog preventing use of British airfields during the War was a major strategic issue. Churchill and his War Cabinet spared no expense in developing the FIDO system to allow British air raids to take place on foggy nights, with aircrews returning safely. Employed at Langhurst in West Sussex, Callendar researched and designed several key components of FIDO, including the trench burners. He was among the patent holders of the system. His FIDO research involved several large-scale experiments, some of which took place indoors at the Empress Hall in Earl's Court, London. There, he used the ice-making machinery at the large skating rink to produce synthetic fog, which he would heat up in a wind tunnel using various arrangements of burners. When eventually perfected in November 1943, the operational FIDO comprised a system of burners arranged on both sides of a runway, which, when turned on, would warm the air above by several degrees, thereby evaporating the fog. The system was massively energy intensive, with some 6,000 gallons of petroleum typically consumed during the four minutes it took to land an aircraft.[94] It was highly successful, being used in the landings of approximately 2,500 British and American airmen. By the end of the war, some thirty million gallons of petroleum had been used by the FIDO system.

Keeling and the Measurement of CO_2

Measurements of CO_2 concentrations in the atmosphere, which had been crudely pieced together by Callendar, reached a higher level of accuracy and consistency through the work of Charles David Keeling (1928–2005). Raised in a Chicago suburb during the 1930s, Keeling studied at the University of Illinois Urbana-Champaign and Northwestern University before starting a post-doc in a new geochemistry department at Caltech in 1953.[95] There, he was challenged by his advisor to develop a method for determining the concentration of carbonates in groundwater, based upon the equilibrium with carbon in limestone and in the atmosphere. A problem with the proposed procedure, however, was that methods of measuring CO_2 in the atmosphere were crude; hence, Keeling set about constructing his own device. His first measurements at the Caltech campus showed high variation, as expected, due to local human activities. So further measurements were made at Pfeiffer Big Sur State Park in California. This location was much to Keeling's liking: he was an outdoorsman who loved geology, mountain climbing, and studying glaciers, and he met his wife Louise during a canoe trip in Canada. The success of Keeling's initial instrumentation caught the attention of Harry Wexler, from the weather bureau in Washington, DC. The two met, and Keeling was able to convince Wexler that even better measurements could be made using an infrared gas analyzer.

In 1956, Keeling obtained a position at the Scripps Institute of Oceanography in San Diego. From there, with Wexler's support, he set about establishing four CO_2 measurement stations: one at the Mauna Loa volcano in Hawaii, one at the South Pole, another back in San Diego, and the last aboard a ship. From his improved measurement devices, Keeling was able to discern a seasonal fluctuation in CO_2 concentrations that could be explained by the natural withdrawal of CO_2 from the air during the summer plant-growing season. After a few years of data collection, moreover, he observed a background trend of gradually increasing concentrations over time. The average annual concentration was increasing at a rate of about 0.7 parts per million per year in 1960, and by 2000, the rate of increase had reached about 2.0 parts per million per year (still being measured with one of Keeling's devices).[96]

Keeling's measurements obtained at the South Pole were pivotal to a landmark paper in which he noted: "[A]t the South Pole the observed rate of increase is nearly that to be expected from the combustion of fossil fuel."[97] Funding cuts prevented continued monitoring of all four stations, but Keeling was able to maintain measurements at the Mauna Loa site continuously from 1958. This became what one colleague described as "the single most important environmental data set taken in the twentieth century."[98] In 2002, Keeling was awarded the National Medal of Science, the highest award in the United States for scientific research.

Malthus Enigma

Captain of Spaceship Earth

With growing evidence from Keeling and others that the greenhouse effect was a problem – and mounting concerns about a host of other environmental stresses – in the early 1970s, the world needed a climate change leader. The person who perhaps more than anyone stepped up to this role was Maurice Strong, who, on the face of it, came from an unlikely background, being a Canadian oil executive.[99] Strong was a staunch socialist who was concerned about human development – and saw the emerging environmental agenda as a means to address a wide range of economic and equity issues faced by the world. The Swedes were worried about slow progress with organization of the first United Nations Conference on the Human Environment – held in Stockholm in 1972 – so Strong, who was a consummate diplomat and a determined and effective leader, was asked to become Secretary-General of the conference, picking up the title Under-Secretary-General of the United Nations responsible for environmental affairs. He did an impressive job, encouraging reluctant developing world leaders to attend, including Indian Prime Minister Indira Gandhi. The conference was considered a great success, producing the Stockholm Declaration of Principles and Action Plan for the Human Environment to deal with global environmental issues. With Strong's leadership, a new era of international diplomacy on the environment began. The *New Yorker* Magazine proclaimed him to be "Captain of Spaceship Earth."[100]

Strong was born in 1929 to a relatively poor family in Oak Lake, a small town in southern Manitoba. Encouraged by an austere school principal and his mother, a former teacher, Strong graduated from high school early and went travelling. Apparently, in the early 1940s, Strong was riding freight trains across Canada, headed for Vancouver, when he heard news that Churchill and Roosevelt were planning to form an organization called the United Nations, once World War II was over. The aim was to bring peace and justice to the world; this impressed upon Strong a desire to one day work for the United Nations. A year or so later, he had gained an apprentice role at a remote northern trading post of the Hudson's Bay Company. There, he met an American prospector, Bill Richardson, whose wife was the heiress to an oil fortune. The well-connected couple somehow managed to arrange for Strong to get a low-level job as a junior security officer at the temporary United Nations headquarters, which were then at Lake Success on Long Island. Strong's goal was thus achieved early, but he aspired to have more influence.

One Big Greenhouse

Realizing that he needed to get some better qualifications, Strong returned to Canada and trained as a specialist in oil and mineral resources. He moved to Calgary and learned the operational side of the business from a charismatic leader of the oil industry named Jack Gallagher. Strong then decided to set up his own business and took over a small, failing natural gas company. He was so successful at turning it around that the Montreal-based Power Corporation of Canada – an investment company specializing in the energy sector – encouraged him to become their executive vice-president and, later, president. In Montreal, Strong grew into influential political circles – including those around Canadian Prime Minister Lester B. Pearson. Liking Strong's view that Canadian foreign policy should focus on development, Pearson appointed Strong to become a deputy minister and eventually the first head of the Canadian International Development Agency. In this role, Strong thus returned to the United Nations.

Following his leadership of the Stockholm conference, Strong took on myriad influential roles internationally and back in Canada, while maintaining his own business interests. He was appointed the first head of the United Nations Environment Programme, established in Nairobi in 1972, before returning to Canada to head the new national oil company Petro-Canada, and later Ontario Hydro. He continued to take assignments for the United Nations, including leading the famine relief program in Africa. Strong was chair of the World Resources Institute, foundation director of the World Economic Forum, a member of the Brundtland Commission, senior advisor to the president of the World Bank, and an international advisor to Toyota Motor Corporation. He was on the board of numerous organizations, including the International Institute for Sustainable Development, the Stockholm Environment Institute, the World Business Council for Sustainable Development, the World Conservation Union (now the International Union for Conservation of Nature), the World Wildlife Fund, and Resources for the Future. Twenty years after Stockholm, the United Nations Conference on Environment and Development, also known as the Earth Summit, was held in Rio de Janeiro, and Strong was again asked to lead – taking on the role of Secretary-General. Attended by numerous heads of government, the Earth Summit produced agreements on biological diversity and climate change. Of particular importance to Strong was an agreement on principles and an action plan toward global sustainable development known as Agenda 21.

Although the world's emissions of CO_2 and other greenhouse gases have continued to increase, the Earth Summit and the Stockholm

Malthus Enigma

conference – both led by Strong – achieved the difficult and necessary step of establishing the political architecture of the United Nations for addressing climate change. This included, for example, the formation of the United Nations Framework Convention on Climate Change, the Intergovernmental Panel on Climate Change (IPCC), and the Kyoto Protocol. Perhaps a sign that his work was done, Strong died in 2015, shortly before the landmark United Nations Climate Change Conference in Paris. The 21st Session of the Conference of the Parties (COP 21)[101] brought further agreement and hope that the world could find a way to substantially reduce global greenhouse emissions.

The Soft Path Forward

Underlying part of the optimism of the United Nations Climate Change Conference in Paris were rapid increases in power generation using renewable sources that had occurred over the previous decade. From 2005 to 2015, the amount of global electricity supplied by wind power increased by a factor of eight, while the amount from solar photovoltaics increased by a factor of sixty-two.[102] Much of this progress was due to feed-in tariffs, which provided financial payments to generators of renewable power who fed back to the electric grid. With the development of feed-in tariffs and other incentives from government, along with technological improvements, the prices for wind turbines and solar cells had continually declined – making renewable power financially competitive with fossil fuel generation. Over the decade, wind and solar together went from generating 0.6 percent of the world's power supply to 4.6 percent. In some countries the penetration was even higher, with Germany, which led the way, having over eighteen percent of its electricity supplied by wind and solar in 2015. By the time of the IPCC's *Sixth Assessment Report* in 2022, prices for wind and solar power had become competitive against fossil fuels, and the design of net-zero electricity systems had become realistic in many places.[103]

The potential to develop energy supply systems dominated by renewable sources was encouraged by American energy guru Amory Lovins in the early 1970s. In one of Lovins' classic articles, "Energy Strategy: The Road Not Taken?" published in *Foreign Affairs* in October 1976, he begins by asking "Where are America's formal or de facto energy policies leading us? Where might we choose to go instead? How can we find out?"[104] The answer Lovins provided to the second question was soft energy technologies, which he argued were not only flexible, resilient, sustainable, and benign, but also more economically stable.

Central to Lovins' soft path were renewable energy sources – such as solar, wind, and biomass. He also placed a strong emphasis on energy-efficiency measures and was open to high-efficiency clean coal as a transitional technology. The hard path to be avoided, in his view, was a continuing upward trend in energy use dominated by oil, natural gas, and especially capital-intensive nuclear-generated electricity or large hydropower.

Looking back at the development of US energy use in 2011, Lovins noted that his projection of total energy use under the soft-path scenario was remarkably close, within four percent of the actual trend.[105] His understanding of the potential of energy-efficiency measures had proven to be correct. Energy efficiency was an endeavour that Lovins himself worked on passionately. He wrote extensively on the great potential and untapped benefits of energy efficiency, especially in buildings and cars, producing influential reports for governments and numerous journal articles. Lovins also authored or contributed to many popular books, including *Factor Four: Doubling Wealth, Halving Resource Use*; *Winning the Oil Endgame*; *Reinventing Fire*; *Factor Ten Engineering*; and *Natural Capitalism*. Moreover, he put his ideas into practice at his Rocky Mountain Institute, housed in Lovins' super energy-efficient home, high in the mountains, complete with a banana tree in the greenhouse. Uptake in the installation of renewable energy technologies was, however, only just beginning in 2011 when Lovins was reflecting. Between 1970 and 2010, there had not been sufficiently aggressive public policy for development of renewables, except perhaps in Germany, which initiated feed-in tariffs in 2004.

By the time of the United Nations Climate Change Conference in Paris, the world's largest greenhouse gas–emitting countries had produced studies showing how renewable sources could provide most of their electricity production by mid-century. First was the *Renewable Electricity Futures Study* by the US National Renewable Energy Laboratory in 2012, which looked at the feasibility of integrating high quantities of biomass, geothermal, hydroelectric, solar, and wind power into the electricity system. The study found:

> *The central conclusion of the analysis is that renewable electricity generation from technologies that are commercially available today, in combination with a more flexible electric system, is more than adequate to supply 80% of total U.S. electricity generation in 2050 while meeting electricity demand on an hourly basis in every region of the United States.*[106]

Malthus Enigma

Three years later, the Chinese government produced a report of similar ambition.[107] The *China 2050 High Renewable Energy Penetration Scenario and Roadmap Study* showed the way forward to supplying over sixty percent of China's total energy needs and over eighty-five percent of its electricity from renewable sources. A striking aspect of the roadmap was that it envisioned China's use of coal to peak in 2020. The potential to meet much of the energy needs in China and the United States from renewables is promising for the whole of the planet. From a systems perspective, though, there is a bit more to say here, learning from an occurrence in Germany.

A Factory in Dresden

In 2009, a short article appeared in the *European Energy Review*, celebrating and promoting advancement in the manufacture of photovoltaic panels in the German state of Saxony.[108] At the time, there were fifteen manufacturers of photovoltaic products in Saxony, employing 2,500 people and generating annual revenues of 1.6 billion euros. Among the companies mentioned in the article was Von Ardenne Anlagen GmbH – a manufacturer of vacuum solar plants for solar cells and modules. Although the company manufactured just one component of the photovoltaic energy supply system, there was something notable about its facility:

> *Von Ardenne Anlagen GmbH, a Dresden-based company, is gaining hands-on experience of thin-film technology. The exterior wall of its assembly building is adorned by a 300 m^2 photovoltaic system which has 420 thin-filmed modules installed, made from the compound semiconductor copper indium diselenide. The system was manufactured by Würth Solar GmbH using coating equipment made by Ardenne Anlagen itself.*[109]

The Ardenne Anlagen factory is probably not the first facility in the photovoltaic manufacturing process to generate electricity from its own photovoltaics. There may well be earlier factories in Germany or another country that has been active in manufacturing photovoltaics. Nonetheless, the occurrence is arguably as significant as Boulton and Watt's first demonstration of the improved steam engine at Bloomfield Colliery back in 1776. If a factory, or a group of factories, can be built with the capacity to generate more electricity on the rooftop than the equivalent energy required in its construction, then a circular, self-reinforcing, net-useful energy-generating process is at play. We can now use energy from the sun to construct further facilities for manufacturing photovoltaic cells without requiring the use of fossil fuels.

Actually, the conditions necessary to create a circular self-sustaining energy system are not as stringent as the Ardenne Anlagen factory example above. All we need is for the photovoltaic cells produced by the factory over its lifetime to generate more energy than is required to produce them – including construction of the factory. To do the calculation of the energy inputs required, we need to study what industrial ecologists call the life cycle of the product, which includes the processes of mining, transporting, and transforming the material inputs to the production process, as well as operating the facility and disposing of or recycling wastes. The analysis is also generalizable to other renewable energy technologies, such as wind turbines. The turbines need not be on the roof of the production facility; they are better located where wind conditions are greatest. Although slightly simplistic, the energy return on investment (EROI) measures that we discussed earlier indicate that self-sustaining energy systems based on renewables are possible. The energy generated by a photovoltaic cell or a wind turbine clearly depends on the local solar conditions or wind speeds, but typical EROI values reported before 2010 were about seven to one for photovoltaics[110] and about nineteen to one for wind.[111] Moreover, EROI values are increasing with improvements in technology.[112] The energy returns are not as high as early-twentieth-century values for coal and oil, which were above eighty to one. This signifies that twenty-first-century society needs to work harder to access the energy it needs to sustain itself. Renewables, nonetheless, can form the backbone of a new sustainable energy system.

For the world to dramatically decrease human-caused greenhouse gas emissions – and save itself from the severe consequences of climate change – it needs to undertake a complete system change. Over the past 200 years, industrialized countries have grown and developed infrastructure systems, technologies, and lifestyles that are locked in to relatively abundant supplies of fossil fuels. Human societies need to go through a transformation of industry equivalent to the one Watt's steam engine began over 200 years ago. Power generation from renewable sources will be central to the new system but wrapped with layers of other technologies, such as electric vehicles and energy storage. The transformation will entail electrification, decarbonization, and a reduction in energy use – occurring on multiple scales, from buildings and communities[113] to cities[114] and entire nations.[115]

Achieving deep reductions in greenhouse gas emissions will require society to transform to a state that Nicholas Georgescu-Roegen referred

Malthus Enigma

to as a third type of Promethean recipe.[116] Georgescu-Roegen (1906–1994) – who was a major influence on ecological economics – borrowed this term from the ancient Greek myth of Prometheus, who gave fire to humans after stealing it from the gods. With control of fire, a Promethean I technology, it was possible for humans to cook food, keep warm, and protect themselves. Georgescu-Roegen considered the development of heat engines, including steam engines, internal combustion engines, and nuclear reactors, to be a Promethean II technology. A world that reverses the greenhouse effect – recognized by the labours of Arrhenius, Callendar, Keeling, and others – will do so only by employing Promethean III renewable energy technologies. Georgescu-Roegen defined Promethean III devices to be solar energy collectors that generated sufficient net energy to provide for the manufacture of another solar collector of the same variety – including wind power and photovoltaics.

Conducting the Science to Understand the World Around Us

In this chapter, I wove the scientific discovery of humankind's role in the greenhouse effect into a wider story about the development of sustainable energy systems. The main take-away with regard to the Malthus Enigma is that humanity needs to keep undertaking the science to understand the Earth system. Scientific learning can, of course, help with technological innovation, too – the first level of Malthusian struggle. The main point, though, is that science is required and must continue to evolve so we can understand the constraints of the planet on which we live. This is the second layer of the Malthus Enigma.

The story told here of discovery of the greenhouse effect, including Eunice Foote's forgotten experiments and Ångström's erroneous rejection of Arrhenius's work, later revived by Callendar, shows how the science has indeed been a struggle. Moreover, despite the critical measurements of CO_2 concentrations by Keeling, at the time of the 1972 Stockholm conference, global cooling was still being seriously considered as likely an outcome as global warming. Only with the advancement of global climate models and the evolving experience of a changing climate – just as *Time* Magazine predicted – are scientists today confident of humankind's role in global climate change. Still, there is much we still do not understand well about the Earth system – as was clearly communicated from the work on planetary boundaries.

Transforming into a new solar society – to Promethean III – humanity will need to continue to learn about the nature of the planet, and the ways

One Big Greenhouse

in which we impact it. Living within the planetary boundaries of carbon, biodiversity, and others may require further technological development, but it will also require more science. In the same way that humanity struggled for decades to understand the causes and consequences of the greenhouse effect, it will have to evolve a new science that understands the Earth as a sustainable system. The roots of this science are already in place in the field of thermodynamics, which is what we turn to next.

In the next chapter, we will continue to add elements of the first two layers of the Malthus Enigma, but we will also start building toward the third layer, that of developing policies for addressing global environmental challenges and resource constraints. We have seen from the current chapter that, at least on the issue of climate change, the science is well established, and much of the technology for reducing greenhouse gas emissions exists. Despite the work of Maurice Strong and the United Nations to establish the political architecture for addressing climate change, policies for mitigation are still seriously lacking in most countries. This, arguably, is symptomatic of a higher-level problem of failure to incorporate ecology into economic decision making. Recall in Chapter 2, we identified the question of how to bring ecological principles into economics. An understanding of ecology is also essential to developing technological systems to be in harmony with Nature – a question identified in Chapter 3.

To bring ecological knowledge into our decision making and technologies, we first have to recognize that ecology is fundamentally based on physics – in particular the laws of thermodynamics. So to develop the third layer of the Malthus Enigma with a socio-ecological systems lens, the next topic we need to tackle is thermodynamics. Hence, in the next chapter, we ask: *How did we discover the laws of thermodynamics, and of what relevance are they for sustaining humanity?*

Chapter 5: The Science of Sustainability

Everyone knows that heat can produce motion. That it possesses vast motive-power no one can doubt, in these days when the steam-engine is everywhere so well known.

To heat also are due the vast movements which take place on the earth. It causes the agitations of the atmosphere, the ascension of clouds, the fall of rain and of meteors, the currents of water which channel the surface of the globe, and of which man has thus far employed but a small portion.

Nicolas Léonard Sadi Carnot, 1824[1]

From the French Revolution in the late eighteenth century through the turbulent decades of war and counter-revolutions of the nineteenth century, the nation of France produced a remarkable number of exceptionally brilliant engineers, mathematicians, and scientists. The names of many of these famous scholars were engraved in gold on the Eiffel Tower upon its opening in 1889. Among them is the name Carnot, but this refers to Lazare Carnot (1753–1823), the mathematician, military strategist, and politician. Strangely missing recognition in the golden letters is Lazare Carnot's son, Nicolas Léonard Sadi Carnot, who is now seen as a key founding figure of the science of thermodynamics. During his relatively short life, Sadi Carnot (1796–1832) wrote just one short manuscript: *Reflections on the Motive Power of Heat*. Yet it contained the first recognized exposition of the second law of thermodynamics – a theory that would aid development of technology from the nineteenth century, be expanded to living systems in the twentieth century, and today provide insights into Malthusian struggle through its application on a global scale. In this chapter, I am going to gently take you through the historical progress of thermodynamics to explain how it became an underlying science for sustainable human development.

Malthus Enigma

Through his father, Lazare Carnot – an influential figure in French society – Sadi Carnot's life and work were intimately intertwined with the successes and failures of France's great emperor, Napoleon Bonaparte. Following the French Revolution, in 1791, Lazare became a member of the Legislative Assembly, and subsequently the National Convention. He was the main architect of the French Revolutionary Army, which found itself embroiled in European wars upon the overthrow of the *ancien régime*. It was his organizational skills that lay behind the resurgence of the French army, which began to win battles against European neighbours after 1793. Indeed, it was Lazare Carnot who promoted his protégé Napoleon from the rank of captain to that of general. When Sadi was born in 1796, he lived the first year of his life at the smaller Luxembourg Palace in Paris, where his father was one of five executive members of the French Directory.[2] At age one, however, Sadi was forced to flee with his mother to Saint-Omer in the north of France, following a coup d'état, from which Napoleon eventually came to power. Sadi's father was exiled to Switzerland and Germany at this time, but by 1800, Lazare was back in Paris as Napoleon's minister of war. In December 1804, when Napoleon crowned himself emperor, Lazare Carnot resigned his post in protest and went into retirement. Yet he returned to serve Napoleon again in various functions, including minister of the interior during Napoleon's Hundred Days resurgence in 1815.

In his youth, Sadi Carnot showed the same intellectual abilities and strong personality as his father. When he was four, Sadi would sometimes accompany his father on visits to Napoleon's residence and headquarters at Château de Malmaison, to the west of Paris. There, Sadi was a favourite of Madame Bonaparte.[3] Apparently, on one occasion, Madame Bonaparte and her ladies were in rowboats on the lake when, for his pleasure, Napoleon started throwing stones into the water to splash them. Deeming this to be improper conduct, the four-year-old Sadi placed himself between Napoleon and the ladies, raised his fists, and gave a scolding to the soon-to-be Emperor, much to everyone's amusement.

At age sixteen, Sadi entered the École polytechnique, the oldest and perhaps most prestigious of France's *Grandes écoles*. Founded in 1794 by his father, along with mathematician Gaspard Monge, the École polytechnique was transformed into a military academy by Napoleon in 1804. Sadi gained his admission on merit, placing twenty-fourth out of many applicants on the entrance exam. The professors at the school included some of the most famous French scientists and mathematicians, men such as: André-Marie Ampère, whose name is used for the scientific measure of electric current; Joseph Louis Gay-Lussac, a physical chemist who worked on gases; and Siméon Denis Poisson, who made numerous contributions to mathematics. One of his classmates was Gaspard-Gustave de Coriolis, who produced an early mathematical description of the Coriolis Effect for objects under rotation. Situated near the Panthéon on the inspiring Montagne Sainte-Geneviève – a gentle hill on Paris's Left Bank – the École polytechnique went on to produce many of France's most brilliant scholars, several French presidents, and CEOs of large French corporations. Upon graduating from the school, Sadi Carnot ranked sixth among those entering the French Army Corps of Engineers.

At first, Sadi pursued a military career, although this had complications. He saw early active service at the Battle of Vincennes in March 1814, before moving to Metz as a cadet sub-lieutenant. His position within the military became difficult, however, following Napoleon's defeat at Waterloo in June 1815 and the return of French monarchy. His father once again took exile in Germany, and opportunities for the younger Carnot were limited by his family name. Sadi therefore took leave of the military, returned to Paris in 1819, and began to engage in academic pursuits. He took interest in a wide range of topics, including mathematics, natural history, manufacturing processes, industrial art, and political economy. He attended lectures and workshops at many of the capital's great venues: the Collège de France, the Sorbonne, the Jardin des plantes, the École des mines, the Louvre Museum, and the Conservatoire national des arts et métiers. With time for other activities too – Italian theatre, piano, gymnastics, fencing, swimming, and other sports – life in his mid-twenties seemed to be quite enjoyable. Moreover, it was during this period, in 1824, that Carnot produced his sole publication on thermodynamics.

In 1826, Carnot was required to return to the Corps of Engineers, and he rose in seniority to the rank of captain, but in 1828, he gave up his uniform. There was a further French Revolution in 1830, with Louis Philippe – the Duke of Orléans – coming to power, but by then, Carnot was engrossed in some profound research on the physical properties of gases and vapours. Sadly, he did not complete this research – it was taken up by Henri Victor Regnault – for in the summer of 1832, Sadi Carnot was taken ill with scarlet fever, and so weakened, he died of cholera on August 24 of that year at the age of just thirty-six.

Decades later, in introducing the 1897 version of Carnot's treatise, the editor Robert Thurston remarked:

> *It is this man who has probably inaugurated the development of the modern science of thermodynamics and the whole range of sciences dependent upon it, and who has thus made it possible to*

Malthus Enigma

construct a science of the energetics of the universe, and read the mysteries of every physical phenomenon of nature.[4]

The significance of Sadi Carnot's work took about three decades to be fully understood, within the works of Lord Kelvin and Rudolf Clausius. This was in part because Carnot had discovered the second law of thermodynamics before the nature of the first law – conservation of energy – had been fully recognized. Today the notion of conservation of energy seems almost trivial. The first law is an idea that all children are taught in high-school physics – and for anyone who goes on to be an engineer or scientist, the conservation of energy is a seemingly intuitive idea that becomes second nature. Like all laws of physics, however, the first law of thermodynamics was at some point in history discovered. As described in this chapter, several eighteenth-century scientists, including Joseph Black and Benjamin Thompson (Count Rumford), had been close to understanding how energy, or at least heat, is conserved, but it was not until the decisive experiments of James Joule in 1843 that the first law was cemented. Prior to Joule, there were two competing theories on the nature of heat: *caloric theory*, which regarded heat as a substance, and *kinetic theory*, which considered heat to be a movement of molecules. The competition between these two paradigms extended over many decades before being essentially settled in the mid-nineteenth century.

The process of scientific discovery about the nature of the Earth is, I have argued, part of our Malthusian struggle – along with the necessity to innovate. These two layers of the Malthus Enigma were discussed in the last two chapters, motivated by the focus on the two great environmental challenges of feeding humanity without destroying global ecosystems and of addressing climate change. The efforts of Callendar to resurrect Arrhenius's theory of human-induced climate change was indeed a struggle. This does not mean to say that all scientific discoveries are substantially part of the Malthus Enigma – just those that enable us to learn about the environmental system we inhabit. In this chapter, we turn to the even more fundamental discovery of the laws of thermodynamics, which exemplifies Malthusian struggle on two accounts. First, the study of thermodynamics has massively assisted technological development and innovation. Second – and key to the overall thesis – application of thermodynamics on a global scale has in recent decades begun to provide a rigorous scientific basis for understanding the environmental limits of sustaining human species on Earth.

The connections between thermodynamic theory and technological development can be demonstrated through several examples, including two that we have discussed in previous chapters. James Watt's improvements to the steam engine and Fritz Haber's process of fixing nitrogen from the air are, as discussed, two of the most important inventions of all time. Each provided humanity massive hands up in our Malthusian struggle. Watt's engine was the lynchpin to a system that provided access to huge reserves of coal – leading, within a few decades, to a revolution in land transportation in the form of the railway locomotive. Later, the Haber-Bosch process provided fertilizers and, hence, food that has fed half the population of the planet. Another example is Rudolf Diesel's invention of the diesel engine in the 1890s. What is intriguing about these inventions is that they were each influenced by theoretical breakthroughs in thermodynamics. To be specific: Watt had a long correspondence with Scottish physicist Joseph Black, who had a primitive understanding of the first law through his writing on the conservation of heat; Haber's work on nitrogen fixation followed from his understanding of the Nernst equation, developed by the brilliant chemist Walther Nernst – who is credited with establishing the third law of thermodynamics; and Rudolf Diesel's invention of the diesel engine explicitly followed from his understanding of the Carnot cycle.

The second – and most significant – way in which thermodynamics is relevant to our inquiry here is in its application beyond technological development to a scientific understanding of the Earth system. Early thermodynamicists such as Carnot, Lord Kelvin, and Rudolf Clausius understood that the science of heat flows applied just as much to the "vast movements which take place on the earth"[5] as to heat engines or cannons. Today, thermodynamics is being applied in ways that help us understand the sustainability of the whole Earth system. As will be discussed, this advanced understanding came about in the twentieth century after Austrian physicist Erwin Schrödinger and Belgian chemist Ilya Prigogine had worked out how thermodynamics applied to living systems that are away from equilibrium. So in this chapter, I am going to recount the tortuous route by which the laws of thermodynamics were established – and explain why they are so important for furthering the science of sustainability.

Beyond the first two layers of the Malthus Enigma, thermodynamics is also important because it underlies ecology. To bring ecological principles into economics and develop technological systems in harmony with Nature, we first need to understand how the laws of thermodynamics apply to living systems. This puts us on our way to developing the third layer of the Malthus Enigma.

Malthus Enigma

Carnot's Treatise

In *Reflections on the Motive Power of Heat,* Sadi Carnot set out to answer an important practical question: What is the greatest possible efficiency of the heat engine? By *heat engine*, he meant a general family of devices that turn thermal energy into mechanical energy – for example, to turn a shaft or wheel axle. A steam engine is just one common case of a heat engine. At the time of Carnot's writing, the Industrial Revolution was in full swing. The earliest steam locomotives had been developed, but it was still a few years before Stephenson's Rocket (1829), and great efforts were being made to build faster and faster locomotives. Gradual improvements had been made to Watt's steam engine, so it was important to ask how much better or more efficient these could be made. Carnot recognized: "The study of these engines is of the greatest interest, their importance is enormous, their use is continually increasing, and they seem destined to produce a great revolution in the civilized world."[6] Carnot had access to data on steam engine performance that assisted his work. He was also influenced by others around him. One of these was perhaps his father, who had an interest in perpetual motion machines. In a sense, what Sadi Carnot set out to do was establish how close to a fully efficient, frictionless, perpetual motion machine a heat engine could be. Another important influence was the knowledge of hydraulics that had been gained by French engineers building canals. Essentially, Carnot treated the movement of heat through a steam engine as analogous to water in a water wheel. This followed from his adherence to the prevalent caloric theory, by which heat was considered a material substance. The theory was later rejected, as will be discussed shortly, but it still proved to be useful in Carnot's work.

Carnot's first important realization was that it is possible to turn heat into mechanical work only when there is a flow of heat from a hot body to a cold body. In the case of a steam engine, the flow of heat is from the boiler to the condenser. The boiler is a vessel, in which water is heated and transformed into high-pressure steam, and the condenser is where low-pressure steam is cooled, turning back into water. Carnot realized that mechanical work could be obtained only with a controlled flow of heat between two temperature vessels, not just by simple heating, as in the case of a stove or home hearth, for example.

Perhaps Carnot's most important insight was that the maximum theoretical efficiency of a heat engine depends upon the perfection of its reversibility. A heat engine can also be designed to be run in reverse – that is, starting with mechanical energy, it can be used to transfer heat

The Science of Sustainability

from a cold body to a hot body. This is essentially how a refrigerator works. In the case of a steam engine, instead of using the temperature difference between the boiler and the condenser to turn a shaft, the process can be reversed, with the shaft being turned to create a temperature difference. Having developed a systematic understanding of the physics of heat engines, Carnot realized that complete reversibility of the processes was the condition that gives the maximum efficiency – that is, the maximum amount of work produced from the flow of heat. All the processes that are not reversible, such as heat loss by conduction through the apparatus or frictional losses, are what diminish the efficiency of the engine. Through this reasoning, Carnot originated the idea of a cycle and its reversibility when perfect.

Carnot proceeded to wrestle with a further question, which proved to be important for generalizing his theory toward the second law. He asked whether it would be possible to obtain more mechanical output from an engine using a different gas from water vapour, while keeping the temperature difference unchanged. Carnot answered this question in the negative by drawing upon the impossibility of perpetual motion. He considered a comparison of two identical engines using different fluids to produce mechanical work. If, by using a different fluid, one engine was more efficient than the other, then part of its mechanical work could in theory be used to drive the other engine backward – thereby eliminating the net heat flow, but still producing some work. This was known to be impossible, and so it was clear that the efficiency of the two engines had to be the same regardless of the fluid used. Thus, Carnot concluded that all perfectly reversible machines acting over the same incremental temperature difference have the same efficiency. He stated:

> *The motive power of heat is independent of the agents employed to realize it; its quantity is fixed solely by the temperatures of the bodies between which is effected, finally, the transfer of the caloric.*[7]

The amount of work delivered, for a given quantity of available heat, depends on the temperature difference, not the fluid. Carnot's insight turned out to be hugely influential toward the end of the nineteenth century for the development of diesel engines, but several decades of scientific struggle were required first.

Malthus Enigma

Caloric Theory Versus Kinetic Theory

While Carnot's treatise eventually became a cornerstone to thermodynamic theory, it included an axiom – known as caloric theory – that was later proven to be erroneous. Remarkably, Carnot's insights into the motive power of heat were developed two decades before the law of conservation of energy had been substantially formalized. It was not until 1843, following decisive experiments by James Joule, that the first law of thermodynamics was fully discovered. There were earlier primitive forms of the first law, but these were somewhat confused by a lack of understanding of the nature of heat. As mentioned, two rival theories existed at the time of Carnot's work – *caloric theory* and *kinetic theory*. The competition between the two is another rich example of the scientific struggle that forms part of humanity's Malthus Enigma.

In the early nineteenth century – when Carnot, and indeed Malthus, was writing – most scientists believed in a theory of heat known as caloric theory. This theory considered heat to be an elemental substance, which, moreover, was conserved. The theory was quite comprehensive and could provide a coherent explanation for phenomena such as the expansion of solids, the boiling of water, and work done by the expansion of steam.[8] Among those who subscribed to the theory was Antoine Lavoisier, who, after identifying oxygen in the late eighteenth century, listed the special substances of light and heat among the first few elements of his primitive periodic table. Lavoisier considered the substance of heat – which he called *calorique* – to be composed of indestructible atoms.[9]

The rival kinetic theory, which considered heat to be a form of motion, had been supported by many earlier philosophers and scientists. Plato had observed that heat and fire were a form of motion that could be obtained by impact or friction. Francis Bacon had clearly declared "the very essence of heat ... is motion, and nothing else."[10] The likes of Galileo, Newton, Locke, and Leibniz had also reached similar conclusions. In 1738, Daniel Bernoulli produced an influential treatise on the flow of fluids known as *Hydrodynamica*. In one chapter, he proposed that gases were composed of rapidly moving particles – and related the pressure of gases to the number of particles and the temperature – but this part of Bernoulli's work still had little impact at the time.[11]

With caloric theory being most generally accepted, understanding of conservation of energy was primitive. The Dutch scientist Herman Boerhaave (1668–1738), professor of medicine, chemistry, and botany at the University of Leiden, had proposed the basic axiom that fire – as he put it – is always conserved.[12] Boerhaave's understanding of conservation was picked up by a similarly broad-minded Scottish doctor, George Martine (1702–1741), who was an early contributor to the Scottish Enlightenment. Following Martine were several notable Scottish scientists, including Joseph Black, James Watt, Lord Kelvin, William Rankine, and James Maxwell, who together made a huge contribution to the study of thermodynamics.

Joseph Black (1728–1799) is of particular interest because of his relationship to Watt. After completing an MD in chemistry at the University of Edinburgh, Black became a university lecturer in medicine and chemistry, first in Glasgow and then, from 1764, back in Edinburgh. Assessing Black's contributions to thermodynamics is tricky, because he had no formal publications, but his lecture notes and other records showed that he had a partially developed understanding of the first law. Black invoked conservation of heat in his work via measurements of temperature, and from it was able to study the phenomena of heat capacity and latent heat before 1760. The heat capacity of a body is the amount of heat required to raise its temperature by 1°C. Latent heat refers to the additional heat needed to undergo a phase change – for example, when turning a liquid into a gas. Black's work on these concepts took place several years before Swedish physicist Johan Wilcke formally published his "discoveries" of them.[13] From his understanding of conservation of heat, Black deduced that latent energy was required in phase transformations. He gave two examples: a kettle of water producing steam and the melting of a snowfield. He noted that, in both cases, if there was no latent heat required, the phase transformations would be explosive, with potentially dangerous consequences. Historian John Cardwell summarized: "Black's great achievements were to set up the fundamental concept of quantity of heat and, by considering its implication, infer the correlative concepts of heat capacity and latent heat."[14]

Other historians have proposed that Watt's major advances to the steam engine were directly derived from Black's discovery of latent heat, although Watt himself denied this, and Cardwell concurred with Watt.[15] Cardwell suggested that, through his tinkering with steam engines, Watt had independently arrived at his key breakthrough of adding a second cylinder to Newcomen's design. Nonetheless, over the four decades or so that he worked on steam engines, Watt clearly learned from Black and the other early pioneers of thermodynamics. Black and Watt exchanged many letters on various properties of steam,[16] including a letter from Black to Watt in 1764 explaining the principle of latent heat. It is also worth noting that Watt's first partner, Dr. John Roebuck, had been a

Malthus Enigma

student under Boerhaave. So even if Watt's invention was not directly attributable to Black, he was influenced by the emerging understanding of thermodynamics in his work.

Also in the late eighteenth century, around the time of Black, caloric theory received a substantial rejection from another notable scientist, Count Rumford of Bavaria, who attempted to disprove the theory through experiments using cannons. The Count of the Holy Roman Empire, who rose to become minister of war for the Duke of Bavaria, came from a remarkably humble and distant upbringing. His real name was Benjamin Thompson (1755–1814), and he was actually born as an American of British descent in rural Woburn, Massachusetts. Thompson came from a relatively poor family and received education at his local village school before becoming an apprentice to a merchant in nearby Salem at the age of thirteen. He later became a teacher and married a rich widower, Sarah Rolfe, moving into her former husband's property in Rumford (now called Concord) in New Hampshire. Thompson joined the New Hampshire Militia, but being a loyalist, he ended up working for the British during the American Revolutionary War – initially as a spy.[17] At one point, he was forced to flee his house as a mob attacked, permanently abandoning his wife as he fled to the British lines. When Boston was besieged in March 1776, Thompson sailed to London, where he became Under-Secretary of State for the Colonies.[18] Rising to the rank of lieutenant colonel in the British Army, he returned to North America to lead several British operations during the war.[19] After the war, he was made a colonel upon retirement and subsequently headed to Europe as a soldier of fortune. His services were taken up by the Duke of Bavaria in 1784, though he was also knighted by King George III the same year, with the expectation that he would continue to provide intelligence for the Foreign Office.[20] In his later years, Thompson split his time between London and Paris, active in scientific societies and briefly marrying Marie-Anne Lavoisier, the widow of the great French chemist.

During his childhood and throughout his military career, Thompson was constantly conducting scientific experiments – and had particular interests in theories of heat. Among his many scientific investigations were experiments on gunpowder and measurements of the thermal conductivity of cloth used in military uniforms.[21] He also claimed to be the first to discover convective currents, which he observed through temperature fluctuations in an alcohol thermometer.[22] Thompson was an avid inventor, too, becoming famous for his advancements in the designs of fireplaces, industrial furnaces, the double boiler, kitchen ranges, and coffee pots. Among his most important scientific experiments were those

The Science of Sustainability

conducted at the munitions workshop in Munich.[23] There, Thompson conducted a series of experiments involving the boring of military cannons. He was skeptical of caloric theory and especially questioning of its ability to explain the heat generated by friction. In one experiment, he arranged for the boring of a cannon to be conducted underwater, observing that after 2.5 hours, the water had warmed to a boil.[24] He argued that the production of heat underwater without fire from a seemingly inexhaustible source was contrary to the idea that heat was a substance. In hindsight, Thompson was correct, but the lack of quantification in the experiment – and a belief that caloric theory could be adjusted to account for Thompson's observations, kept the paradigm alive.[25]

Caloric theory eventually met its end in the middle of the nineteenth century with the experiments of James Joule and the formalization of the law of conservation of energy. Several scientists have claims on determining the conservation of energy: Hermann von Helmholtz first definitively declared the law in 1847, but Julius Robert von Mayer had earlier described it – and Joule's experiments of 1843 provided the fundamental evidence. Several key scientific developments had preceded Joule's work: Alessandro Volta had discovered the electric battery in 1800; André-Marie Ampère – one of Sadi Carnot's instructors – had invented the electromagnet; and, in 1831, Michael Faraday had found that an electric current could be produced by a change in a magnetic field – leading to the invention of the electric generator. The quest was on to find some unifying way of connecting mechanics, electricity, magnetism, chemical reactions, light, and heat. Joule, who was the son of a wealthy Lancashire brewer and was privately trained by the scientist John Dalton, performed several key experiments. In one of these, Joule placed a closed wire loop inside a container surrounded by water, between the poles of a magnet. The container was arranged so it could rotate, driven by a falling weight. Rotations of the coil induced an electric current to flow in the loop, with the heat produced in the electric circuit measured by a temperature change recorded by a thermometer in the water. Thus, the work done by the falling weight was converted into heat via an electric current. The nature of the experiment – including the electric current – was such that the heat could not be a material quantity transferred from elsewhere. Joule was able to measure the mechanical work done in terms of the heat produced.[26] Most important, though, was that Joule's experiment established conservation of energy – the first law of thermodynamics.

Malthus Enigma

The scientific struggle to establish the law of conservation of energy had taken decades, or perhaps centuries, to resolve. What is quite striking about the timing of Joule's key experiment is that it occurred almost ten years after the death of Malthus. So Malthus's writings on the population problem were done at a time when our understanding of energetics was still primitive. He failed to see not only how innovation could help overcome the population trap, but scientific discovery too. The significance of this becomes quite profound as we continue on with the story of the development of thermodynamics.

Formalization of the Second Law

Technically, the term *thermodynamics* is thought to have first been used by the Scottish–Irish scientist Lord Kelvin in an 1849 paper describing and expanding upon Carnot's work.[27] Born in Belfast in 1824 as William Thomson, Lord Kelvin played a pivotal role, along with Rudolf Clausius, in founding the formal field of thermodynamics. Lord Kelvin was a brilliant scientist who also made contributions to electricity, geology, mathematics, engineering, and mechanics. He is perhaps best known for the temperature scale beginning at absolute zero (−273.15 degrees Celsius or 0 Kelvin), which bears his name. Lord Kelvin was knighted, in 1866, for providing scientific and engineering leadership on the eventually successful laying of the first transatlantic cable. Later, in 1892, he became the first British scientist to be made a Lord – for his work on thermodynamics. That he was buried in Westminster Abbey following his death in 1907, next to Sir Isaac Newton, speaks volumes of his influence and achievements.

An important part of Lord Kelvin's leadership in thermodynamics was that he rediscovered and then promoted Carnot's treatise. The way this came about was quite intriguing. After graduating from the University of Cambridge in 1845, Lord Kelvin went to Paris for several months and was able to obtain a position in the physical laboratory of Henri Victor Regnault at the Collège de France. Through access to the books in Regnault's cabinets, Lord Kelvin became familiar with a paper by Émile Clapeyron: "Memoir on the Motive Power of Heat." Ten years after Carnot's tragic death, Clapeyron – another graduate of École polytechnique – had produced a graphical representation of Carnot's cycle using Watt's indicator diagram.[28] Moreover, Clapeyron had also expressed Carnot's cycle in a mathematical form. Inspired by this, Lord Kelvin went in search of Carnot's original treatise, first inquiring in vain at the libraries of the Collège de France – and then attempting to purchase

The Science of Sustainability

a copy from the old booksellers by the banks of the Seine. Evidently, Paris booksellers were not aware of the book, and the only Carnots they had heard of were Lazare Carnot, "the organizer of victory," and Sadi's brother Hippolyte – a social and political writer. Lord Kelvin later recounted his frustrating search:[29]

> *I went to every book shop I could think of, asking for the* Puissance motrice du feu, *by Carnot.* "Caino ? Je ne connais pas cet auteur." *With much difficulty I was able to explain that it was "r" not "i" I meant.* "Ah! Ca-rr-not! Oui, voici son ouvrage," *producing a volume on some social question by Hippolyte Carnot; but the* Puissance motrice du feu *was quite unknown.*

Amazingly, after three more years of searching, Lord Kelvin eventually obtained a copy of Sadi Carnot's rare treatise from his colleague Lewis Gordon, a professor of civil engineering at the University of Glasgow.[30] In 1849, Lord Kelvin then published a key paper, "An Account of Carnot's Theory of the Motive Power of Heat; with Numerical Results Deduced from Regnault's Experiments on Steam." The paper not only brought attention to Carnot's treatise, but also used Clapeyron's equations and experimental results from Regnault to support Carnot's ideas. In the paper, Lord Kelvin also mentioned the important experimental work of Joule, with whom he was in communication. In this 1849 paper, however, Lord Kelvin was unsure and inconsistent in addressing Joule's great insights. Essentially, Lord Kelvin rejected Joule's ideas, but then later, in contradiction, stated, "Nothing can be lost in the operation of nature – no energy can be destroyed."[31] At this point, Lord Kelvin was struggling to reconcile Joule's work demonstrating conservation of energy with Carnot's insights that had been arrived at using caloric theory. His biographer, Silvanus Thompson, noted: "The apparent conflict took possession of [Lord Kelvin's] mind and dominated his thoughts."[32]

While Lord Kelvin was trying to reconcile the insights of Joule and Carnot, it was German scientist Rudolf Clausius (1822–1888) who was able to assimilate the two.[33] Cardwell noted that Clausius could almost be considered a disciple of Lord Kelvin – and he had learned about Carnot's treatise through Lord Kelvin's writings. Clausius was able to see that with rejection of caloric theory (and hence conservation of heat), it was possible to accept Joule's conservation of energy and Carnot's core observation that the work done by a heat engine is determined by the heat transmitted and the temperature difference. Moreover, Clausius recognized that Carnot's treatise provided fundamental insights that lay

Malthus Enigma

beyond the idea of conservation of energy. He realized that the common experience "that it is impossible for heat to flow on its own accord from a cold to a hot body"[34] could become an axiom of science – in other words, the second law of thermodynamics.

The interplay between Lord Kelvin and Clausius remained important, though. Within two years, Lord Kelvin had accepted Clausius's reasoning and abandoned caloric theory. Lord Kelvin then began to produce a number of hugely important papers, including, in 1853, "On a Universal Tendency in Nature to the Dissipation of Mechanical Energy," in which he began to recognize heat as the grand moving agent of the universe. Cardwell summarized Lord Kelvin's contributions during this period as follows:[35]

> *Once Kelvin had accepted the new theory, however, he produced in the course of a few years an abundance of new ideas, which were to prove immensely important: many of them in the hands of other men, such as Clausius. Prominent among these ideas were the concept of the "energy" of a body, the insight into the process of dissipation of energy, the establishment of an absolute scale of temperature, the awareness of the importance of "transformation" and "equivalence" and, finally, the expression of the second law of thermodynamics in a mathematical form.*

More should be said on Clausius' contribution here, because what he achieved – influenced by Lord Kelvin – was to define a set of universal laws for heat flow. This was a similar achievement to the seventeenth-century establishment of the laws of motion by Newton. In doing so, Clausius added a third type of transformation to the mechanical energy transformation to heat (Joule) and the flow of heat from hot to cold bodies (Carnot).[36] The third transformation is a change in the arrangement of molecules in a body, which is observed, for example, when ice is transformed into water. Clausius argued that when heat is added to a body, some of it is used to: (1) do work against external forces, (2) increase the thermal content of the body, and (3) perform internal work against forces of mutual attraction between constituent molecules.[37] Clausius then combined the second and third of these terms together – and divided by temperature for equivalence – to give a measure of the heat that is used for internal transformation of the body. He called this transformational content *entropy*, from the Greek word *entropia* – meaning "transformation." Similar to Lord Kelvin, Clausius then took his theorems for the nature of heat and generalized them to fundamental laws of the universe:[38]

The Science of Sustainability

1. The energy of the universe is constant.

2. The entropy of the universe tends to a maximum – meaning that it moves toward a disordered dead state.

Impacts of the Second Law

Clausius's conception of entropy and its tendency toward a maximum poses some challenges for sustaining the human species – indeed any species – which we will turn to shortly. First, however, we can recognize through three examples how formalization of the second law of thermodynamics has been useful for engineers – and has enabled them to develop increasingly efficient technologies in the course of Malthusian struggle.

Indeed, the impacts of uncovering the second law of thermodynamics on the development of technology have been outstanding. Malthus witnessed incredible improvements to the efficiency and applications of steam engines during his lifetime. The impacts of the second law, however, came later, toward the end of the nineteenth century. One example of this was Rudolf Diesel's internal combustion engine. In developing internal combustion engines, Diesel – and others, such as Nicolaus Otto, Gottlieb Daimler, and Karl Benz – changed to using air as the working fluid rather than steam. As Cardwell noted, Diesel drew directly from Carnot's observation that the motive power of heat was independent of the working fluid. Thermodynamically, air has a major advantage over steam in that its pressure does not increase as dramatically with temperature, as is the case with steam. This means it can be used over a wider range of temperatures "without blowing the engine to bits."[39] Development of the internal combustion engine required advancements in metallurgy, fuel technology, and engineering standards.[40] These occurred over the second half of the nineteenth century, thus enabling Diesel to draw upon the amazing insights of Sadi Carnot in inventing his high-efficiency internal combustion engine.

Increasing the energy efficiency of technology is a constant part of our Malthusian struggle – and our progress in this regard has been substantially boosted through understanding of the second law. A further example of this occurred in the United States following the Oil Crisis of October 1973, when interest in energy efficiency rose to prominence. America's domestic oil production had peaked in 1970, as M. King Hubbert had predicted,[41] and it had become dependent upon exports from Middle East oil producers. When the United States found itself on the

Malthus Enigma

opposite side as the Organization of Arab Petroleum Exporting Countries over the Yom Kippur War, the Arab nations began an oil embargo. Within six months, the price of oil rose from US$3 per barrel to US$12 per barrel. The impacts in the US were significant: the stock market crashed, gasoline sales became limited at service stations, and a national speed limit of 55 miles per hour was imposed. Other countries, such as Canada, Japan, the UK, and parts of Europe, were also impacted. The oil shock had a variety of long-term effects: countries began developing other sources of supply, strategic oil reserves were increased, and research in solar and wind power was spurred. From an energy demand perspective, however, the response was an accelerated effort to improve the energy efficiency of transportation vehicles, homes, and industry.

In the midst of the Oil Crisis, the American Physical Society held an intensive four-week summer stay at Princeton University entitled "Technical Aspects of Efficient Energy Utilization." Co-sponsored by the National Science Foundation, the Federal Energy Administration, and the Electric Power Research Institute, the aim of the summer meeting was to identify physics research topics that could beneficially impact technologies governing energy efficiency.[42] With some of America's brightest physicists and engineers gathered to work on a unified framework for research on energy efficiency, naturally the laws of thermodynamics were again a centrepiece. Notably, the group put particular emphasis on the use of the second law of thermodynamics in addressing energy efficiency. They developed the second law efficiency for a device or system based on the ratio of the useful heat or work done to its maximum theoretical value using any device. The second law efficiency has many advantages over the efficiency of a device based solely on the first law of thermodynamics. First law efficiency is the ratio of useful work done to the energy input to the system. The second law efficiency is broader and captures how much room there is for improving energy use for a desired outcome. Subsequent to the American Physical Society meeting, second law efficiencies have become the experts' way of assessing improvements to the energy use of technologies.

A third example of technological innovation stemming from the laws of thermodynamics is the development of photovoltaic panels for capturing radiation from the sun. The path to this invention has a bit more to it, but it essentially stemmed from the statistical interpretation of entropy by Austrian physicist Ludwig Boltzmann (1844–1906). Going beyond thermodynamics as a study of heat, Boltzmann provided a more general, statistical interpretation of Clausius's concept of entropy. German physicist Max Planck (1858–1947) then drew upon Boltzmann's

The Science of Sustainability

statistical interpretation of entropy in his studies of radiation, which started the revolution in quantum physics. One practical outcome of our subsequent understanding of the quantized nature of energy and the photoelectric effect was photovoltaic technology. A remarkable feature of photovoltaic panels is that they generate useful energy at an efficiency that is ten to twenty times greater than the photosynthesis that occurs in plants.[43] Thus, as discussed in the last chapter, they are an essential Promethean III technology for humanity's transformation to a sustainable solar society.

Thermodynamics of Life

While the laws of thermodynamics have been fantastic for the development of technology, their uncovering also makes them potentially useful for understanding natural living systems. Recognition of their applicability to living systems has been philosophically challenging and has taken a long time to develop. To understand the challenge, we have to go back to Clausius's interpretation of the second law, that the entropy of the universe is ever increasing. Essentially, this means that the universe as a whole tends to move toward a disordered "dead" state – which is quite a predicament for the sustainability of the human species. Modern-day German thermodynamicist Axel Kleidon, from the Max Planck Institute for Biogeochemistry in Jena, describes the predicament particularly well:

> *Matter mixes, water flows downhill and wood burns into ashes. If nothing else were to take place, sooner or later all matter would end up in a uniform mix of everything, water would collect in the world's oceans and all biomass would be burnt to ashes. All processes would lead to a "dead" Earth state with no gradients present to drive fluxes and no free energy available to run life.*[44]

If the second law gives that the universe tends toward a state of disorder, how, then, does life exist? One of the earliest to wrestle with the significance of the laws of thermodynamics on the existence of life was the Austrian physicist Erwin Schrödinger. Born in Vienna in 1887, Schrödinger studied physics at the University of Vienna and taught at Jena, Stuttgart, Breslau, and Zurich before, in 1927, succeeding Max Planck as professor of physics at the Royal Friedrich Wilhelm University of Berlin. Although best known for his Nobel Prize–winning work on quantum mechanics, Schrödinger was a polymath, making contributions

Malthus Enigma

to cosmology, general relativity, statistical mechanics, and thermodynamics. He left Germany in 1934, with the rise of the Nazis, eventually settling in Dublin for seventeen years, where he became an Irish citizen. In these middle to late years of his research, he wrote more broadly on philosophy, theoretical biology, and genetics.[45]

In 1944, Schrödinger published a short philosophical text entitled *What Is Life?* In it, he contemplated that one of the fundamental processes underlying life involves the creation of order from disorder. Schrödinger observed that, on first glance, living organisms seemingly fly against the second law of thermodynamics. Rather than follow the natural tendency to go into a state of disorder, living organisms develop complex, orderly structures. He wrote: "Life seems to be orderly and lawful behavior of matter, not based exclusively on its tendency to go over from order to disorder, but based partly on existing order that it kept up."[46] In recognizing that living organisms evade, at least temporarily, the tendency to decay into an equilibrium disordered state, Schrödinger was touching upon the essence of contemporary concerns over sustainability. In asking what are the characteristics by which a piece of matter can be said to be living, he answered: "When it goes on 'doing something', moving, exchanging material with its environment, and so forth, and that for a much longer period than we would expect of an inanimate piece of matter to 'keep going' under similar circumstances."[47] The challenge of sustaining the human species – which is arguably synonymous with Malthusian struggle – can thus be interpreted in Schrödinger's terms as keeping the species alive for longer against the inevitable tendency toward collapse into a disordered dead state.

How is it that a living organism is able to sustain itself against the tendency toward disorder that the second law dictates? The answer that Schrödinger gave was that living organisms are able to maintain themselves only at the expense of increasing the entropy – that is, the disorder – of the environment around them. Through eating, drinking, breathing, or assimilating matter, organisms can maintain and even grow their internal ordered structure but necessarily must increase the disorder of their external environment. Thus, the second law still holds. The entropy of the overall system is increasing, while organisms maintain themselves at a state away from thermodynamic equilibrium.

The theory of non-equilibrium thermodynamics was later developed by the Nobel Prize–winning Belgian chemist Ilya Prigogine. As a professor at the Université libre de Bruxelles, and then the University of Texas at Austin, Prigogine discovered how the import and dissipation of energy into chemical systems could locally reverse the maximization of entropy implied by the second law of thermodynamics. He mathematically formulated an extended version of the second law that applied to open systems, away from equilibrium, as well as isolated systems.[48] In his writings, he identified several systems that could develop complex ordered structures away from equilibrium – including lasers, chemical networks, and weather systems.[49]

Gaia and the Thermodynamics of the Earth

To recap so far, during the nineteenth century, through the works of Carnot, Lord Kelvin, Clausius, and others, humankind uncovered the scientific laws of thermodynamics. These laws were fantastic for understanding and furthering some of the amazing technologies that engineers have developed, such as steam engines, fertilizers, and automobiles. Over the twentieth century, Schrödinger, Prigogine, and others began to understand how the laws of thermodynamics apply to living systems – life forms that have ordered structure and exist away from a dead state of equilibrium. Implicit in this work on living systems is an understanding that the laws of thermodynamics apply everywhere to everything – such are the laws of physics.

So next we turn to the scale of the planet – to the dynamics of the Earth – its atmospheric system, hydrological cycle, geological processes, etc. As briefly discussed before, the thermodynamics of the Earth had been considered by Clausius and Lord Kelvin, but only in recent decades have thermodynamicists begun to understand the Earth as a living system. The inspiration for this has been, in part, the controversial Gaia Hypothesis, developed by James Lovelock in the early 1970s. The central idea of the Gaia Hypothesis is that the Earth and life on it act as one system. As discussed further below, German researcher Axel Kleidon has begun to examine the Gaia Hypothesis in the context of the thermodynamics of the Earth – and this potentially leads to deeper understanding of some absolute constraints on Malthusian struggle.

James Lovelock – the originator of the Gaia Hypothesis – was a British scientist who spent several years working for NASA in the United States. Lovelock was born in Hertfordshire, England, and studied chemistry at the University of Manchester, before obtaining a PhD in medicine at the London School of Hygiene & Tropical Medicine. During the 1960s, Lovelock worked as a consultant for NASA, involved in the development of scientific instruments. Among his inventions was an electron capture device, with which he was the first person to measure the presence of chlorofluorocarbons (CFCs) in the Earth's atmosphere. CFCs were later linked with ozone depletion and found to be a significant cause of the hole in the ozone layer. Lovelock's ideas about Gaia were also first developed when consulting for NASA. He was employed to create instruments for analyzing the atmospheric composition of extraterrestrial planets. Prior to NASA's Viking Project that went searching for life on Mars in the late 1970s, Lovelock noted that the composition of Mars's atmosphere was close to a stable equilibrium. Unlike the Earth's atmosphere, it was dominated by CO_2 (over ninety-five percent), with only small amounts of nitrogen, argon, oxygen, and other trace elements. Lovelock's intuition was that the stable state of Mars's atmosphere and its lack of chemical diversity were strong indications that no life would be found there.

The Gaia Hypothesis, which Lovelock first published in 1972, aims to explain the existence of life on Earth for billions of years. The planet and life on it are considered to be part of a single system that has the ability to both regulate and repair itself.[50] The regulation, achieved through changing the composition of the atmosphere, is able to maintain temperatures suitable for life on Earth. Moreover, even in the event of a major shock – such as a meteorite hit – the system is able to repair itself. Reflecting these abilities, Lovelock also named the system *Gaia*, after the ancient Greek goddess of the Earth.

The Gaia Hypothesis has been controversial. Some used to see it as more of a philosophy or religion rather than a scientific theory. Yet the idea that the Earth and the life on it can be seen as one system is not so far-fetched – and is somewhat consistent with modern approaches to modelling global climate change. The contention has been that the Gaia Hypothesis is untestable and therefore not scientific, but there are counter-arguments here, too. Perhaps the Gaia Hypothesis is hard to test using conventional reductionist science, but this applies to most Earth systems science – which is a complex type of science, as will be discussed in the next chapter. In other respects, some aspects of the Gaia Hypothesis are seen as cutting against other scientific doctrines. For example, the Gaia notion that species will evolve on the Earth as part of regulation and repair is altruistic, and – some argue – counter to evolution through competition and survival of the fittest.[51]

Axel Kleidon provides a rigorous description of the thermodynamics of the Earth system, from which the concept of Gaia can be further assessed. In an impressive series of academic papers, including some published in *Philosophical Transactions of the Royal Society*, Kleidon explains and, moreover, quantifies the major Earth systems in terms of thermodynamics. In masterful diagrams, he shows how the flux of solar

The Science of Sustainability

radiation arriving at the Earth's atmosphere drives the dynamics of the atmosphere and propels the hydrological cycle, the carbon cycle, and other geochemical cycles. Subtracting radiation that is immediately reflected by the Earth's atmosphere and adding a small amount of heat generated from the Earth's cooling core, the solar radiation that penetrates the atmosphere is eventually reflected back into space as long-wave radiation.[52]

One of the most important components in thermodynamic analysis of the Earth system is primary production. As we discussed in Chapter 2, primary production is the transformation of sunlight into useful chemical energy by the biota on the Earth's surface, which involves an uptake of carbon. It can be expressed as gross primary production, which is the total uptake of carbon by living organisms, or net primary production, which excludes the carbon used up in meeting the energy needs of plants. Recall that Steven Running proposed that net primary production could be seen as an overarching planetary boundary that limits the food, fibre, and fuel that humanity depends upon.[53] We will pick up the discussion of planetary boundaries again below, but first let us continue with Kleidon's work.

Through his research, Kleidon has begun to test and tentatively corroborate the Gaia Hypothesis.[54] To do so, he first formulated the hypothesis in a precise manner. Specifically, he used gross primary production – the total global uptake of carbon by all biota – as a measure of how well environmental conditions serve life on Earth. In early work, he then ran numerical climate models with drastically different levels of vegetation on the Earth – one a "desert world," the other a "green planet." Results of these simulations showed that the presence of vegetation led to a climate that was more conducive to the uptake of carbon. He concluded that "life has a strong tendency to affect Earth in a way which enhances the overall benefit (that is, carbon uptake)."[55] The approach received some criticism, so Kleidon devised a second test of the hypothesis using thermodynamics. In his thermodynamic calculations, Kleidon invoked the theory of maximum entropy production. His analysis suggested that the presence of living biota on the Earth's surface occurs in a way that maximizes the rate at which disorder of the Earth system is produced. The flip side to this, following the second law of thermodynamics, is that the biota exists such that the amount of ordered structure of life on Earth is maximized, constrained only by available solar radiation.

The work of Kleidon and others, applying thermodynamics on a global scale, provides deeper scientific insights into the significance of

Malthus Enigma

biota – the living stuff that occupies the surface of the Earth. Biota is remarkable for the amount of free energy that is generated through photosynthesis. Only a small fraction of the solar radiation that penetrates the Earth's surface is transformed or captured through photosynthesis. The rest is reflected back, used to drive the hydrological cycle, propel wind currents, etc. Biotic activity, however, produces approximately 215 terawatts of free energy[56] – which means it is useful energy available to do work. As Kleidon observed, the free energy generated by biota is about ten times greater than that generated from human processes in the biosphere; the quantity of fossil fuels mined by humans amounts to only approximately seventeen terawatts. To give some other comparisons, the power produced by the gravitational effects of the Moon, which drives ocean tides, is only five terawatts, and the convection emanating from the Earth's cooling core is about forty terawatts. So biota generate a relatively large amount of free energy. Kleidon, furthermore, noted that this large amount of free energy generated by biota "substantiates the suggestion given by Lovelock that the Earth's planetary geochemical disequilibrium is mostly attributable to the presence of widespread life on the planet."[57]

The free energy photosynthesized by biota is responsible for net primary production, which has been recognized as a potential hard, absolute planetary boundary for human life on Earth.[58] Recall that Johan Rockström and his colleagues have attempted to quantify nine planetary thresholds, including CO_2 concentration and species extinction rates, that humanity needs to live within to avoid catastrophic environmental change. Net primary production by biota could be considered as a universal planetary boundary – because it integrates over the nine Rockström boundaries. For example, recall from Chapter 3 that the diversity of biota provides a whole range of ecosystem services to humans, and, of course, biota play a critical role in the carbon cycle, mediating some of humanity's role in climate change. The observation that humans already exploit thirty to forty percent of the fifty-four petagrams of carbon that life on Earth produces each year, with approximately fifty-three percent considered unharvestable, leaves humanity with little room to grow. If, as we discussed in Chapter 2, humans can increase their exploitation of net primary production by only about another five petagrams of carbon, then we are almost at our limits.[59] This is a really small breathing space – and makes Tilman's projection of a doubling of global food demand by 2050 seem particularly daunting.

One Earth System

The placement of biota within the framework of thermodynamics notably brings together multiple environmental challenges of our modern-day Malthus Enigma. Indeed, through biota and other components, our thermodynamic model of the Earth system integrates, or relates to, all of the environmental concerns expressed in the planetary boundaries work of Rockström and colleagues.[60] Destruction of biota through land-use change exacerbates the problem of biodiversity loss, undermining ecosystem services essential to human welfare. The biota require carbon, water, and nutrients (nitrogen and phosphorus) to function, and, moreover, their destruction has impacts on the carbon cycle, the hydrological cycle, and biogeochemical cycles. Human use of fossil fuel energy, with its related problems of air pollution, greenhouse gas emissions, and subsequent ocean acidification, clearly fits within our thermodynamic model of the Earth, too. Thermodynamics thus provides a framework of understanding – at least from a biophysical perspective – of the connections between a range of critical environmental challenges.

Wrapping up this chapter, we see that the discipline of thermodynamics has the potential to become the bedrock upon which humanity's Malthusian struggle can be scientifically studied. The discipline has evolved from early pioneers such as Sadi Carnot wrestling with the efficiency of steam engines and James Joule experimenting on energy conservation. Clausius and Lord Kelvin masterfully formulated the fundamental laws, which were initially largely grasped only in the context of equilibrium dead states but have become understood over the twentieth century for living systems away from equilibrium. Now we have a greater understanding of the thermodynamics of the Earth system as a whole.[61]

We saw in the discovery and development of thermodynamics plenty of the scientific struggle, which I have proposed is a second layer of the Malthus Enigma – and spillovers into human innovation, which is the first layer. There is, moreover, a greater role for thermodynamics in the overall story of sustainable human development – it plays a central part as a science of sustainability. Schrödinger's realization of how Carnot's theory of the energetics of the universe applies to living systems was pivotal. This provided understanding – at least in biophysical terms – of how complex living organisms can develop, and the limits to which they can be sustained.

The laws of thermodynamics apply to both the natural world and the human creations within the natural world – including economies and industries. As such, thermodynamics has a role in the third layer of the Malthus Enigma, which is about policy-making. The sustainability-focused disciplines of ecological economics and industrial ecology – which we turn to in Chapters 7 and 8 – are both built upon thermodynamic principles. In these later chapters, we will see that thermodynamic principles also apply to "human" socio-economic systems – for which values matter too – and not just the biophysical world.

First, though, we turn to the broader endeavour of humans wrestling with interconnected problems using systems approaches. To develop policies that address critical global environmental challenges, those challenges have to be systematically framed within a broader set of global issues. Use of thermodynamic principles will continue to be important as we examine systems approaches and ask: *How do we conceptualize the interrelationships between global problems?*

Chapter 6: The Global Problématique

That night at dinner – at Peccei's house – about six of us closed ranks and decided to carry on. A small study group, mainly for our own education, was formed and decided that the function that was most important and would be the central theme of the group was what we called the global problématique – namely the interconnection of problems.

Alexander King[1]

On April 7–8, 1968, a meeting of about thirty industrialists, financiers, civil servants, and academics took place at the Villa Farnesina in Rome, to discuss global challenges of pollution, poverty, and security – and the systematic connections between such issues. The event was hosted by the Accademia Nazionale dei Lincei, Italy's venerable National Academy; founded in 1603, it had supported Galileo Galilei against the Roman Inquisition with his views that the Earth revolved around the sun. The April 1968 meeting at first turned out to be an "abject failure,"[2] as the largely European audience was highly skeptical of co-opting American approaches to anything at the time of the Vietnam War. Undeterred, however, six men[3] met later at night for dinner at the house of Aurelio Peccei and began a quest to understand connections between global problems. Thus, the Club of Rome was born. Led by Peccei and Scotsman Alexander King, the informal group would unleash the power of computer systems models to address humanity's greatest challenges. Most notably, it supported the highly publicized and controversial *Limits to Growth* study,[4] led by MIT researcher Dennis Meadows, using methods developed by his advisor, Jay Forrester. With strong Malthusian undertones, *The Limits to Growth* provoked a huge scientific debate about the *predicament of mankind*.[5] But through this study, the Club of Rome helped establish systems modelling more broadly as an essential approach to studying the human condition on Earth.

Malthus Enigma

The principal architect of the Club of Rome was the Italian industrialist Aurelio Peccei. The preface to a short biography on Peccei notes that he "played a historical role in bringing scientists and statesmen together, and motivating them to think about how to tackle problems of humankind."[6] As a young man, Peccei lived an adventurous life. Born in Torino in 1908, Peccei was a great traveller and a talented linguist and communicator. After completing a degree in economics at the University of Torino, he studied for six months in Paris at the Sorbonne before travelling to the Soviet Union, where he worked at a foundry near Moscow owned by Fiat. In December 1935, while only twenty-seven years old, Peccei was moved to a role with Fiat's special affairs division in China. There, he oversaw the manufacture of Fiat warplanes for the Chinese military while also finding time to complete a doctoral thesis on Lenin's New Economic Policy.[7] When World War II broke out, Peccei was remarkably able to fly back to Italy with a suitcase full of money – China's payments for the planes – despite Italy and China ending up on different sides. Successful delivery of Fiat's Chinese earnings put Peccei in good favour with Vittorio Valletta, the director general of the Fiat group. Remaining in Italy as a Fiat employee, Peccei actively participated in the anti-Fascist movement. This led to him being arrested, imprisoned, and tortured during the war – an experience he was able to draw upon, learning about the inner strength of humans and their incredible potential. At one point, Peccei was condemned to be executed by his captors, but Valletta got wind and intervened, with counter-threats against the Fascist militia holding Peccei.[8]

After the war, Peccei worked on the rebuilding of Fiat, including a ten-year position in Argentina, where he presided over the reopening of the company's operations. While he was continuously employed by Fiat, either as a director in South America or elsewhere as a special team member, Peccei also split his time between many roles. From the late 1950s, he headed the non-profit firm Italconsult, which provided engineering and economic consulting services to developing countries. Then, in the early 1960s, he co-founded and was a board member of the Atlantic Community Development Group for Latin America – a private sector institution that invested in medium-size businesses. In 1964, Peccei became the managing director of Olivetti and its twenty-eight associated companies, which employed 54,000 people with operations in 117 countries.[9] Following the formation of the Club of Rome, Peccei's stature as a global statesman grew. He became a board member of numerous organizations, including the World Wildlife Fund, Friends of the Earth, and the Population Institute.[10] Peccei became so influential

The Global Problématique

that, in February 1974, he was able to encourage ten heads of state to attend a meeting in Salzburg, Austria, to discuss global challenges.[11]

From the late 1960s, Peccei also began to turn his hand to writing, combining his Malthusian leanings with his rich, broad, and sophisticated view of global economic affairs. He published several books, including *The Chasm Ahead* (1969), *The Human Quality* (1977), and *One Hundred Pages for the Future* (1981). The first of these – which became a source of inspiration for Dennis Meadows at MIT – was an elaboration on a short discussion paper that Peccei wrote, "The Challenges for the 1970s for the World of Today." Peccei had first delivered the paper as a lecture at the National Military College in Buenos Aires on September 27, 1965.[12] In it, he drew upon his wealth of experience in Europe, the USSR, China, and Latin America to address a variety of global issues: poor cooperation between the First and Second Worlds (i.e., West and East), the need for Europe to keep abreast of rapid technological change in North America, and the need for the developing world not to miss out entirely on the technological transformation. His understanding of technological progress was quite profound – he correctly diagnosed the failure of the Soviet Union to innovate, which would lead to its eventual collapse in 1990. He recognized that the development of computers and the increased automation of production would change not only man's way of working, but also his "socio-cultural environment, his relations with others, his view of himself, his philosophy of life."[13] Peccei also wrote that "it is difficult to find in history any other period in which whole peoples have looked to the future with such lively concern."[14] He was, of course, writing during the years of the Cold War, at a time when the spectre of nuclear holocaust loomed. The solution, as he saw it, was to find alternatives to the military stimulus that was guiding and regulating technological development. Peccei proposed:

> *[T]he technological revolution must be guided in the attack on the real problems of the next decade: survival in the nuclear age, ... overpopulation, hunger in large parts of the world, education in the broadest sense, justice in liberty, better circulation and distribution of wealth produced inclusive of the technological patrimony itself, that is well-being.*[15]

These were the challenges the Club of Rome went on to wrestle with.

The other essential instigator of the Club of Rome was Dr. Alexander King, the director of scientific affairs at the Organisation for Economic Co-operation and Development (OECD). Born in Glasgow, King had studied and lectured in chemistry at Imperial College London; he then

Malthus Enigma

had a distinguished career as a British overseas science advisor before joining the OECD. Howard Brabyn, in the *New Scientist*, described King as follows: "Much more than just a good scientist turned competent international administrator, he is a trader in thought, a cool catalyst with the golden knack of transforming ideas into action."[16] King was instrumental in the founding of the Club of Rome. He and others at the OECD, including then Secretary-General Thorkil Kristensen, were concerned that national governments were not forward-looking enough in their ability to act upon a variety of contemporary problems, including environmental destruction. In King's words, "The Club of Rome was actually born in the OECD, around such concerns."[17] Copies of Peccei's speech in Buenos Aires had been circulating around a few United Nations meetings.[18] King received a copy from Dzhermen Gvishiani, a member of the Soviet State Committee for Science and Technology. Impressed, King wrote to Peccei – and about a week later, they had lunch together in Paris. This led to a dinner in Rome and several further meetings in quick succession, during which they decided "to do something about the world."[19]

Together, King and Peccei arranged the meeting at the Accademia Nazionale dei Lincei. Peccei convinced the Fondazione Agnelli to fund the gathering, while King encouraged a highly respected consultant at the OECD, Erich Jantsch, to produce a background paper. Unfortunately, while academically strong, the paper was too technical and, moreover, had an overly ambitious title: "A Tentative Framework for Initiating Systems Wide Planning of World Scope."[20] It did not succeed in inspiring the assembled discussants. The French attendees at the meeting, for example, argued that the word "system" had no meaning in the French language.[21] When the small group met at Peccei's house for dinner later that night, however, systems thinking was foremost in their minds. They decided to continue working on what they called the *global problématique* – in other words, the interconnection of problems.

Many talented, influential, and famous people joined the Club of Rome.[22] Early members included Hugo Thiemann, director of the Battelle Institute in Geneva; Saburō Ōkita, president of the Japan Center for Economic Research and later minister of foreign affairs; Eduard Pestel, minister of science and technology for the state of Hanover; and Carroll Wilson, a professor at the MIT Sloan School of Management.[23] Along with King and Peccei, these four became the club's early executive committee. The club was not a formal organization, however; it had no core funding, and each member paid their own way to meetings. Also, the club did not seek consensus among its members, limited to a maximum

The Global Problématique

of one hundred, but rather had the aim of catalyzing debate on key issues. The flexibility of the club has since enabled many powerful industrialists, bankers, diplomats, senior civil servants, elected officials, and even royalty to join. Among the members in the past forty years or so have been Queen Beatrix of the Netherlands; Prince Hassan bin Talal of Jordan; and several former national leaders, including Mikhail Gorbachev (Soviet Union), Horst Köhler (Germany), and Pierre Elliott Trudeau (Canada). Several Nobel Prize winners have also joined the club, including Ilya Prigogine and Joseph Stiglitz. In early days, many members were encouraged to join the Club of Rome due not only to Peccei's vision, but also to his warmth, intelligence, and personal charm.

Through Carroll Wilson, some members of the Club of Rome were first introduced to Jay Forrester and his group at MIT, who were making advancements in systems dynamics models. The club had begun meeting about every six to eight weeks in Geneva, where Hugo Thiemann provided facilities at the Battelle Institute. Peccei was keen for them to go beyond discussion of problems and find systematic methods for analysis of the global issues. At first, they were keen on the work of Professor Hasan Özbekhan, whom King described as "extraordinarily, unusually brilliant, a genius type."[24] Özbekhan, along with Erich Jantsch and Alexander Christakis, had drafted the original prospectus of the Club of Rome, entitled *The Predicament of Mankind*,[25] but there was doubt as to whether the proposed methods were achievable. Forrester attended the club's general assembly in Berne in June 1970, where he described the methods he had developed for analyzing industrial systems. The following month, he hosted a ten-day seminar in Boston for six members of the Club of Rome, at which details of a possible global model were discussed. There was considerable debate between members of the club as to whether systems modelling fulfilled their objectives. Peccei thought it missed some of Özbekhan's humanistic elements. Eduard Pestel, however, was able to encourage the Volkswagen Foundation to commit $250,000 to the project.[26] Forrester passed the project to his postgraduate associate Dennis Meadows, who, with his wife Donella and an international team of about fifteen researchers,[27] built a systems model of the world!

The World3 computer model that Meadows and his team developed used 150 mathematical equations to describe the functioning and interrelations of five attributes of the world: population, agriculture, industry, resources, and pollution.[28] Philosophically, the idea that our complex, evolving, uncertain world can be suitably described by a few equations may be difficult for some people to comprehend. But this is the

Malthus Enigma

nature of systems modelling – to create a simplified, abstract representation of the real world including just sufficient detail to describe essential functioning. Some of the equations in the World3 model have a solid physical basis; these are the balance equations, which, for example, add births and subtract deaths in calculating total human population and similarly keep track of balances of agricultural land, capital stock, and non-renewable resources. The other equations are more difficult to establish; they are essentially empirical equations – determined from data, where available – and based on knowledge of human and/or environmental interactions, but not scientific laws. All of the equations are solved together, using a computer, to produce simulations of possible global trends under various assumptions and scenarios.

The final report for the project, written primarily by Donella Meadows and released in 1972, had a startling principal conclusion:

If the present growth trends in world population, industrialization, pollution, food production, and resource depletion continue unchanged, the limits to growth on this planet will be met sometime within the next one hundred years. The most probable result will be a rather sudden and uncontrollable decline in both population and industrial capacity.[29]

The authors were, however, careful not to be prophetic in their findings. They went on to further conclude that it was possible for growth trends to be altered toward more stable ecological and economic conditions, and it was best to start soon.[30] The mechanism of collapse determined by the model began with a growing population and industrial demand overshooting a limited global supply of resources. As greater efforts are required to obtain resources from hard-to-reach locations, there is less capital available to maintain industrial capacity. After the industrial base collapses along with agriculture and services, population then drops sharply due to declining food supply and medical services. Meadows and his team ran other simulations with different assumptions, but these mostly still led to collapse. For example, when resources were increased, the collapse still occurred due to overwhelming pollution; when pollution was controlled and agricultural productivity increased, the model still reached a point where population had to decline due to insufficient food. The findings were like "Malthus with a computer."[31] The way to avoid collapse demonstrated in one scenario required that population and industrial output become stationary; technology was used to sharply decrease pollution and increase the recycling and efficiency of resources;

The Global Problématique

and a shift occur away from manufacturing and toward the service sector, or health and education.

Prior to the official publication of *The Limits to Growth*, Peccei worked tirelessly to raise awareness of the forthcoming report. Approximately one hundred preliminary versions of the report were shared with heads of state, the media, and university professors.[32] In many countries, Peccei also managed to convince publishers to produce the book at cost. The Club of Rome waived royalty rights in several countries at first, although the Meadowses did receive a personal cheque from Peccei for writing the report.[33] The book was translated into thirty languages, selling over ten million copies.[34] Sales in some countries were particularly spectacular. In September 1971, about six months before the official release, Peccei arranged for Dennis Meadows to promote the study during a lecture tour in the Netherlands. The book ended up selling one million copies in the Dutch language, making it the second-highest-selling non-fiction book ever, after the Bible.[35] The Dutch royal family was so concerned about the book's findings that they initiated – and attended – a gathering of Dutch scientists, politicians, and others at the Royal Palace of Amsterdam to discuss its implications.[36] Queen Juliana provided personal signed copies to every secondary school library in the country. Peccei's efforts to promote the book clearly worked well. *Time* Magazine ran a substantial article on the forthcoming publication on January 24, 1972 – and Anthony Lewis referred to the draft report in the *New York Times* on January 29, stating that *The Limits to Growth* was "likely to be one of the most important documents of our age."[37]

When *The Limits to Growth* was officially released in English – on February 26, 1972 – it was featured on the front page of the *New York Times*. The article, "Mankind Warned of Perils in Growth," began:

> *CAMBRIDGE, Mass., Feb. 26—A major computer study of world trends has concluded, as many have feared, that mankind probably faces an uncontrollable and disastrous collapse of its society within 100 years unless it moves speedily to establish a "global equilibrium" in which growth of population and industrial output are halted.*[38]

The correspondent, Robert Reinhold, went into some detail explaining how the sophisticated method of systems analysis worked. He described how birth rates, death rates, food production, industrial output, pollution, and natural resource use were "all part of a great inter-locking web in which a change in any one factor will have some impact on the others."[39] Reinhold noted how the study bolstered the intuitive warnings of environmentalists but then went on to raise the skeptical views of many economists. One anonymous economist exclaimed, "It's just utter nonsense." Nobel Prize–winning economist Simon Kuznets questioned the wisdom of stopping growth, while Yale economist Henry Wallich was quoted in the same article: "I get some solace from the fact that these scares have happened many times before – this is Malthus again."

Controversy around the *Limits to Growth* study intensified further after its release. On March 2, 1972, Dennis Meadows presented the report at an invitation-only event to 250 high-level guests and media outlets at the Smithsonian Institution in Washington, DC.[40] One week later, *Nature* published a deeply questioning article, "Another Whiff of Doomsday," stating: "On balance, in spite of its provenance, the book is at once over simple and confusing."[41] Though praising the impressive use of computer simulation, the article suggested that "Dr. Meadows and his colleagues would have had a more convincing tale to tell if they had been seen to be more aware of where their predecessors went astray"[42] and concluded that "Mr Peccei and his colleagues in the club should try to do better next time."[43] *The Economist* Magazine followed this up the next day – March 11 – with a scathing article entitled "Limits to Misconception." A month later, in *The New York Times Book Review*, three economists, Peter Passell, Marc Roberts, and Leonard Ross, weighed in further: "*The Limits to Growth* in our view is an empty and misleading work."[44] They went on to note that, while simulations were useful for testing engineering designs and had helped, for example, the Apollo rocket make thousands of trips to the Moon on an IBM 300 before it was built, computer models unfortunately had a poor record of simulating economies even just a year or two ahead.

The Limits to Growth was an extremely controversial study that gave rise to a huge debate, but perhaps more significantly, the Club of Rome's creation helped establish systems modelling techniques more broadly as a tool for humankind's Malthusian struggles. There was a more careful, considered debate about the merits of the *Limits to Growth* study that played out primarily in academic literature for several more years – which will be discussed later in this chapter. There were also further studies by the Club of Rome and a more recent revitalization of studies that looked back at progress since 1972. Whether it is simplistically viewed as right or wrong, the *Limits to Growth* project did advance the use of computer systems models for studying the challenges of human population growth and the global environment. This important accomplishment – and other developments in Earth systems modelling – is the focus of this chapter.

The Global Problématique

Before going further, let me pause and recap how the development of Earth systems models fits within the broader storyline of this book. We are on the way to adding appropriate policy response as a third layer of the Malthus Enigma. So far, we have seen that the need to innovate and doing the science to understand the world around us are the first two dimensions of the Enigma. These are difficult, but critical, activities that humanity has to undertake to sustain itself within the carrying capacity of the Earth. In the previous chapter, I also argued that the science of thermodynamics is particularly important in this regard. Earth systems modelling techniques are a necessary ingredient for supporting policy to address our great environmental challenges of biodiversity loss and climate change. A systems perspective frames our understanding of the interconnectedness of things, helping support wise choices. In earlier chapters, I noted several examples where some form of systems model or framing was used to make judgments on the state of the world. These ranged from simple models such as Malthus's population model and Ehrlich and Holdren's IPAT equation, to more complex models such as those used in the *Global 2000* study of the 1980s. In this chapter, we will explore a variety of Earth system models, which have important differences in their approaches.

In parallel to the development of systems dynamics models, such as the World3 model, there has been advancement of other types of systems approaches to global environmental challenges, which have made notable progress. The first of these are the general circulation models (GCMs), which evolved from early toy models of the global climate to become sophisticated models of the Earth's systems, including the entire global carbon cycle. Having a more solid biophysical basis – relying on thermodynamics – these models made the important progression from being mere scenario tools to being a legitimate part of a holistic Earth science. On a smaller scale, the development of systems approaches for studying the Earth's ecosystems – led by the brothers Eugene and Howard Odum – was also important. Despite the complexity of natural ecosystems, the Odums were able to bring the rigour of physical conservation laws to bear in describing how they function. Herein lies a lesson for the integrated assessment models that evolved from the early systems dynamics models of Forrester and Meadows and continue to be an essential part of our Malthusian struggle to this day.

Malthus Enigma

Forrester: Pioneer of Digital Computing and Systems Dynamics

The founder of the field of systems dynamics is widely recognized to be Jay Forrester (1918–2016), who initially undertook pioneering work on digital computing.[45] Forrester was a farm boy from Nebraska who gained early interest in electrical engineering as a schoolboy when building a wind turbine to power the family cattle ranch. After completing undergraduate studies in electrical engineering at the University of Nebraska, he went to work with Professor Gordon Brown, an expert on feedback control systems at the Massachusetts Institute of Technology. Through graduate work with Brown, Forrester engaged in a range of practical World War II–related projects, such as the control of radar antennas and gun mounts. At one point, he was on board the US Navy's USS Lexington aircraft carrier repairing an experimental radar control unit when it was hit by Japanese fighter planes in the Pacific. After World War II, Forrester headed an MIT project for the US military to build an aircraft flight simulator, from which he led the development of the Whirlwind digital computer. This was an experimental military combat information system, which eventually became the Semi-Automatic Ground Environment (SAGE) air defence system for North America. By 1947, Forrester and his collaborator, Robert Everett, had designed the Whirlwind to be a high-speed stored program computer (i.e., a computer that stores program instructions in its memory). In the process of developing the Whirlwind, Forrester also invented the basics of magnetic core memory – the predominant form of random access memory (RAM) used in computers from approximately 1955 to 1975, after which it was superseded by semiconductors. Having been a pioneer of digital computing, however, in the mid-1950s, Forrester took a change of direction.

In 1956, Forrester was asked by MIT President James Killian to join the MIT Sloan School of Management.[46] MIT Sloan had been founded just four years earlier, with the intent of including substantial technical engineering content within an established business program. Forrester considered the potential to use his computer engineering skills to further fields such as operations research and management information systems, but he instead began to develop the field of systems dynamics. Essentially, he was able to do this because he had developed the necessary tool – a digital computer – to do so. Systems dynamics involves developing mathematical models of complex systems exhibiting non-linear behaviour through internal feedback loops and time delays. The mathematical equations are typically too difficult – or too laborious – to solve without a computer. At first, Forrester focused on industrial applications of systems dynamics, producing a seminal book, *Industrial Dynamics*, in 1961.

One of Forrester's earliest applications of the methods of systems dynamics was at a household appliance factory in Kentucky.[47] For several years, General Electric had been wrestling with a production issue that saw fluctuations between the facility running extra shifts and then half the workforce being laid off. Although partially caused by cyclical variations in demand, Forrester was able to show that the instability could be largely explained by internal decision-making processes. He developed a mathematical simulation model – at this point using pencil and paper – that tracked orders, inventories, and employees. The model showed how employment at the factory became unstable due to the way that hiring and inventory decisions were made. The General Electric factory was essentially behaving like a complex, non-linear system, exhibiting behaviour that was hard to understand without a formal model. Such learning about non-linear systems became central to the teaching of systems dynamics at MIT – and later became incorporated into an exercise known as the "beer game" that students undertake.[48]

Adding to his work on industrial systems, Forrester next moved on to urban systems. The inspiration for this came from the former mayor of Boston, John Collins, who took a year's sabbatical at MIT in the same office area as Forrester.[49] Collins was able to draw upon an extensive network of urban practitioners who aided Forrester's research on understanding the inner workings of cities – and eventually to a book, *Urban Dynamics*.

Even though Forrester was a colleague of Carroll Wilson, a member of the Club of Rome, it was through his work on *Urban Dynamics* that he first met Aurelio Peccei. The two became acquainted at a meeting on urban challenges held at Lake Como in Italy in 1968. So Forrester's work was already known to Peccei when Forrester boldly took the floor at the Club of Rome meeting in Berne in June 1970 and made strong claims that systems dynamics could help with understanding the *global problématique*. Indeed, having received an encouraging response, Forrester made a first sketch of a world model when he took a Pan Am flight back to Boston on July 28, 1970.[50] Forrester's sketch included similar elements as the World3 model that Dennis Meadows and his team later constructed. But Forrester went even further, developing his own model – ahead of his research associate – and published the results in a book, *World Dynamics*, in June 1971. While overshadowed by *The Limits*

Malthus Enigma

to Growth, Forrester's book also gained some media attention, with a review on the front page of *The London Observer* – and articles in *The Straits Times*, *The Christian Science Monitor*, *The Wall Street Journal*, and *Fortune* Magazine.[51]

Forrester took the findings of his and Meadows' modelling work very seriously. Testifying to a congressional committee in the 1970s, he recommended that investments in industry and food production be reduced to slow population growth.[52] In a sense, he had arrived at difficult conclusions similar to those Malthus had reached about feeding the poor over a century before. Recall from Chapter 2 that Malthus took very controversial positions against the English Poor Laws and against repeal of the Corn Laws, in both cases because he believed that they tended to heighten population growth. Forrester's recommendations were nowhere near as wild as Ehrlich's scenarios in *The Population Bomb*, but then, Ehrlich could be dismissed as being sensationalist. Forrester, like Malthus, came to difficult conclusions after careful, considerate study using the best available mathematical tools of the day.

Criticisms and Revival of *The Limits to Growth*

Naturally, the Malthusian warnings of Forrester and Meadows did not go unchallenged. The conclusions to *The Limits to Growth* sparked an immediate response in the media that was largely critical. Beyond the media, moreover, the work provoked a massive response from various strands of academia. Lasting several years – at least until a 1978 paper by Francis Sandbach, "The Rise and Fall of the Limits to Growth Debate" – the academic critiques tended to be more thoughtful and more carefully studied than the reactionary articles in the media. Surveying the debate over forty years later is intriguing, for several reasons. First, some of the debate was seemingly encouraged by members of the Club of Rome themselves – with differences addressed through further studies. Second, generally clear battlegrounds were formed between the economists and the systems dynamists – among which the exchange between Forrester and Yale economist William Nordhaus was most prominent. This battle was full of misunderstanding – and confused by notions of what qualified as being "scientific" on both sides. Third, an interesting twist in the debate on *The Limits to Growth* occurred in 2000, when Matthew Simmons, an investment banker and peak oil writer, published a retrospective paper, "Revisiting *The Limits to Growth*: Could the Club of Rome Have Been Correct, After All?" This was the first of several works

The Global Problématique

that looked back at simulations of the World3 model, to see how they stood up after several decades.

Given that the Club of Rome did not seek consensus among its members – and indeed encouraged debate – then it is perhaps not surprising that some of the critique of *The Limits to Growth* can be traced to club members. Recall that the awarding of the project to Forrester's team had been a hard decision. Both Peccei and King thought highly of Özbekhan's work, even if it lacked a fundamental framework and was considered difficult to achieve.[53] The decision was fractious. Özbekhan had proposed a budget of $900,000,[54] nearly four times what Forrester received. He and Jantsch quit the club when Forrester got the project.[55] King's reflections on *The Limits to Growth* are also quite telling. He stated that the club was very unhappy at first about some of the misinformed discussion around *The Limits to Growth*, especially the insinuation that the club itself was in favour of a zero-growth economy. He remarked that "to be labelled as prophets of zero growth very much annoyed us."[56] Later, he added: "So many people in the Club disliked the *Limits to Growth*. I personally thought it was the best that could be done at the time. It had its faults, but it would have been difficult to better them."[57]

Although King's personal perspective is diplomatic here, there is some speculative evidence that he may well have encouraged one of the most substantial criticisms of *The Limits to Growth*, written by researchers at the University of Sussex in England. An interdisciplinary team from the university's Social Policy Research Unit produced a thorough evaluation of the World3 model, published as a series of fourteen essays. Their book was released in 1973, under the title *Thinking About the Future* in the UK, and under the more provocative title *Models of Doom* in the United States. Yet one of the editors of *Models of Doom*, Keith Pavitt, was a former assistant to King at the OECD. Indeed, King more blatantly remarked: "Everyone at the Social Policy Research Unit (SPDU) at Sussex is a former member of my section at OECD."[58] Perhaps the connection to King was just a coincidence, but that seems unlikely. The main criticism of the Sussex team distilled down to: (1) *The Limits to Growth* being too pessimistic in some of its assumptions and (2) there being fundamental flaws in the model. Looking back at this critique with hindsight, recent Club of Rome member Ugo Bardi justifiably argues that some of the counter-assumptions proposed by the Sussex group turned out to be somewhat naive and that the fundamental flaws were perhaps more like minor flaws.[59]

Malthus Enigma

Meadows and his team were, however, well aware of the limitations of the World3 model – and communicated them reasonably well. The limitations were echoed by members of the club. For example, King noted the model was "highly aggregated and takes no cognisance of the huge disparities between major factors in different parts of the world."[60] It was because of such deficiencies that the club later supported the development of a more detailed multi-regional model by Eduard Pestel and Mihajlo Mesarović.[61] Along similar lines, the World3 model also used a singular index for generically representing resources and again for pollution. In *The Limits to Growth*, Meadows' team provides estimates of the known stocks of about twenty non-renewable natural resources, ranging from eleven years for gold to 2,300 years for coal (at then current usage rates).[62] Such high variation is not reflected in the actual model, for which under the base scenario, the stock of available resources is aggregated into a single variable. Pollution was similarly represented by a single index, thereby not recognizing the huge differences between local and global – and solid, aqueous, and gaseous – forms of pollution. Interestingly, climate change has since emerged as such a critical issue – with CO_2 emissions also interlinked with many local forms of air pollution – that CO_2 or aggregate greenhouse gas emissions could perhaps now be used as a single proxy measure of global pollution. To their credit, Meadows et al. did provide a figure showing data on CO_2 concentrations at Mauna Loa in Hawaii,[63] although they did not explicitly include climate change in the World3 model.

King's later synopsis of the limitations of the model, made in 2006, hence seems quite reasonable:

> *The model had many shortcomings; the imminence of material shortage was overstated, while the magnitude of dangers from pollution could not be fully taken into account until many years later when global environmental effects like global warming and depletion of the ozone layer began to be understood. The criticisms concerning lack of appreciation of technological change, the power of the market and the degree of aggregation were, of course, valid.*[64]

Among the many economists who were critical of *The Limits to Growth*, perhaps the most vociferous was Yale economist William Nordhaus. He had a heated, protracted exchange with Jay Forrester over the methods of systems dynamics used in the World2 and World3 models. This all began with a twenty-seven-page critique, "World Dynamics: Measurement Without Data," which Nordhaus published in 1973. In examining the

The Global Problématique

exchange between Nordhaus and Forrester, Ugo Bardi summarized Nordhaus's criticism in three categories: personal accusations, unsubstantiated statements of disbelief, and quantifiable criticism.[65] Under the first category, Forrester was accused of lacking humility – essentially because he conducted simulations to the end of the twenty-first century. This was ambitious but not surprising, given the question being addressed. Nordhaus's most central criticism of "measurements without data" turned out to be a gross misunderstanding. In both "World Dynamics" and *The Limits to Growth*, historical data is absent from figures showing the calculated scenarios, but Bardi noted that the models were calibrated using historical data; moreover, the full World3 model and data sources were later made fully available by Meadows and his team.[66] Nordhaus also presented some data on birth rates and material standards of living and showed that the relationship was drastically different from Forrester's model assumptions. In his twenty-one-page response, however, Forrester pushed back, explaining that Nordhaus had incorrectly done the comparison in a single dimension and had failed to account for other factors influencing the birth relationship. It seemed that Nordhaus had not really understood the methods used in systems dynamics modelling.

This became further apparent eighteen years later, when the two Meadowses, along with Jørgen Randers, published an updated book, *Beyond the Limits: Confronting Global Collapse, Envisioning a Sustainable Future*. Nordhaus responded to this with a forty-three-page paper, in which he essentially confessed to the difficulty that critics had in understanding the systems modelling technique.[67] Nordhaus then focused on showing that the World3 model was missing technological progress and was therefore flawed. He also went to some length to compare the model to a standard economic approach – a neoclassical production function – which was an exercise of frankly dubious merit. As Bardi noted, the debate essentially died out at this point, as both sides lost interest.

Interest in the *Limits to Growth* study resurfaced again in 2000, however, with Matthew Simmons's article speculating whether the study might have been correct after all.[68] This was followed by several retrospective studies – including one by Meadows and colleagues – that looked back at original World3 model results and compared them against actual data.[69] In some respects, the World3 standard model run was remarkably accurate; it predicted global population, birth rates, and death rates fairly well – in addition to industrial output.[70] Some of the underlying details were far off, however: the model substantially overestimated population in the over-sixty-five age group and significantly missed the rise of the service-sector economy. The model also overestimated land conversion to arable land but underestimated land yields, so its estimation of food production was not too bad. Some of the variables in the World3 model, such as resource use and pollution, were difficult to compare against real data, as the model used indexed variables. If fossil fuel use and particulate pollution were used as proxy measures, then the model was found to greatly overestimate both variables. While hindsight shows the World3 model predicted some things well and others poorly, the comparisons in the revivalist literature were being made before the "collapse" – which occurred roughly around 2020 in the standard World3 model. This, of course, was the year the coronavirus dramatically spread throughout the world, killing several million people and causing a massive global recession. Thus, ironically, there was a collapse, of sorts, though the mechanism was completely different. So let us move on from the *Limits to Growth* study and discuss another type of global systems model – those used to understand global climate change.

Global Climate Models

Early in the morning of June 4, 1944, Dwight D. Eisenhower, Supreme Commander of the Allied Expeditionary Force, made one of the most difficult and important decisions in the course of World War II – to delay the launch of the D-Day landings by twenty-four hours.[71] With a full Moon for visibility and a low dawn tide to expose German underwater defences, June 5 was the first day in a narrow three-day window that would be perfect for the Allies' Normandy landings. A team of weather forecasters from the Royal Navy, the British Meteorological Office, and the US Air Force – led by Group Captain James Stagg – had warned of a storm moving into the English Channel that would have hampered the beach landings. The Luftwaffe's chief meteorologist also registered the storm, reporting that rough seas and gale-force winds would likely continue until mid-June, prompting the German commanders to move their coastal defences to a state of low readiness. The Allies, however, had access to a much more robust network of weather stations and weather ships in the North Atlantic and forecast that there would be a temporary improvement in the weather on June 6. Eisenhower took Stagg's advice – which proved to be correct – catching the Germans by surprise in a critical turning point for the war.

The Global Problématique

Weather forecasting proved to be so important for D-Day, and for the military in general, that following World War II, computer pioneer John von Neumann chose weather prediction as a showcase application for the development of electronic computers.[72] A talented mathematician and physicist, von Neumann had worked on the Manhattan Project, and through the Electronic Discrete Variable Automatic Computer (EDVAC) project, he developed the architecture for one of the earliest digital computers. As an influential member of Princeton's Institute for Advanced Study, von Neumann set up a research program in meteorological systems modelling at the institute.[73] In the post-war period, the military had a continuing interest in meteorology, in part because of a belief that future combatants might seek "control" of the weather as a strategic advantage in warfare.[74] Von Neumann tried to persuade leading meteorologist Carl-Gustaf Rossby of the University of Chicago to come to Princeton to lead the program, though Rossby declined. Rossby, who had pioneered the use of fluid mechanics to describe large-scale motions of the atmosphere, continued, nonetheless, to advise von Neumann.[75] Indeed, after Rossby returned to his native Sweden to a position with the Swedish Meteorological and Hydrological Institute in Stockholm, he led the first consortium to achieve real-time weather forecasting using numerical computer models.[76] Since this breakthrough in the mid-1950s, meteorology has been one of the largest consumers of supercomputer power.

Back at Princeton, researchers had their sights on a bigger goal – to model the world's climate system. Weather forecasting is concerned with understanding the weather over a period of hours to days, while climate modelling is about changes over years to decades. In 1948, Jule Charney, who had spent two years studying with Rossby in Chicago, took over management of von Neumann's meteorological project at Princeton. In planning out the research program, Charney envisioned a hierarchy of models with increasing complexity for describing the atmosphere. The pinnacle of these was a general circulation model (GCM) that would mathematically describe the climate system for the entire Earth over periods of decades – all based on thermodynamics.[77] To push the research, Charney led teams of researchers to code atmospheric models during intensive multi-week-long work sessions on the Electronic Numerical Integrator and Computer (ENIAC) at the University of Pennsylvania. The pinnacle of Charney's hierarchy of models – the programming of a GCM – was achieved in 1955, the year before he left Princeton to join MIT. The researcher who developed the first GCM was Norman Phillips, a graduate of the University of Chicago, who had also

Malthus Enigma

studied with Rossby before moving to Princeton's Institute for Advanced Study. Publication of his paper, "The General Circulation of the Atmosphere: A Numerical Experiment,"[78] was a major breakthrough for climatology and received much attention from fellow climatologists.

To understand the basics of how GCMs work, it is necessary to return to the Earth's radiative budget, discussed briefly in the last chapter. Near the equator, the Earth receives more heat than it can re-radiate back out to space, while conversely, at the poles, it receives relatively less heat. Thus, the major phenomenon underlying the Earth's climate system is a transport of heat through the atmosphere from the equator to the poles,[79] just as Sadi Carnot had described. GCMs – starting from that of Phillips and colleagues – represent this heat flow using energy balance equations, which are solved on computers for spatial grids over the Earth's surface and simulated through multiple time steps. In Phillips' original paper, the heat flows were resolved on a coarse spatial grid of just sixteen by seventeen zones horizontally, with upper and lower atmosphere zones on the vertical.[80] Of course, that was in the 1950s, when computer power was a tiny fraction of what it is today. In subsequent decades, the spatial grids used in GCMs have become more granular. More significantly, though, the GCMs have incorporated more and more essential physical details as computing speeds and memory have increased. These developments in the physical details are well described in Paul Edwards' book, *A Vast Machine: Computer Models, Climate Data, and The Politics of Global Warming*.[81] Major developments prior to the 1980s included model representation of radiative transfer, cloud formation, ocean circulation, the reflectivity of the Earth's surface, and – importantly – sulphate emissions and particulate aerosols. In the 1980s, modellers added representation of sea ice, vegetation, snow, and agriculture as the models became more complete Earth systems models. Since the mid-1990s, greater attention has been paid to representing the entire carbon cycle, including CO_2 uptake and release by plants.

With the addition of so many physical processes – pushed on by a global community of modelling groups – GCMs evolved from being simple play models to become realistic models of the Earth system central to the science of global climate change. As well as the modelling group originating at Princeton, early GCMs were developed by the UCLA Department of Meteorology and the National Center for Atmospheric Research.[82] These groups spurned offshoots so that, by the mid-1990s, thirty-three of them could participate in model inter-comparison projects.[83] Already by the 1970s, GCMs had become sophisticated enough to become reasonable predictive tools. At the 1972 United

The Global Problématique

Nations Conference on the Human Environment in Stockholm, which Maurice Strong had led, there was a split on whether greenhouse gas emissions caused global warming or global cooling. The split occurred because the cooling effects of sulphate and particulate matter emissions, relative to the warming of greenhouse gases, had not been fully resolved.[84] Over the 1970s, modellers were able to incorporate the countering effects sufficiently well to establish that warming was dominating. In 1975, Princeton-based researchers Syukuro Manabe and Richard Wetherald were the first to use a GCM to estimate the long-term effects on the climate of a doubling of CO_2. A further watershed mark occurred in 1979 with a report produced by Jule Charney for the Carter Administration.[85] Based on results from six GCMs, Charney concluded that the most likely impact of a doubling of CO_2 levels would be a mean global temperature rise of between 2°C and 3.5°C. Edwards noted that the Charney report was the "first policy-oriented assessment to claim a concrete, quantifiable estimate of likely global warming."[86] What's more, he noted that Charney's estimate had largely stuck and had been consistently verified in reports of the Intergovernmental Panel on Climate Change.

Another climate modeller who had substantial impacts beyond the advancement of GCMs was James Hansen. Following education in physics in his native Iowa, Hansen joined NASA's Goddard Institute for Space Studies in New York City in 1967. There, he made numerous contributions to global climate modelling, including radiative transfer models and climate forcing mechanisms – for which he was elected to the National Academy of Sciences in 1996. Between 1981 and 2013, Hansen was head of NASA's Goddard Institute, and it was during this time that he gave an influential testimony to the US Senate. On June 23, 1988, Hansen testified before the Standing Senate Committee on Energy and Natural Resources: "Global warming has reached a level such that we can ascribe with a high degree of confidence a cause and effect relationship between the greenhouse effect and observed warming. ... It is already happening now."[87] Several science writers and historians have recognized that Hansen's testimony was a pivotal moment when global warming was elevated to a high level of attention among the public, the media, and policy-makers.[88]

Some politicians were hesitant at first to accept conclusions based on GCMs, because they did not conform to simplistic ideals about the scientific process. A few decision makers rejected the modelling results outright, arguing that proof of human-induced climate change based on raw experimental data was required. As Edwards argued, however, the

Malthus Enigma

battle between models and raw data was incorrectly posed. The global environmental system cannot simply be examined on a laboratory bench. "You can't study global systems experimentally; they are too huge and complex."[89] This challenge of coming to terms with global climate change exemplifies the second dimension of the Malthus Enigma – to scientifically understand the global environment around us. As Edwards further explained: "There is no 'control Earth' that you can hold constant while twisting the dials on a different, experimental Earth."[90] Occasionally there may be opportunities to observe natural experiments like a major volcanic eruption, but still there is no control, and we do not know with certainty what would have happened without the eruption.

Developing simulation models of the Earth system has essentially become a major part of the process of understanding the environment around us. It involves a different type of science from a pure reductionist approach, though it still incorporates scientific understanding of components established by conventional reductionist means. In the case of GCMs, they have been developed over many decades, continually being tested by and adapting to other scientific evidence of the mechanisms of global climate change – from Keeling-like measurements of CO_2 concentrations to reconstruction of past climates through ice cores and similar geological findings. Indeed, with decades of work using the best available computing power and continual examination, GCMs have crossed over from being simulation tools to being hypotheses of a legitimate holistic science.

The Odum Brothers

Two scholars who fully appreciated that science should include synthetic understanding of functional wholes as well as reductionist explanation of components were the brothers Eugene Pleasants and Howard Thomas Odum. Widely recognized as the founding figures of systems ecology, they helped create, as Eugene Odum put it,[91] "one of the few academic disciplines dedicated to holism." Sir Arthur Tansley established the term *ecosystem* in 1935, and other scientists had contributed key ideas to the growing discipline of ecology; however, it was Eugene Odum's textbook *Fundamentals of Ecology* – first published in 1953, with contributions from his brother – that became the bible for education on the topic. Eugene Odum considered the text to be revolutionary in two respects: First, it was arranged in chapters starting with the principles of the whole and then getting into the details of the component parts. Second, the text used energy as an integrating concept linking all elements in ecosystems, whether they be animal, mineral, or vegetable. In other words, it was based on thermodynamics. So although the methods and contexts are quite different, the Odums took an integrated, systems approach to studying ecosystems, as did Forrester, Meadows, and the climate modellers for their inquiries.

The Odum brothers spent much of their lives in the Old South of the United States. Their father, Howard Washington Odum, moved to Chapel Hill, North Carolina, in 1920, to become a professor of sociology. This is where the younger Howard was born. Eugene was born in Newport, New Hampshire, in 1913, and his sister, Mary Francis, was born six years later. Eugene studied zoology at the University of North Carolina and then completed a PhD from the University of Illinois before taking a faculty position at the University of Georgia in 1948, where he stayed for the whole of his career. Howard was more transient; he also completed a bachelor's degree in zoology at North Carolina, albeit interrupted by three years of World War II service in Puerto Rico and the Panama Canal Zone. Howard earned his PhD in zoology from Yale University in 1950, although the focus of his dissertation on the global biogeochemistry of strontium revealed his early interests in systems ecology and understanding of the Earth as a single system. Following four years of intensive research work at the University of Florida, Howard spent two years at Duke University and seven years as the director of the University of Texas Marine Science Institute. After further positions at the University of North Carolina and the University of Puerto Rico, Howard eventually returned to the University of Florida in 1970. There, until his retirement in 1996, he taught in the Department of Environmental Engineering Sciences and founded centres for environmental policy and wetlands. Although they were ten years apart and worked at different universities, Eugene's and Howard's overlapping interests in ecology produced many fruitful collaborations.

In addition to his book, Eugene's greatest contributions to ecology revolved around the Institute of Ecology at the University of Georgia, which he founded in 1961 and directed until 1984. The institute provided teaching for both undergraduate and graduate degree programs in ecology and has been the training grounds for many graduates who went on to leadership positions in the field.[92] Eugene received a host of awards for his work as an educator, from organizations such as the National Wildlife Federation, the Society of Environmental Toxicology and Chemistry, and the International Association of Landscape Ecology.[93] His research was equally impactful. In 1955, Eugene and Howard published a paper on coral reef ecology[94] that received the Mercer Award from the Ecological

Malthus Enigma

Society of America. Eugene followed this with other classic papers, such as "The Strategy of Ecosystem Development,"[95] and a study of the Georgia salt marsh ecosystem.[96] In later years, Eugene began to concentrate on the role of ecology as a life-support system, publishing books such as *Ecology and Our Endangered Life-Support Systems*[97] and *Ecological Vignettes: Ecological Approaches to Dealing with Human Predicaments*.[98] In 1970, Eugene was made a member of the National Academy of Sciences.

In introducing a festschrift celebrating Howard Odum's career, his former graduate student Charles Hall picks him out as a modern equivalent to Leonardo da Vinci – for his contributions were so innovative and wide-reaching. Hall supports this by recognizing at least six scientific journals in which Howard published "either the first, the first significant, or the first systems oriented paper."[99] Generally speaking, Howard's work progressed from ecology to ecological engineering to ecological economics, though there were many twists along the way.[100] In his early days at the University of Florida, he produced a landmark study of the energetics of the Silver Springs ecosystem. This was followed by further whole ecosystem studies of marine systems, such as the Texas Gulf Coast and tropical rainforests. Howard's work on ecological engineering largely began in 1966, upon his return to North Carolina, when he began to research ways to make use of wastewater effluent. In 1971, he published one of his most influential books, *Environment, Power and Society for the Twenty-First Century: The Hierarchy of Energy*, in which he recognized the importance of energy and the environment in supporting human well-being. Paralleling developments by Forrester and others, Howard's book also emphasized the use of quantitative, systems-oriented methods. Following his move back to the University of Florida, Howard's ideas on ecological economics took further shape, and his systems-oriented work led to a further significant book, *Systems Ecology*, in 1983, which was republished in 1994 under the broader title *Ecological and General Systems*. Throughout his career, Howard also took an interest in thermodynamics, writing extensively on maximum power theorem. Relating to Malthusian concerns in the late 1960s, he also produced a study of the energetics of food production for the President's Scientific Advisory Committee.[101] Scientific historian Joel Hagan summarized: "More than any other ecologist, Howard Odum has shaped the way biologists think about energy."[102]

A couple examples, out of many possible candidates, help demonstrate the systematic approach the Odum brothers took in exploring ecosystems. Their award-winning work on coral reef ecology was the result of an intensive six-week study trip with a team of researchers. They went to a large coral reef known as Enewetak Atoll, located in the Marshall Islands near the equator in the Pacific Ocean.[103] There, they mapped and measured the entire trophic structure of part of the reef, from the producers (e.g., algae) to the herbivores (e.g., snails and anemones) and carnivores (e.g., larger fish and crabs) to the decomposers. Howard later conducted a similar, but longer, study along a three-quarter-mile stretch of the Silver Springs ecosystem in Marion County, Florida.[104] The aquatic ecosystem was particularly convenient for scientific analysis, as it had a constant temperature and flow rate from the underground spring throughout the year. As the ecosystem was at a steady state, Howard and his team were able to undertake an extensive sampling program of the water, the light intensity, and all living species, lasting over two years. They characterized the whole community structure of species in the ecosystem, measuring rates of production and assessing mechanisms by which the community metabolism was self-regulated. Anchoring the study all together, Howard produced an energy flow diagram showing kilocalories per year from solar radiation flowing through producers, herbivores, carnivores, and decomposers and then exiting the system. This Silver Springs study became the standard example of energy and material flow through ecosystems used in most ecology textbooks.[105]

Overall, the work of the Odum brothers was highly impactful, not only within the field of ecology, but also in inspiring disciplines such as ecological economics and industrial ecology. Their observance of the physical laws of conservation of energy and mass will be important to further development of global systems models. They showed how thermodynamics can be applied in the complex context of natural ecosystems, so why not human socio-economic systems too?

From Systems Modelling to Holistic Science

Earth systems modelling has come a long way since the controversial *Limits to Growth* study of the 1970s. Systems analysis was still in its infancy in the early 1970s. Following conceptual and mathematical advancement of general systems theory in the 1950s and '60s – by the likes of Ludwig von Bertalanffy and Kenneth Boulding[106] – there were several efforts to analyze global systems. These included the modelling works of Forrester, Meadows, Pestel, Mesarović, and others. Among these, the *Limits to Growth* study had by far the greatest impacts, largely

Malthus Enigma

because of the huge political influence of Peccei and others from the Club of Rome. The *Limits to Growth* model was not much more than a toy model for demonstrating possible global collapse mechanisms. This was recognized by Meadows and his team – who were clear that the model could not make predictions, even if the revivalist literature sought to cling somewhat to the outputs of some scenarios. From a long-term view, however, perhaps the most important contribution of the *Limits to Growth* effort was that it established systems modelling as an important technique for addressing global environmental challenges.

Notable among the organizations that conduct large-scale global systems modelling today is the International Institute for Applied Systems Analysis (IIASA), located in Laxenburg, Austria, near Vienna. Funded by a substantial consortium of national governments, among the activities that IIASA supports is integrated assessment modelling, which simulates global environmental changes under possible evolutions of the global economy and society.[107] Included in its modelling applications are scenarios of future greenhouse gas emissions used in reports of the Intergovernmental Panel on Climate Change. Intriguingly, the founding of IIASA has some strong ties back to the Club of Rome. IIASA was formally started in 1972, after six years of negotiations, as a project to bring scientists together from either side of the Cold War divide to address complex, growing environmental problems. The key actor on the Russian side was the deputy minister of the Soviet State Committee for Science and Technology, Dzhermen Gvishiani, who had been instrumental in introducing Peccei to Alexander King. In his memoirs, King later described Gvishiani as a "hidden member of the Club,"[108] and King also discusses his own role in helping choose Vienna as the location for IIASA.[109] So in the same year the Club of Rome was rolling out *The Limits to Growth*, it also established a permanent institute to continue similar work. That said, while IIASA's integrated assessment techniques offer significant advancement on the *Limits to Growth* work, they essentially remain scenario tools.

In the case of global climate change models, however, which have a much more rigorous physical basis – using thermodynamics – the modelling has moved into another domain – that of holistic science. In this new form of science, the data, the interpretation of the data, the conceptual model, and the mathematical model all together essentially become a hypothesis about the workings of the Earth. The hypothesis cannot be tested on a laboratory bench or with a control experiment. But continued observations and testing of the model can help to corroborate it, or not. Such holistic science does not replace reductionist science, but

The Global Problématique

builds upon it. A good systems model incorporates physical laws and principles that have been tested by reductionist science, but there will also be model parameters that can be established only by testing the model as a whole.

Continued development of global systems models – such as integrated assessment models – has much to learn from the transformation of global climate modelling into a legitimate form of science. Of course, developing systems models of complex socio-economic phenomena including technological change is different from modelling biophysical processes. But perhaps this is where inspiration from the Odum brothers comes in. Howard Odum, in particular, learned how to apply conservation of energy and mass to complex ecological systems – and, over time, began to see socio-economic systems in terms of the same physical laws. In doing so, Howard Odum was a forerunner and influencer of the fields of ecological economics and industrial ecology, which we turn to next. The rigour of applying physical conservation laws in the continued development of global systems models will be important in our continued Malthusian struggle to understand the world around us and guide appropriate policy responses.

In the next chapter, we address the question: *How do we bring ecological principles into economics?* In doing so, we fully arrive at the third layer of the Malthus Enigma, concerned with developing policies to tackle pressing global environmental challenges. The next chapter covers the development of ecological economics, which necessarily adds human value theory to the understanding of thermodynamics, systems thinking, biodiversity, and climate change that we have covered so far.

Chapter 7: Ecological Economics

> *The closed economy of the future might ... be called the "spaceman economy," in which the Earth has become a single spaceship, without unlimited reservoirs of anything, either for extraction or for pollution, and in which, therefore, man must find himself in a cyclical ecological system.*
>
> Kenneth E. Boulding, "The Economics of the Coming Spaceship Earth," 1966

The Apollo program of the late 1960s stirred an environmental consciousness that brought a resurgence of Malthusian sentiment – expressed by the likes of Paul Ehrlich and Dennis Meadows. A consequence of these environmental concerns was the emergence of a school of thought that called for a re-examination of the world's economic system. Elements of this direction were apparent in the motives of the Club of Rome and in the work of others discussed earlier, such as Howard Odum. The calling for a new set of economic principles was strongly made by Kenneth Boulding, an English-born American economist, who penned an influential essay, "The Economics of the Coming Spaceship Earth." Boulding noted: "Economists in particular, for the most part, have failed to come to grips with the ultimate consequences of the transition from the open to the closed earth.[1]" The essay recognized that economies of previous centuries included expansion into undiscovered parts of the Earth, but now the whole Earth was known. Boulding raised the emerging challenges of managing resources and pollution in a thermodynamically closed system. Such ideas were taken further by the "frustrated genius" of Nicholas Georgescu-Roegen (1906–1994), a Romanian-born statistician and economist. Georgescu-Roegen threw a formidable challenge to conventional economic theory by formally defining the thermodynamic basis of the economic system. Although others were important too, the work of Georgescu-Roegen

Malthus Enigma

became highly influential in the development of the revolutionary discipline of ecological economics, which tackles the codependence of economic and natural systems.

Georgescu-Roegen was born with a humble background in the Romanian city of Constanța on the shores of the Black Sea. His parents – an army captain and a sewing teacher – encouraged a hard-work ethic in him. He excelled in mathematics at school, and despite spending two years as a refugee in German-occupied Bucharest during World War I, Georgescu-Roegen was accepted into the University of Bucharest. In 1926, he graduated in mathematics with such a high standing that he obtained a national scholarship to study for a PhD at the Sorbonne in Paris. He completed his doctorate at the Institut de Statistique, writing a highly acclaimed thesis, "On the Problem of Finding Out the Cyclical Components of a Phenomenon."[2]

Following his PhD, Georgescu-Roegen had remarkable opportunities to work with some of the most brilliant scholars of the mid-twentieth century. The first of these, from 1930 to 1932, was Karl Pearson at University College London. Pearson was the founder of modern statistics, developing statistical methods that continue to be used by most of science today.[3] After working together on statistical methods, it was Pearson who encouraged Georgescu-Roegen to apply his mathematical prowess to the field of economics. Subsequently, Georgescu-Roegen obtained a Rockefeller scholarship, which he used to study economics at Harvard with Joseph Schumpeter and Wassily Leontief. Schumpeter, perhaps best known for the concept of creative destruction,[4] and Leontief, who developed input–output methods, were highly influential economists. Georgescu-Roegen worked with them applying statistical analysis to business cycles.

Georgescu-Roegen later wrote that one of his biggest regrets in life was not staying longer at Harvard.[5] Apparently the Harvard economists were impressed by his abilities, but Georgescu-Roegen felt some obligation to his native Romania, so he returned there in 1937. For the next eleven years, Georgescu-Roegen held several government positions back in Bucharest, but he and his family fled for their lives following the communist revolution at the end of World War II. Having been a prominent member of the pro-monarchy National Peasants' Party, Georgescu-Roegen was likely to have been imprisoned by the communists, so using a counterfeit passport, he was smuggled out of Romania on a freighter headed for Turkey. Returning briefly to Harvard, Georgescu-Roegen accepted a faculty position in economics at

Ecological Economics

Vanderbilt University in Nashville, Tennessee, where he stayed from 1949 until his retirement in 1976.

Throughout his career, Georgescu-Roegen worked on a variety of topics, but it was in 1971, just five years before his retirement, that he published his most notable book, *The Entropy Law and the Economic Process*. Up to this point, Georgescu-Roegen had been a relatively mainstream economist – he worked on subjects such as utility theory, production, and monetary theory.[6] From his practical experience in Romania, he was deeply questioning of some of the concepts of neoclassical economics – especially when applied in countries with excess redundant labour. In *The Entropy Law*, however, he made a fundamental departure from mainstream economics. He rejected the macroeconomic concept of circularity of economic processes, instead arguing that "the economic process consists of a continuous transformation of low entropy into high entropy, that is into irreversible waste, or ... pollution."[7] He demonstrated his ideas about biophysical economics by showing how natural factors could be incorporated into a conventional equation used by economists to describe production by industry.[8] He also presented equations showing how laws of conservation of matter and energy should be included in economic input–output analysis.[9]

Through his wrestling with thermodynamic principles, Georgescu-Roegen came to hold extreme Malthusian views about the fate of humankind. He joined the Club of Rome for a while, but left it because his opinion on the need for negative growth was not accepted.[10] In *The Entropy Law*, Georgescu-Roegen noted that it was fashionable to estimate the potential maximum population of the Earth, but he considered a greater question: How long could the Earth support such a population?[11] Given finite resources, he argued that a more relevant metric was *life quantity*, which he defined as the sum of the years lived by all present and future individuals. He proposed: "The population problem, stripped of all value considerations, concerns not the parochial maximum, but the maximum of life quantity that can be supported by man's natural dowry until complete exhaustion."[12] Georgescu-Roegen took the dismal view that the carrying capacity of the Earth was in continual decline as its finite stock of mineral resources was continually being used up. He wrote: "If we abstract from other causes that may knell the death bell of the human species, it is clear that natural resources represent the limitative factor as concerns the life span of that species."[13] Georgescu-Roegen's pessimism here is perhaps due, however, to his misinterpretation of thermodynamics. Physicist Robert Ayres argued that

Malthus Enigma

Georgescu-Roegen incorrectly applied the entropy law to material resources, whereas technically it holds only for energy. Georgescu-Roegen had theorized that materials would ultimately reach a high entropy state in which they were homogeneously mixed – like a well-stirred cake mix – but the continual flux of solar energy received by the Earth could be used to separate out minerals where required. Thus, Ayres argued a spaceship economy was theoretically obtainable.[14] Nonetheless, Georgescu-Roegen's insight that the first and second laws of thermodynamics applied to economies was still profound.

In a paper entitled "Energy and Economic Myths," Georgescu-Roegen drew upon thermodynamics to attack a whole range of economic misconceptions – including conventional neoclassical growth theory.[15] The paper was wide-ranging, addressing topics including the mechanistic nature of economic theory, wastes produced by economies, forms of technological progress, fallacies of exponential growth, and the difference between growth and development. He commented on contributions from Sadi Carnot to Rachel Carson and Paul Ehrlich and provided a relatively strong defence of the Earth systems modelling approaches used in the *Limits to Growth* study. Georgescu-Roegen criticized the modern agricultural practices devised by Borlaug as being against the long-term interests of the human species, in that they support too large of a population. His greatest criticism, however, was reserved for economist Robert Solow – the primary architect of neoclassical growth theory. Georgescu-Roegen picked on a statement Solow had made at a lecture of the American Economic Association in 1974:

> *If it is very easy to substitute other factors for natural resources, then there is in principle no 'problem'. The world can, in effect, get along without natural resources.*[16]

This dismissal of the role of resources – of energy, materials, and food – in the functioning of economies was not a passing joke; the reality was that Solow's Nobel Prize–winning model for growth of economic output relied solely on inputs of capital and labour.[17] Georgescu-Roegen's protégé Herman Daly explained the situation further:

> *Solow's recipe calls for making a cake with only the cook and his kitchen. We do not need flour, eggs, sugar, etc., nor electricity or natural gas, nor even firewood. If we want a bigger cake, the cook simply stirs faster in a bigger bowl and cooks the empty bowl in a bigger oven that somehow heats itself. Nor does the cook have any cleaning up to do, because the production recipe produces no wastes. There are no rinds, peelings, husks, shells,*

or residues, nor is there any waste heat from the oven to be vented.[18]

Solow did not reply to Georgescu-Roegen's original critique but was provoked by Daly's follow-up over twenty years later. His response to the last of five questions posed by Daly was particularly telling. Daly asked:

Do you believe that the matter/energy transformations required by economic activity are constrained by the entropy law?[19]

Solow replied:

No doubt everything is subject to the entropy law, but this is of no immediate practical importance for modelling what is after all a brief instant of time in a small corner of the universe.[20]

Solow, like most economists, failed to understand that the laws of physics provided a huge amount of information about how every sector of the economy works and, in time, would provide insights into phenomena such as the growth of cities,[21] the genesis of the Great Depression,[22] and the whole process of capital formation.[23] The genius of Georgescu-Roegen was the recognition that thermodynamics could be applied to human economies and not just the biophysical world or, indeed, the universe.

Georgescu-Roegen's broader work was held in high esteem. MIT economist Paul Samuelson, who won the Nobel Prize for economic sciences in 1970, considered Georgescu-Roegen to be "an economist's economist."[24] It is said that Georgescu-Roegen was also nominated for the Nobel Prize for economics but never received it.[25] He was hugely influential in inspiring the field of ecological economics – as explained further in this chapter – but *The Entropy Law*, despite having potentially deep implications for the field of economics, was not well received by other economists. An obvious reason for this is that Georgescu-Roegen was calling for a paradigm shift, which would be counter to the paradigm of the prize committee.[26]

There are a couple of further broad reasons that could explain the lack of recognition in his lifetime. The first – to be blunt – was that Georgescu-Roegen had a difficult personality.[27] In exploring his character, Duke University researcher Samuel Iglesias noted that Georgescu-Roegen was somewhat of a lonely mathematical recluse as a child, but that his difficult personality traits came more to fruition later in life, after he became established as a professor. His relationship with one of his few PhD students, Herman Daly, is telling in this regard. Through engaging with

Malthus Enigma

Georgescu-Roegen, Daly, as will be discussed, went on to be a key figure in the establishment of ecological economics. In an interview with Iglesias, Daly recounted his "long and stormy relationship"[28] with Georgescu-Roegen. Daly was drawn to work with Georgescu-Roegen because he was a terrific teacher, but he observed that Georgescu-Roegen could flip from being "kind and considerate on occasion" to being "extremely harsh and vindictive" – and that "[h]e could change his opinion of you on the basis of a chance remark."[29] Daly saw Georgescu-Roegen's personality traits as a "small price to pay," yet when it came to his thesis, Daly successfully defended it, in spite of heavy opposition from Georgescu-Roegen. This left some bitterness, but Daly and Georgescu-Roegen still managed to maintain some form of friendship afterward. Later, however, after Daly had joined the World Bank, he learned that Georgescu-Roegen had given a speech at the Romanian Embassy criticizing him and suggesting that he ought to be fired. Perplexed by news of the speech, Daly wrote to Georgescu-Roegen, but he never heard back.[30]

Perhaps the second reason *The Entropy Law* was not well received by economists was that it was broadly philosophical, hard to read, and arguably not far enough developed to provide practical applications. *The Entropy Law* is over 400 pages long, including the appendices. There are eleven chapters, but only two of these, Chapters 9 and 10, might be considered as pertaining to economics – and the reader is required to wade through several chapters of heavy philosophy before arriving at the economics content.

In reviewing the contributions of Georgescu-Roegen to ecological economics, scholars Cutler Cleveland and Matthias Ruth recognized that he created the "pre-analytic vision" of the field, but it still needed work to be developed. They remarked at "the vagueness of some of Georgescu-Roegen's arguments and to instances where he over-extended his interpretations of thermodynamics and its role in economic systems."[31] Georgescu-Roegen had not developed the theory far enough for readers to fully understand its implications. He did, however, inspire a generation of ecological economists that followed. Cleveland and Ruth noted that Georgescu-Roegen made fundamental contributions to the conceptual framework of ecological economics: he exposed shortcomings of conventional economics and drew attention to the constraints imposed by thermodynamics.[32] Building upon Georgescu-Roegen's core insights, ecological economists have gone on to wrestle with concepts such as the substitutability of natural and human-made capital, and to begin to understand how biophysical constraints play out at the level of industries

Ecological Economics

and economies and on a global scale. Ecological economics has begun to create the tools and concepts – and the right kinds of systems thinking – that are necessary to develop policies in the context of Malthusian struggle.

This chapter describes the development of the field of ecological economics and some of its most important insights for the policy layer of the Malthus Enigma. The philosophy of Georgescu-Roegen, Boulding, and Howard Odum had a huge influence on the field,[33] but it was the next generation of scholars that came together to make it happen. Notable among them was Georgescu-Roegen's student Herman Daly, whose strongly neo-Malthusian treatise, *Steady-State Economics*, was a landmark text that proposed controversial market-based mechanisms for controlling growth of population and the capital stock. Daly became part of a core group of economists and ecologists that established ecological economics as a discipline with its own journal and society.

The field of ecological economics builds upon many of the concepts already discussed, so a brief recap may be useful. Earlier chapters on feeding humanity within the Earth's carrying capacity and global climate change were concerned primarily with the development of technology and science – the first two layers of the Malthus Enigma. With ecological economics, we turn now to the third layer of the Enigma, of finding appropriate policy responses to the human predicament. Ecological economists take a whole systems perspective – discussed in the last chapter – in that they place the workings of society and the economy within the biosphere.[34] Following Georgescu-Roegen, ecological economists see the laws of thermodynamics – discussed in Chapter 5 – as a fundamental starting point. In adding the human economy, though, they necessarily add human values, desires, and needs to the biophysical world.

Ecological economics spans between the physical sciences and the social sciences. As such, the thermodynamic principles that it draws upon are necessary, but not sufficient to understand the economy. The limitations of thermodynamics for representing economic processes can be seen in two respects. First, raw thermodynamic quantities such as energy and entropy are insufficient to represent metabolic processes in society. As ecological economist Kozo Mayumi put it: "Humans cannot be fed gasoline, nor can they power a refrigerator with pizzas."[35] So the form of energy matters. Second, just as Georgescu-Roegen recognized, economic processes cannot be understood without invoking a human value system. Let us pick up on Mayumi's example, and suggest that we feed the pizza to the humans. Whether we choose meat toppings or insist

Malthus Enigma

on vegetarian pizza of similar caloric input will clearly depend on human values – in particular those regarding other species on the planet.

According to Clive Spash of the Vienna University of Economics and Business, ecological economics could "aspire to becoming a meaningful critical social science embedded in a good understanding of biophysical reality."[36] There are, however, many social issues that ecological economists care about, making the field of study pluralistic. To Spash, ecological economics is important for leading a radical social and ecological transformation that is required to address the huge environmental challenges such as biodiversity loss and climate change.[37] Ecological economists wrestle with issues such as decoupling economic growth from energy and material use, exploring notions of de-growth in economies, and developing sustainable consumption – including decreases in absolute levels of consumption.[38] All this, while also still aiming to achieve the goal of meeting basic human needs for everyone on the planet, addressing large inequalities in wealth, and recognizing the ethical significance of other species, too. The philosophy and internal wrestling of ecological economics becomes more apparent when we reflect upon some of the market-based approaches that have emerged since 1970 for addressing environmental challenges.

Working outside of ecological economics, the field of finance and other more mainstream economists have developed market trading mechanisms aimed at reducing environmental pollution. Early ideas about trading of pollutants from economist Ronald Coase began to creep into US policies in the 1970s; following the Reagan–Thatcher era of market liberalization, full-blown experiments on markets for local air pollution began. Through the efforts of folks such as Chicago futures pioneer Richard Sandor, the earliest carbon markets were established – and some wild ideas were even hatched that we could develop markets for biodiversity. In parallel to this, ecological economists had been wrestling with concerns of how to suitably value the Earth's natural ecosystems, which are so seriously under threat. But the notion that we might actually trade biodiversity on global markets strongly goes against fundamental principles of ecological economics, as this chapter will reveal. Before we get to the trials and tribulations of markets for carbon and biodiversity, however, let us return to Georgescu-Roegen's student Herman Daly.

Daly and *Steady-State Economics*

In Daly's influential 1970s text, *Steady-State Economics*, he wrestles with the modern Malthusian challenge of sustaining human population within ecological constraints. Daly identifies with Malthus on many levels, from recognizing the limited carrying capacity of the Earth to advanced exploration of similarities between Malthusian and Marxist theories of poverty.[39] Daly's neo-Malthusian perspective strongly motivates his arguments against economic growth and his calling for a different economic paradigm. He begins: "Growth of the human household within a finite physical environment is eventually bound to result in both a food crisis and an energy crisis and in increasingly severe problems of depletion and pollution."[40] Daly's concerns extend beyond the question of how to feed a growing population to doing so within the environmental boundaries of the planet. He draws upon Lester Brown, founder of the Worldwatch Institute, in noting that "the question is not can we produce more food, but what are the ecological consequences of doing so?"[41] Daly also recognizes that Malthus had himself identified this more nuanced higher-level struggle. Malthus postulated that an unlimited supply of food would "plunge the human race in irrecoverable misery"[42] The reason was that unlimited food, or unlimited energy supply, for that matter, would make it easier for the human population to grow and destroy the surrounding ecosystems. Thus, Daly pursues the ambitious challenge of developing a political economy that is bounded by both ecological scarcity and human existential concerns over equity and justice.

Daly argues that the necessary solution to this human predicament is the establishment of a steady-state economy. He defines this as "an economy with constant stocks of people or artifacts, maintained at some desired, sufficient levels by low rates of maintenance throughputs, that is, by the lowest feasible flows of matter and energy from the first stage of production (depletion of low-entropy materials from the environment) to the last stage of consumption (pollution of the environment with high-entropy wastes and exotic materials)."[43] In developing this definition, Daly builds upon the thermodynamics of his mentor, Georgescu-Roegen. He clearly draws upon *The Entropy Law* in expressing the first stage of production and last stage of consumption in thermodynamic terms. To live within the thermodynamic constraints, Daly proposes that the human population and the stock of physical wealth need to be held constant. He also stresses things that can change, including culture, genetic inheritance, and knowledge, as well as technology, design, and the

Malthus Enigma

distribution of artifacts.[44] In short, Daly's notion of a steady-state economy entails development without growth.

Daly proceeds to boldly describe three types of institutions that would be necessary for a steady-state economy to come to fruition, while recognizing the difficult ethical dilemmas in establishing them. In his steady-state economy, there are still a price system and property rights, but three essentially new forms of institutions are required: institutions for stabilizing population, institutions for stabilizing the stock of physical capital, and institutions for limiting inequality. All are extremely ambitious and controversial goals.

On the matter of stabilizing population, Daly draws upon an idea from Kenneth Boulding for establishing a system of transferrable birth licences.[45] Under the proposed arrangement, each couple, or female, is freely awarded a fixed number of birth permits; these can be used or otherwise bought or sold on a market, thereby allowing family sizes to be larger or smaller than the standard. Daly, like Boulding, suggests 2.1 children as the permitted quota given to each woman – as this is the fertility rate for a stable population. Recall from Chapter 3 that fertility rates among women in developed countries are below 2.1 children on average, but the rate in Africa is almost double. The permits are tradeable in units of one-tenth of a child, so a couple desiring a third child could purchase nine more deci-child permits or buy a whole child permit outright, for example. The birth licence system is, of course, hard to accept, because it is such a direct way to stabilize population. Daly recognizes that societies tend to prefer indirect approaches that lower fertility rates through means such as liberalization of women in the workforce, changing tax laws, limiting the size of public housing, and encouraging celibacy or late marriage.[46] He understands the reluctance to combine reproduction and financial means but argues that "life is physically coupled to increasingly scarce resources, and resources are coupled to money."[47]

Daly's second institution is considered quite separate from the first – without any dependency between the two. The purpose is to stabilize the stock of physical human artifacts – that is, the stock of physical capital assets – to keep the throughput of materials within ecological limits. While the objective of the institution is to both conserve resources and minimize pollution, Daly argues that both can be controlled simultaneously, as they are linked by material throughput – that is, conservation of mass. He considers it easier to monitor and control the depletion of resources compared with emissions of pollutants – and suggests that depletion quotas are the key tool for the second institution.

Ecological Economics

Daly's reasoning is that there is a smaller number of resource sources, such as mines and wells, than pollution sources, such as smoke stacks, vehicle exhausts, and a wide variety of diffuse emissive sources. By controlling the supply of raw materials, he argues that society will become more efficient at controlling the stock it has. Daly, moreover, is in favour of the public sector conducting geological exploration for resources and having full control of the quota system. He has a strong preference for using quotas rather than discouraging exploitation through taxes, since quotas will definitely limit resource inputs as desired. Taxes, he argues, may be used as a secondary mechanism but have indirect control with uncertain results. He sees taxes as a means for fine-tuning resource markets; they are a means for pricing externalities (e.g., pollution or resource depletion), but they do not enable markets to align with ecosystem limits as quotas can. In Daly's view: "The market is not allowed to set its own boundaries."[48]

The third institution comes about because Daly recognizes that using market mechanisms to control population and capital stocks will lead to equity issues. His third institution has responsibility for managing the distribution of wealth in society. He argues there is a need to place minimum and maximum limits on income, as well as a maximum limit on the wealth of individuals. Minimum income levels are recognized in many societies – for example, through social assistance and minimum wages – but putting direct maximum limits on income is uncommon, at least in contemporary market economies. Daly's concern is that exchange in free markets between super rich and super poor can too easily turn into a form of exploitation. He draws upon scholars such as John Stuart Mill and John Locke in arguing that private property rights are a "bastion against exploitation,"[49] but only if inequality of wealth is kept within justifiable limits. Thus, deep social considerations enter Daly's philosophy for a steady-state economy.

Daly's *Steady-State Economics* and later works have inspired many ecological economists, as will be discussed shortly, but in particular, they have encouraged the development of ecological macroeconomics. Several grand, big-picture treatises on how the macroeconomy can sustainably function within planetary boundaries and without growth have followed. Examples include works by authors such as Peter Victor from York University in Toronto and Tim Jackson from the University of Surrey in the UK.[50] There has also been the recent development of a de-growth movement, but let's go back to the 1970s again, to the formative years of ecological economics.

Malthus Enigma

Development of Ecological Economics

Daly was just one of a handful of economists who realized in the early 1970s that mainstream economics was seriously lacking in its ability to address critical environmental concerns. After a gestation period of almost twenty years, these few breakaway economists had teamed up with ecologists and other scientists to form a society of ecological economists. The story of how the development of ecological economics occurred is well told by Danish professor Inge Røpke, whom I substantially draw upon here.[51] Before recounting the early history of ecological economics, it is worthwhile to explain why it was needed.

Mainstream economics had broadly developed some understanding on three environment-related issues. First, there was work on optimal use of renewable and non-renewable resources. Second, starting with the work of English economist Arthur Pigou (1877–1959), methods of taxing the external costs of environmental pollution had been developed. Pigouvian taxes are applied to correct for the negative externalities of an economic activity – that is, societal costs that are not borne by the market. Third, economists had prescribed ways to value the enjoyment gained through the access, or simply existence, of unspoiled natural environments. These three issues largely became the focus of the field of environmental economics, which became fully established in the 1970s. Environmental economics, however, gave little or no attention to Georgescu-Roegen's ideas related to entropy, consideration of material and energy flows, or deeper Malthusian concerns expressed by Daly. Resource use, pollution, and the value of amenities were largely treated as three separate issues, essentially using microeconomic principles. There was no recognition that the macroeconomy was dependent upon severely stressed global ecosystems. Røpke summarized the state of affairs well, noting: "The message that externalities are pervasive and potentially threatening for the life support of the human economy was nearly invisible in environmental economics at the time."[52]

There were several important theoretical developments that influenced the gestation period of ecological economics. Though difficult to read, Georgescu-Roegen's book *The Entropy Law* was highly influential – becoming the focus of study groups for the emerging discipline. Daly's work, especially his book *Steady-State Economics*, was also important – and easier to read. Howard Odum's large body of work on studying energy and material flows through ecosystems also had a lot of followers. Although Odum was not directly involved with the formation of the International Society for Ecological Economics, he

Ecological Economics

played a role in connecting some of those who were. Further theories that developed in or around the 1970s that influenced ecological economics included Ilya Prigogine's work on non-equilibrium thermodynamics,[53] Buzz Holling's work on the resilience of ecosystems, and related ideas dubbed chaos theory.

The broader context of the energy crisis also had an impact on ecological economics, as this encouraged research on connections between energy and the economy. Studies on energy return on investment (EROI), which we discussed in Chapter 4, were relevant to ecological economics. Research by Charles Hall, Cutler Cleveland, Robert Herendeen, and others suggested that EROI should tend to fall over time as the most accessible non-renewable energy sources get used up. Bruce Hannon at the University of Illinois Urbana-Champaign developed detailed economic models of the US economy, showing its links to energy use. Others wrestled with questions of energy quality, implications of increased labour productivity on fossil fuel consumption, and the efficiency of energy in food production. Among the most intriguing research was work by Robert Costanza on whether the energy embodied in goods and services provides a good measure of their value.

Costanza came to be one of the most central actors in the formation of the field of ecological economics – along with Daly, Swedish ecologist AnnMari Jansson, and Spanish economist Joan Martínez-Alier. Costanza joined Howard Odum's South Florida wetlands project, working on land-use history as part of a master of architecture degree in the early 1970s, and went on to complete a PhD in systems ecology with Howard Odum. Through his work on energy flow accounting, he became acquainted with research by Bruce Hannon, whom he visited for a while, as well as Daly's work. In 1980, soon after completing his PhD, Costanza attended a symposium organized by Daly, "Energy, Economics, and the Environment," at a meeting of the American Association for the Advancement of Science. Daly encouraged Costanza to publish his work on embodied energy in *Science* – and this was the start of a long association between the two. Costanza applied for a position at the Center for Wetland Resources at Louisiana State University, motivated by Daly's presence at the university, in a different department. He sat in on Daly's economics seminars and began a study group – including his first graduate student, Cutler Cleveland – to understand books by Georgescu-Roegen and others. Costanza and Daly would go on to achieve many things together. They produced a special issue on ecological economics in the journal *Ecological Modelling*. Then, in 1987–1988, they negotiated with potential publishers over the establishment of a new journal,

Malthus Enigma

Ecological Economics. It was the journal's eventual publisher, Elsevier, that encouraged them to establish a society to support the journal. Costanza became the chief editor of the journal and the first president of the society.

The person who was first responsible for initiating gatherings of the emerging ecological economists was Swedish ecologist AnnMari Jansson. Again there is a link with Howard Odum here. Jansson was a zoologist turned marine ecologist who gained insights into human impacts on the environment through studies of algal blooms in the Baltic Sea in the 1960s. In 1971–1972, she and her husband Bengt-Owe Jansson – who was also an ecologist – went to join Odum as visiting scholars at the University of Florida. They were inspired by Odum's studies of energy flows in ecosystems. So, upon returning to Sweden, AnnMari Jansson set out to conduct similar studies for Gotland, Sweden. Later, Bengt-Owe Jansson, from his position on the board of the Marcus Wallenberg Foundation, was able to encourage the foundation (which supported international cooperation in science) to sponsor a workshop bringing ecologists and economists together. When AnnMari Jansson asked Odum whom to invite, he replied with a list topped by Daly. The symposium was held at a luxury hotel in Saltsjöbaden, Sweden, in 1982, with Daly making an impressive contribution: "Alternative Strategies for Integrating Economics and Ecology." Several economists and ecologists who became active in ecological economics met for the first time at this meeting. AnnMari Jansson then hosted a second meeting on ecological economics in 1986, under the European Coordination Centre for Research and Documentation in Social Sciences.

The second of these meetings was attended by Spanish economist Joan Martínez-Alier, who proceeded to join Jansson as an early instigator for meetings of ecological economists. Martínez-Alier was an agricultural economist who studied and worked in Oxford from 1966 to 1973. He had been introduced to Georgescu-Roegen's research in 1974 – and had met and corresponded with him, as well as with Daly.[54] At the 1986 meeting in Sweden, Martínez-Alier offered to host the next meeting in Barcelona. This became a legendary meeting in the history of ecological economics. Martínez-Alier was able to encourage many more attendees, especially from North America, in addition to those who had been at the Swedish meeting. Barcelona attracted a critical mass of people with a like-minded perspective, providing the ingredients and impetus to launch a new journal, with the consensus that it should be called *Ecological Economics*.

The Value of the Earth's Ecosystem Services

One of the most important – and also most controversial – questions that ecological economists have wrestled with is the valuation of ecosystem services. Lying beneath the destruction of ecosystems, threats to biodiversity, and the potential for the Earth to witness a sixth mass extinction is an inability of human economic systems to properly value natural ecosystems and the services they provide. This comes about because of the proverbial problem that human economic transactions fail to price in the external costs of damage to the environment. One way of understanding why this happens is Garrett Hardin's notion of the "tragedy of the commons," whereby each of our rational economic actions as independent free agents leads to pollution of our environment or destruction of ecosystems, to the detriment of us all.[55] The hope of some ecological economists was that if they could find a way of expressing the value of ecosystems in monetary terms, it might go some way toward increasing their protection.

Value theory is one of the most difficult topics in economics. As I reviewed in my previous book,[56] there are three perspectives that can be taken in determining the value of an entity: the utility – or satisfaction – that it provides, the labour required to produce it, and its scarcity.[57] All three apply simultaneously and are interlinked, but in practice, the way in which the value of something is actually determined is through buying and selling on an open market. This is where humanity runs into difficulties with ecosystem services, though – as aside from agriculture, fisheries, lumber, etc., there are no markets for the broader, richer set of services that Nature provides humankind. One of the philosophical contributions of Georgescu-Roegen's *Entropy Law* was that he produced a quasi-equation of value based on thermodynamics. Georgescu-Roegen theorized that "low entropy" was a "necessary condition for a thing to have value."[58] This can perhaps be considered as a restatement of the scarcity perspective of value. Yet, drawing upon his background as a neoclassical economist, Georgescu-Roegen realized that low entropy was an insufficient reason for a thing to have value. The entity also had to provide enjoyment of life – that is, utility – to humans. He wrote:

> *[W]e cannot arrive at a completely intelligible description of the economic process as long as we limit ourselves to purely physical concepts. Without the concepts of purposive activity and enjoyment of life we cannot be in the economic world. And neither of these concepts corresponds to an attribute of*

Malthus Enigma

elementary matter or is expressible in terms of physical variables.[59]

Ecological economists have continued to work on the challenge of placing value on Nature in the absence of markets. Daly, for example, co-edited a book, *Valuing the Earth: Economics, Ecology, Ethics*, containing a collection of twenty classic essays on the topic.[60] Most frequently, to sidestep the challenge of having no real market, ecological economists have resorted to a "willingness-to-pay" approach to evaluation. Under this approach, people's valuation of ecosystems is revealed through carefully constructed survey questions.

Substantial efforts have also been undertaken to collect and synthesize evaluations of ecosystem services. An important early meeting in the study of ecosystem services was a gathering of Pew Scholars in Conservation and the Environment that was held in New Hampshire in October 1995.[61] The purpose of the meeting was to produce an authoritative book on ecosystem services, which became *Nature's Services: Societal Dependence on Natural Ecosystems*. The lead editor was Gretchen Daily – an ecologist from Stanford – who went on to be awarded a Blue Planet Prize for her work. Among other contributors present at the meeting were Paul Ehrlich and Robert Costanza. At the meeting, Costanza had the idea that the data being collected could be synthesized into a quantitative assessment of the value of all of the Earth's ecosystem services. It was an ambitious idea but received support from the new National Science Foundation–funded National Center for Ecological Analysis and Synthesis in Santa Barbara, California. Subsequently, on June 17 of the following year, Costanza and twelve colleagues convened a workshop entitled "The Total Value of the World's Ecosystem Services and Natural Capital."

At the Santa Barbara workshop, Costanza and the fellow ecological economists pulled together a disparate collection of previous studies to provide an estimate of the total value of the Earth's ecosystem services. They assembled evaluations largely from willingness-to-pay estimates for seventeen types of ecosystem services over sixteen categories of biomes.[62] The ecosystem services included some that are already captured by markets, such as food production and extraction of renewable raw materials. The majority of them lay beyond markets, however, including services such as climate regulation, erosion control, soil formation, nutrient recycling, and pollination. The workshop attendees determined the combined value of ecosystem services per unit area of each biome type and then, somewhat crudely, multiplied by the area of biome over the surface of the Earth. Recognizing there were many

Ecological Economics

uncertainties in their methods, the ecological economists determined a best estimate for global ecosystem services of $33 trillion (in 1994 US dollars). Just under two-thirds of the value was for marine systems – especially coastal waters. Of the $12.3 trillion valuation of terrestrial ecosystems services, substantial components were calculated for forests ($4.7 trillion) and wetlands ($4.9 trillion). Costanza and colleagues went at some length to describe twelve sources of error and limitations of their study. They added that the uncertainty put the estimate for the total value of ecosystem services between $16 trillion and $54 trillion. To put this value into context, they noted that global GDP in the year of study was $18 trillion.

An article from the workshop, "The Value of the World's Ecosystem Services and Natural Capital," was published in *Nature* – receiving substantial press coverage and stirring academic debate. It was picked up by *The New York Times*, *Newsweek* Magazine, National Public Radio, and the BBC.[63] Academic responses ranged from considering the evaluation to be far too high, to it being far too low. Some thought it was impossible for the value of the Earth's ecosystems to be greater than GDP; others saw the value of ecosystem services to be infinite and considered the exercise to be misguided. The general consensus was, however, more accepting; Costanza and colleagues reported that "most people understood our point in making this admittedly crude estimate: to demonstrate that ecosystem services were much more important to human wellbeing than conventional economic thinking had given them credit for."[64]

Some progress on evaluating and protecting ecosystem services has been made since the Santa Barbara workshop.[65] There has been a proliferation of research on valuation of ecosystem services, covering a variety of biomes on different continents – as well as the development of integrated computer models of ecosystem services. Several large institutional programs have been established, including The Economics of Ecosystems and Biodiversity, an initiative of the United Nations Environment Programme, and the Intergovernmental Science-Policy Platform on Biodiversity and Ecosystem Services, a large global effort to synthesize knowledge. Some institutional mechanisms for paying for ecosystem services do exist – so that landowners receive compensation in return for providing or protecting natural assets. Within existing legal systems, common asset trusts can be used to assign property rights to the commons – with boards of trustees protecting the asset for the community. There is some evidence that payments for ecosystem services under the United Nations' REDD+ framework is beginning to slow

Malthus Enigma

deforestation.[66] Despite the progress, ecological economists remain somewhat pessimistic and worry that valuing of ecosystem services will continue to be a marginal idea unless there is a revamping of the current economic system to better include natural and social capital. They maintain that we need a "new economic paradigm that puts 'nature' at the core."[67] We need measures beyond GDP – ones that recognize the complex non-linear connection between natural and human systems. One of the key challenges remains the lack of general knowledge on how ecosystems function and provide support for human well-being, which can be overcome only through better education. There is one other possible approach to valuing ecosystem services, though it is mighty controversial, and that is through the development of markets.

Experiments with Markets

Ecological economists are concerned primarily with what I have called the third layer of the Malthus Enigma – that is, developing appropriate policy approaches to sustain the human population within the carrying capacity of the planet. This was apparent in Daly's *Steady-State Economics* and was an underlying motive for Costanza and colleagues to put a value on the world's ecosystem services. Among the most innovative – but most difficult – policy approaches under this layer of the Enigma has been the advent of market approaches for managing pollution, and potentially other environmental impacts. Humans have used markets to develop and distribute limited resources since antiquity.[68] On the face of it, using markets to share resources is far better than theft or war, although sometimes markets can fail, and more direct regulations are required. With appropriate rules in place and systems to enforce them, markets can be designed as very effective means for distributing the goods and services that humans desire. Flipping this around and using market mechanisms to do the converse – prevent environmental damage that we don't want – has many challenges, as will be discussed further below. While some ecological economists – such as Daly[69] – see an essential role for markets in sustainable development, others are more skeptical.[70] One reason for the divided opinion, perhaps, is that using market mechanisms to control environmental ills is still a relatively new idea. The first formal markets for pollution started about thirty years ago, in cities such as Chicago and Los Angeles, but we have yet to really move beyond the experimentation stage. Using markets to reduce pollution, control carbon, and perhaps even prevent biodiversity loss has some potential, but we have very limited experience of doing it – and we are struggling to get it right.

The development of pollution trading markets went from theory to implementation in just a few decades, beginning primarily in the United States. The idea that markets could be used to determine rights to pollute is generally attributed to Nobel Prize–winning economist Ronald Coase,[71] although he was skeptical that such a scheme would work. Coase wrote:

> *But if many people are harmed and there are several sources of pollution, it is more difficult to reach a satisfactory solution through the market. When the transfer of rights has to come about as a result of market transactions carried out between large numbers of people or organizations acting jointly, the process of negotiation may be so difficult and time-consuming as to make such transfers a practical impossibility.*[72]

In the late 1960s, however, economists Thomas Crocker from the University of Wyoming and John Dales from the University of Toronto argued for systems of tradeable permits of pollution – and further academic studies followed.[73]

In practice, the first pollution markets developed in the United States in a somewhat evolutionary manner, which has been described as an "unintended consequence of the *Clean Air Act of 1970.*"[74] The Act required states to meet new air quality standards for criteria pollutants such as sulphur dioxide, nitrogen oxides, and carbon monoxide, as established by the Environmental Protection Agency (EPA). A practical challenge emerged where there was demand for a new industrial facility in a region that was already failing to meet air quality standards. The political will to prevent such industrial expansion simply did not exist. A creative solution was devised in 1976 by the regional administrator in the EPA's San Francisco office, Paul DeFalco.[75] His interpretative ruling for California allowed for new stationary pollution sources in non-attainment areas, provided that the lowest technologically achievable emissions rates were met and that emissions were offset at a greater rate elsewhere within the airshed.[76] This ruling – which became codified by the EPA in 1977 – essentially gave recognition to tradeable offsets, although without formally creating a market.

From 1979 to 1987, the EPA used a primitive form of emissions trading during the phase-out of leaded gasoline. Technically, the EPA allowed oil refineries to trade rights to use lead in their products over the period of transformation – rather than the right to emit lead pollution.[77]

175

Malthus Enigma

A report in 1985 by economist Thomas Tietenberg of Resources for the Future called out the EPA for running a *de facto* emissions trading scheme.[78] The EPA was repeatedly challenged by environmental interest groups over its emerging processes of managing pollution, but its provisions for trading were upheld by the US Supreme Court in 1986.

Throughout the 1980s, the US – like other developed countries – wrestled with the problem of reducing emissions of sulphur dioxide and nitrogen oxides that caused acid rain. The means to use market mechanisms were by then established – and were applied in a small way to reduce chlorofluorocarbons under the Montreal Protocol – but there was resistance to using them mainstream. An important political breakthrough in this regard was a bipartisan study known as *Project 88*, led by Senators Timothy Wirth and John Heinz.[79] Written by Harvard professor Robert Stavins, with a strong supporting cast, the *Project 88* report persuasively argued for using market forces to tackle environmental pollution. As well as recommending the use of tradeable permits for stationary sources of air pollution, the report also called for international trading of greenhouse gases and a variety of other measures on energy efficiency, water, solid waste, and contaminated land. *Project 88* was well received by the administration of incoming president George H. W. Bush, and the EPA's use of emissions trading was subsequently bolstered in the 1989 reform of the *Clean Air Act*.[80]

One of the first places in the world to attempt a formal pollution trading market was Los Angeles, California. The city of Los Angeles and the surrounding metropolitan area had been infamous for its smog for many decades. Occurrences of photochemical smog were a prominent issue at least as far back as 1943.[81] By the 1990s, an estimated 6,000 premature mortalities per year were occurring in the Los Angeles region due to particulate air pollution – this being about one-tenth of the total for the entire United States.[82] Between 1989 and 1993, emissions of nitrogen oxides from industrial facilities were reduced by thirty-seven percent through technology-based regulation.[83] Industry was against command and control approaches,[84] though, and inspired by the EPA and the Clinton Administration, companies argued strongly for market-based approaches to pollution abatement in the region.

Responding to the new political mood, the South Coast Air Quality Management District[85] began developing market mechanisms for managing pollution in the South Coast Air Basin. This began with an old-vehicle scrapping program – known as Rule 1610 – that was launched in January 1993. Under the program, stationary emitters of pollution, such as petrochemical and other industrial facilities, could earn permits to pollute by paying for old cars to be removed from the road. The system involved licensed car scrappers selling emissions permits to the stationary polluters. In theory, Rule 1610 required the stationary polluters to purchase emissions permits that would achieve twenty percent greater reductions in emissions than would be achieved using technology-based regulation. The program was designed so that net emissions in the air basin could be reduced by greater amounts at lower costs by the stationary industrial emitters paying for the removal of mobile emissions sources. The following year, the South Coast Air Quality Management District started a second program – known as the Regional Clean Air Incentives Market, or RECLAIM – which established the world's first urban smog trading market.[86] Approximately 400 industrial facilities were potentially included in the market. These facilities had to purchase credits to pollute, with the number of credits available set to decline each year. The aim was to reduce emissions of nitrogen oxides by seventy-five percent and emissions of sulphur oxides by sixty percent, both by 2003.[87]

An evaluation of the performance of RECLAIM and the mobile source trading program was published by Richard Drury and colleagues from Communities for a Better Environment in 1999 – but they were severely disappointed. The analysts reported three major problems, among several others, with how the programs had been designed and implemented. The first was the formation of toxic hotspots and concerns over social injustice.[88] Cars removed from the road under Rule 1610 were regionally distributed, but stationary polluters buying emissions credits were spatially concentrated. In particular, Drury and colleagues highlighted that most of the emissions permits were purchased by four oil companies, and three of these were located close together in the San Pedro and Wilmington area. Moreover, as this neighbourhood was primarily home to non-whites, this raised serious questions about environmental injustice.

The second issue raised was one of fraud and manipulation by companies. Drury and colleagues observed that the car-scrapping program had been "plagued by a history of under-reporting of actual emissions by industry and an over-reporting of claimed emissions reduction from cars."[89] Part of the problem was that the program relied on companies self-reporting. Another aspect was that old polluting engines from the scrapped cars were being sold for reuse.[90]

The third major concern was that, over the first two years of RECLAIM, the net emissions of nitrogen oxides and sulphur oxides actually increased compared with 1993. The underlying problem was that the initial, free allocations of trading credits were based on historical peak production levels, which were some forty to sixty percent above the actual emissions in the year the program was started.[91] With such an over-allocation of trading credits, the initial market costs for pollution were very low. Drury and colleagues speculated that this amounted to "at least a ten-year free ride of avoided emission reductions for the four hundred largest polluters in the Los Angeles area."[92]

In wrapping up their review, Drury and colleagues also hit upon some deeper moral issues relating to pollution trading. Based on their analysis, they concluded: "Pollution trading, as practiced in Los Angeles, has produced immoral, unjust, and ineffective outcomes."[93] They continued: "What once was wrong – polluting – is now a 'right.' The immorality of pollution trading lies in its treatment of a public resource, pollution-free air, as a private commodity. Instead of people having the right to breathe free, businesses have the right to pollute as much as they can afford." Moreover, on the question of whether companies should reduce pollution or purchase permits and continue to pollute: "Profit, not public health, becomes the deciding factor."[94]

A more recent analysis, co-authored by Stavins in 2015, of the design and performance of seven notable emissions trading schemes included RECLAIM in Southern California.[95] Stavins noted that the overall goal of a seventy percent reduction in emissions by 2003 under RECLAIM was achieved, albeit starting from initial allocations that were forty to sixty percent too high. He also pointed to analysis showing that emissions fell by twenty percent more, on average, over the first ten years of the program than in comparable facilities in California under command and control approaches – moreover, with no evidence of systemic spatial differences in emissions reductions between neighbourhoods, income groups, or racial groups.[96] RECLAIM did run into problems in 2000–2001, when several power plants were temporarily closed during California's electricity crisis, causing emissions to exceed permits at some participating RECLAIM facilities, with prices dramatically spiking at over \$60,000/ton.[97] Some emissions sources had to be temporarily excluded from the program, but prices did return to more normal levels, below \$2,000/ton, in 2002. The experience of the electricity crisis exposed the inability to bank allowances between years as a possible design flaw in the market system. Provision for banking of annual allowances as well as avoiding initial over-allocation of emissions were just two of several key lessons that were learned during the first thirty years of pollution trading markets.[98] These lessons came not only from experiments with markets for local air pollutants, but also increasingly from experiments with carbon markets.

Carbon Markets

The development of markets for greenhouse gas emissions was centrally woven into the prolonged negotiations and eventual failure of the Kyoto Protocol.[99] The whole process involved a confusing dance between the self-interests of nation-states. Despite James Hansen's testimony to the US Senate in the hot summer of 1988 and the backing of global carbon markets from *Project 88*, at the Earth Summit in 1992, the United States bowed to the fossil fuel industry and opposed mandatory emissions-reduction targets.[100] The potential for a market-based approach, nonetheless, came about when the Intergovernmental Panel on Climate Change concluded that "for a global treaty, a tradeable quota system is the only potentially cost-effective arrangement where an agreed level of emissions is attained with certainty."[101] Three years later, at the first Conference of the Parties (COP 1), with European oil companies BP and Shell breaking the industry's hard stance, countries agreed to reach mandatory greenhouse gas–reduction targets within the next two years. A pilot emissions offset program was also launched. The United States went into COP 2 willing to make binding commitments, provided that carbon trading was included in a new international agreement. The European Union was reluctant, preferring uniform emissions reductions, but came to agreement with the United States at COP 3 in Kyoto in 1997. Details of the market mechanism in the Kyoto Protocol took a further year to negotiate, until a compromise was reached with China and developing nations. The result was the Clean Development Mechanism, by which developed countries could get credits for paying for emissions reductions in developing countries. Rules for the full international carbon market were then delayed further until 2001, with Japan opposed to the strong enforcement desired by the Europeans. That year, however, the United States abandoned the Kyoto Protocol, ironically leaving the Europeans to push on with their own emissions trading scheme, which they initially had been opposed to.

Before the opening of the European Union emissions trading system, however, came a couple of other carbon markets. One of these was the UK emissions trading scheme, opened in London in 2002; it had only a few participants, many of which essentially received subsidies for voluntarily reducing emissions.[102] Perhaps more notable was the establishment of the Chicago Climate Exchange by the American academic turned finance entrepreneur Richard Sandor. In the 1970s, Sandor took a sabbatical leave from the University of California, Berkeley, to become the chief economist and vice-president of the

Chicago Board of Trade. During this time, he did pioneering work on financial futures markets that revolutionized the world of finance. Turning to environmental concerns, in the early 1990s, Sandor chaired the Chicago Board of Trade Clean Air Committee, which developed early markets for sulphur dioxide. Through a company founded by Sandor, Climate Exchange PLC, the Chicago Climate Exchange opened in October 2003, providing trading in the suite of six Kyoto greenhouse gases. The exchange was a voluntary, legally binding system for offsetting emissions from projects in North America and Brazil. Among the over 400 members of the exchange were power generation companies; a variety of other industrial firms; and several municipalities, states, and universities.[103] By 2007, Sandor was recognized as one of *Time* Magazine's Heroes of the Environment – as the father of carbon trading.[104] In 2010, however, the Chicago Climate Exchange closed down. There had been declining activity on the exchange following the 2008 financial crisis, and by February 2010, trading volumes ground to a halt; the exchange was sold and closed by year-end. Nonetheless, Climate Exchange PLC continues to operate, or be affiliated with, several carbon markets around the world, including the European Climate Exchange.

The European Union's emissions trading system has arguably been the biggest pollution trading experiment of them all. First opened in 2005, the European system is now into its fourth phase of operation, with a lot of learning and no shortage of controversy as it progressed. As of 2015, approximately 12,000 power supply and manufacturing facilities from thirty-one countries were required to participate in the system.[105] All combined, their emissions are about forty-five percent of the EU's total greenhouse gas emissions[106] and account for close to eighty-five percent of global emissions under carbon markets.[107] When it first opened, the system suffered from some of the same challenges as previous market trading experiments. Phase 1 started without good emissions data – and with EU member states free to distribute initial emissions credits relatively liberally. The result was an over-allocation of permits, which led within a short time to a collapse in the price of carbon, disincentivizing activities to reduce emissions further.[108] There was no provision for banking emissions credits between phases 1 and 2, leading the price to fall to zero in 2007. Phase 2 opened in 2008, with a tighter cap on the total allocation. Prices increased at first but then fell by fifty percent when the financial crisis occurred. There was a partial recovery, but the market collapsed again after the summer of 2011.[109] Heavy purchasing of offsets under the Clean Development Mechanism was implicated as a possible explanation for the declining demand,[110] although other factors, such as the declining costs of renewables and natural gas, might also have been the cause. The third phase, 2013–2020, opened with a stricter, centrally determined cap; greater use of auctions in allocation of allowances; limits on purchasing overseas offsets; and unlimited carry-over of allowance from phase 2.[111] The target for phase 3 was to reduce emissions to twenty percent below 1990 levels by 2020, with target reductions of forty percent agreed for 2030.[112]

While the European Union has persevered with its emissions trading system and seems to be committed to it, opinions on its success differ widely. Some have argued that the small emissions reductions achieved can all be attributed to the development of feed-in tariffs, the economic malaise following the financial crisis of 2008, and energy-efficiency improvements and fuel switching that resulted from other policies. Moreover, the trading system detracts from other policies that could reduce emissions faster.[113] The trading system is perceived by some to be a sham, through which some companies realized huge windfalls. This was even recognized by the European Parliament, which noted that "some industries have generated windfall profits by passing on the cost of free allowances to consumers."[114] Other appraisals are more positive, suggesting that the trading system has reduced emissions without any detrimental effects on economic competitiveness – and has partly helped stimulate innovation in the clean technology sector.[115] Nonetheless, it seems that about twenty years after opening, the European carbon market is still finding its way.

Markets for Biodiversity!

While the jury is still out on whether our experiments with carbon markets can seriously reduce greenhouse gas emissions, an even more ambitious idea has been posited of creating markets for biodiversity. Wrestling with the question of whether such a concept is worth seriously considering touches upon some of the key ideas and most difficult issues in the discipline of ecological economics. This is a question that is exemplary of the third layer of the Malthus Enigma – a policy question, which, frankly, as I will conclude here, is probably not a good idea. The question pits ecological economist against ecological economist.[116] Attempting to put a monetary value on Nature, or, rather, ecosystem services, has been a central activity in ecological economics. Led by Costanza, as we have seen, some progress has been made in synthesizing estimates for various types of biomes, largely based on willingness-to-pay studies. Like all goods and services, however, the only real way we

Malthus Enigma

have of establishing prices for ecosystem services is through a market. Whether market prices reflect true value, however, is a deeper question – because all markets are human creations and are subject to the policy regimes, taxes, subsidies, and market rules and imperfections we create. The use of markets clearly lies within ecological economics, as is evident from the work of Daly, but when it comes to considering markets for biodiversity, many ecological economists understandably become uneasy.

One of the earliest explorations on the potential to develop markets for biodiversity was a book published in 2003, *Capturing Carbon and Conserving Biodiversity: The Market Approach*, with support from the UK's Royal Society.[117] The lead editor of the collection of papers was conservation biologist Ian Swingland, with Richard Sandor notably among the contributors. In introducing the book, Swingland and co-editors recount the challenge of avoiding the sixth mass extinction of the Earth's species and question the adequacy of conventional conservation approaches to prevent it. With Malthusian sentiments, they point to increasing population and high levels of consumption as the underlying cause. Then, with a twist on the evaluation of ecosystem services of the ecological economists, they declare: "It is now widely recognized that given the lack of public funding, biodiversity conservation must start to pay for itself, otherwise biodiversity, and perhaps even the human race, are in jeopardy."[118] Consequently, the authors arrive at the concept that biodiversity can be conserved through its carbon content, as valued on carbon markets.

If carbon markets work well – and if carbon sequestration is permitted in such markets – then it is argued that carbon trading provides a means of converting forests into financial capital – for protection of biodiversity, rather than logging. The authors lament that land-use activities that sequester carbon – such as growing trees – were not suitably incorporated into the market mechanisms under the Kyoto Protocol. If they were, then carbon markets could at least be indirectly beneficial to forest conservation. Further consideration of how to incorporate biodiversity and ecosystem services into the world of finance have since been produced by the UN Environment Programme, in partnership with several large banks and multinational corporations.[119] Nonetheless, enthusiastic concluding comments from Swingland and his co-editors reveal some of the frictions underlying the development of markets for biodiversity:

> *The case made here for the maximum use of terrestrial carbon sinks, particularly in the developing world, is overwhelming. The*

> *benefits of such a strategy to the rural poor, indigenous people, habitat preservation, biodiversity, watershed protection, and the climate as a whole is revealed here in persuasive detail. That such a strategy is also, particularly when combined with emission trading, the lowest-cost approach makes it hard to understand why some environmentalists and policy-makers involved in the negotiations have chosen to do their utmost to prevent it.*[120]

The development of markets for biodiversity is intriguing, but there are many reasons to be cautious. A first big reason is that the whole concept assumes that carbon markets can be made to successfully work over a long period. Despite the enthusiasm of the European Parliament for its emissions trading scheme and the more recent opening of carbon markets in Asia and elsewhere, they still remain unproven. When it comes to the creation of a global carbon market, Coase's warning that it would be practically impossible with so many actors has, at least so far, proven correct. A well-balanced review of the history of carbon markets also encourages a tempering of expectations:

> *Experience teaches us that there are certain things that carbon markets can potentially do very well, but it is unreasonable to expect them to cut global carbon emissions, to minimize the short-run cost of abatement, to bring about a low-carbon technological revolution, and to be a tool for achieving a more equitable global income distribution, all at once. In fact, these objectives often conflict, so we must leverage our historical experience to better understand what kind of success is reasonable to expect.*[121]

The notion that carbon markets are the only way to reduce emissions is weakly grounded. Even Stavins, the author of *Project 88*, recognized that the market approach to reducing emissions may be a "brief departure from ... command and control regulation."[122] The development of pollution markets may be just a reflection of the dominant political philosophy of the times. In the 1990s, in the wake of Reagan and Thatcher, it seemed like the market could do no wrong. Further to this are the ethical questions raised by Drury and colleagues about pollution markets in general. The idea that companies have rights to pollute, that take precedence over human rights to breathe clean air or to a stable climate, is one that a future generation with a more advanced sense of morality might find unacceptable.

Malthus Enigma

Even if a suitably functioning global carbon market helping reduce greenhouse emissions could somehow be established, a host of other concerns about the translation from carbon into biodiversity arise. Imagine for a moment that Coase's pessimism could be overcome. Imagine that the countries of the world come together and develop a global carbon market that caps the global budget to be net-zero emissions by 2050. Such a scheme would likely have to give credits for carbon sequestration by forests and other vegetation – so let's imagine that too. Even if such a market worked and saved us from average global climate change of more than 1.5°C, it does not necessarily follow that global threats to declining biodiversity will have been solved. It could be that the response to the market is a predominantly technological path, rather than one heavily reliant on forest preservation. A cap on carbon, designed to address our deepest climate change challenges, does not necessarily suitably address global biodiversity loss. Another big assumption in the discussion of market-based approaches to biodiversity is that the climate change benefits of stored carbon serve as a useful proxy for all other ecosystem services.[123] This essentially means that biodiversity becomes measured only in units of stored carbon, which are then translated into dollars through the carbon market. The notion that biodiversity can be packaged into a single unit for exchange on global markets is very unsettling for ecological economists.

In spite of their work on the valuation of ecosystem services, most ecological economists strongly oppose the idea that such "natural capital" can be traded for other forms of capital.[124] It is one thing to place a value on a wetland swamp, but that does not mean we can then blindly drain it and use the land for some investment opportunity of higher financial value. In rejecting the substitutability of capital, the ecological economists take a strong sustainability approach, a philosophy that distinguishes them from mainstream economists, including environmental economists. This perspective is well articulated by ecological economist Clive Spash in his paper titled "Bulldozing Biodiversity."[125] Spash recounts an abhorrent calculation from the 1970s, which showed – based on conventional economic logic – that blue whales should be hunted to extinction and the revenues invested in growth industries.[126] The problem here, just as Georgescu-Roegen recognized, is that economics inherently invokes human value judgments, but if people don't understand how much they rely on ecosystem services and don't care about Nature, then Nature loses – and so do we. After all, Spash teases, "Who cares about soil microbes, insects, spiders, stinging plants and ugly snakes?"[127] Spash argues further that the design of any financial

Ecological Economics

offsets to biodiversity inherently involves a "political battle over human-Nature relationships."[128] Markets are not natural; they are designed by humans – and they are often flawed, as we have seen from the history of pollution markets. Ultimately, Spash argues, markets for biodiversity "use economic logic to legitimise, rather than prevent, ongoing habitat destruction"[129] and "the approach does not promise to protect biodiversity, and in fact is consistent with the optimal extinction of species."[130]

In summing up, this brief exploration of ecological economics shows just how hard it is for humanity to develop sensible policies in the face of Malthusian struggle. But it is hugely beneficial to humanity that this smart and ethical discipline has developed the language, concepts, and skill to rigorously assess sustainable paths forward. Ecological economics was formed through the confluence of economists who saw the importance of thermodynamics, including Daly and Georgescu-Roegen, and ecologists inspired by Howard Odum. It has the ingredients to lead the policy response – our third layer of the Malthus Enigma – on the formidable challenges of tackling climate change and biodiversity loss that we discussed in Chapters 3 and 4. With that, we move from the interdisciplinary marriage of ecology and economics to another field with similar environmental motives – that is, industrial ecology. Beginning with inspiration to learn lessons from ecology in the design of industrial systems, industrial ecology developed into an interdisciplinary applied science with contributions to all layers of the Malthus Enigma.

Chapter 8: Industrial Ecology

Why would not our industrial system behave like an ecosystem, where the wastes of a species may be resource to another species? Why would not the outputs of an industry be the inputs of another, thus reducing use of raw materials, pollution, and saving on waste treatment?

Robert A. Frosch and Nicholas E. Gallopoulos,
Scientific American, 1989

Two men were deeply engaged in conversation as they walked briskly across Central Park on a crisp morning. It was January 6, 2000 – the first Tuesday of a new millennium. The two made their way through the park from the Empire Hotel to the New York Academy of Sciences on 63rd Street and 5th Avenue. There, they would join with a select group of twenty collaborators to discuss the founding of a new organization – the International Society for Industrial Ecology – concerned with addressing complex environmental challenges by applying ecological concepts to the organization and operation of industry. The older of the two was John Ehrenfeld – a stocky, powerful man with great intellect and a generous character. He was emeritus director of the Technology, Business and Environment Program at the Massachusetts Institute of Technology. The second man was much younger – a researcher from Leiden University in the Netherlands named René Kleijn. His attendance at the meeting was partly fortuitous, standing in for another Leiden professor who was ill at the time. Kleijn, nonetheless, ended up chairing the first international conference of the new society, in 2001, while Ehrenfeld became its first executive director.

Among Ehrenfeld's scholarly works was a classic paper on the evolution of industrial symbiosis in the Kalundborg industrial area in Denmark. As an example of how waste energy and material flows from industry could be made useful through exchange at co-located factories,

Malthus Enigma

Kalundborg exemplified a concept that had been promoted in a seminal article by Robert Frosch and Nicholas Gallopoulos in *Scientific American*. Ehrenfeld conducted a scientific analysis of the factors that led to the evolution of interdependencies in the Kalundborg industrial ecosystem.[1] The system started back in the 1960s, when a coal-fired electric power plant began sending excess heat – in the form of steam – to a nearby Statoil refinery. Later, the steam was also supplied to a pharmaceutical plant, greenhouses, local homes, and a fish factory – all close by. Such an arrangement, known as co-generation, was not novel, but the Kalundborg site went further. The Statoil refinery was producing an excess of high-sulphur gas, from which the sulphur was recovered and sold to a neighbourhood manufacturer of sulphuric acid. The cleaner gas was then sent to the electric power plant, reducing its demand for coal – as well as a nearby wall-board manufacturer. The electric power plant also found other means of making use of its waste products. It added a scrubber to remove the sulphur from its waste emissions and turned this into calcium sulphide, which the wall-board manufacturer could use instead of purchasing gypsum. Ash from the power plant was also used for road building and cement manufacture. Completing the web of material flows, sterile wastes from a fermentation process at the pharmaceutical plant and fish-processing waste from the fish farm were sold as fertilizers to local farmers. Through the symbiotic exchange of energy and materials, the Kalundborg site was able to save annually on the following use of natural resources: 1.2 million cubic metres of water; 30,000 tonnes of coal; 19,000 tonnes of oil, plus the energy requirements of 3,500 oil-fired residential home furnaces; 800 tonnes of nitrogen; 1,200 tonnes of phosphorus; 2,800 tonnes of sulphur; and 80,000 tonnes of gypsum.[2] With large quantities of waste also avoided,[3] Kalundborg showed that it was possible for industry to begin to mimic natural systems through recycling of waste materials and energy.

Following Ehrenfeld's study of Kalundborg, many more cases of industrial symbiosis were initiated or simply discovered around the world. In a 2007 review, Yale industrial ecologist Marian Chertow reported on examples from countries such as Australia, Austria, Canada, China, Finland, Germany, the United Kingdom, and the United States.[4] She noted how some of the industrial networks were planned, while others were self-organized, with pairs of companies making agreements to exchange waste materials without cognition of a wider industrial ecosystem. Often in free-market economies, the self-organized networks were more successful, though they could be enhanced with planned approaches once an initial cluster had formed. In the 1990s, several

Industrial Ecology

attempts at planned industrial symbiosis had been started by the President's Council on Sustainable Development, under the Clinton Administration. These had mixed results. Later, though, the UK's highly successful National Industrial Symbiosis Programme managed to generate £1 billion in new sales while diverting forty-seven million tonnes of industrial waste away from the landfill and reducing greenhouse gas emissions by forty-two million tonnes.[5] The UK program also managed to create or maintain 10,000 jobs while reducing costs of business by £1 billion, saving sixty million tonnes of virgin material and seventy-three million tonnes of industrial water. China – with its planned economy – also began to experiment with industrial symbiosis and wider applications of industrial ecology through its Circular Economy Development Strategies Action Plan. A thousand demonstration projects were initiated in a hundred cities to spur the development of eco-industrial parks, recycling technologies, remanufacturing, and other technological developments to avoid waste and other environmental impacts.[6] Concurrent to these applications, at the industrial-park scale that Ehrenfeld studied, methods of industrial ecology have increasingly been employed in other contexts, from company supply chains to eco-cities and entire economies.[7]

Ehrenfeld was one of several influential thought leaders at the New York meeting whose work was at the heart of industrial ecology. The other attendees at the January 6 meeting included an impressive cast of academics, industrialists, and policy experts. Yale University had a particularly strong contingent from its School of Forestry and Environmental Studies. This included Gus Speth, an environmental lawyer who had worked on the *Global 2000* report as chair of President Carter's Council on Environmental Quality, before becoming the founder and first president of the World Resources Institute. Others from Yale were Tom Graedel, the first professor of industrial ecology, who later became the first president of the society; Marian Chertow; and Reid Lifset, editor of the *Journal of Industrial Ecology*. Lifset had a key operational role as the source of emails communicating organizational details of the meeting, as well as the follow-on correspondence. Among the other sixteen attendees at the meeting were academics from the US, Europe, and Japan; several people from industry; a former vice-president of the World Bank Group; and representatives from the United States Environmental Protection Agency, the World Resources Institute, and the New York Academy of Sciences.[8]

The meeting at the New York Academy of Sciences set the wheels in motion for the formation of the society. There was naturally much discussion and debate. Stefan Bringezu, from the influential German Wuppertal Institute for Climate, Environment and Energy, provided a discussion document on formation of the society from a European perspective.[9] Among the issues were the competition and relationship of the society to other groups, such as the ecological economists; a mainly European community studying material flows under the acronym ConAccount;[10] and the large but disorganized array of life-cycle assessment practitioners, who study the cradle-to-grave environmental impacts of products. Should the society have continental branches or topical sections reflecting subdisciplines within industrial ecology? Other deliberations were the services provided in return for membership dues, the management structure, the balance between research and practice, and the importance of engaging developing countries. Toward the end of the meeting, two "breakout" groups were formed – one drafting a preliminary mission statement and goals, and the other undertaking a roadmapping exercise. An executive steering committee was formed consisting of Brad Allenby of AT&T; Jesse Ausubel, an environmental scientist from The Rockefeller University; and Bringezu, Graedel, and Lifset. Significantly, Allenby declared that AT&T would provide $15,000 toward start-up costs, to which Graedel added a further $5,000. Within four months, over $150,000 had been raised from philanthropic sources.

Over the coming months, the society took shape. A series of scoping memos were prepared by various authors on topics such as governance, awards, advocacy, conferences, and education. An email from Lifset on May 17, 2000, indicated that Ehrenfeld had agreed to be interim director of the society and that a website had been established on a Yale server. About a year after the New York gathering, Yale University issued a press release, which began:

INDUSTRY MEETS ENVIRONMENT IN NEW INTERNATIONAL SOCIETY

New Haven, CT – A community of researchers, policy makers, industrial strategists, and environmental advocates announces the launch of the International Society of Industrial Ecology. The new field of industrial ecology applies ecological concepts in the organization and operation of industry.

The Society will encourage communication among scientists, engineers, policymakers, managers and others who are interested in how environmental and economic concerns can be better integrated. Industrial ecology is a powerful way of finding innovative solutions to complicated environmental problems.[11]

Industrial Ecology

The release went on to note that the society would open its doors to members in February 2001, with an inaugural society conference, "The Science and Culture of Industrial Ecology," to be held in the Netherlands on November 12–14 of that year.

The discipline of industrial ecology was influenced by many of the scholars and topics we covered in earlier chapters, paralleling – and sometimes overlapping with – ecological economics. In its formative years, industrial ecology was motivated by the environmental concerns that drew attention in the 1960s and '70s – especially the problems of waste and pollution that had been highlighted by Barry Commoner, Paul Ehrlich, and others. In some respects, industrial ecology was a technological response to Boulding's Spaceship Earth – a discipline aiming to avoid waste by closing material loops. This was the focus as industrial ecology began to coalesce around 1989, under the influence of Robert Frosch – a former administrator of NASA. The field of industrial ecology grew, however, to address the full suite of environmental challenges. We discussed Lenzen's work on the impacts of trade on biodiversity loss in Chapter 3 – and many other industrial ecologists have worked on strategies for responding to climate change, too. In its initial conception, then, industrial ecology was largely concerned with the first layer of the Malthus Enigma – as I have called it – being about innovation in technology and management to reduce environmental burdens. As this chapter will explore, however, industrial ecology also evolved as a science and a policy paradigm – contributing to the second and third layers of the Malthus Enigma, too.

The science of industrial ecology developed from its use of systems approaches – based on the laws of physics – to study industrial society. Industrial ecologists drew upon principles of conservation of mass and thermodynamics to understand the environmental impacts of industrial systems. But it soon became apparent that the "system" was not just "industry" but the whole of "industrial society" – that is, both producers and consumers matter. A major influence on the science of industrial ecology was physicist Robert Ayres, who conceived of the notion of *industrial metabolism*. The concept borrowed language from – and was inspired by – ecology, but the notion was not an analogy; there really are energy and material flows associated with industrial systems. The study of industrial metabolism, as well as societal metabolism[12] and urban metabolism,[13] became central to industrial ecology. The collection of data on energy and material use in industrial society brought a heightened degree of scientific rigour to analyzing complex resource and environmental issues, which had sometimes been absent in earlier

Malthus Enigma

emotional discourses. Study of the world's stocks and uses of the metals that Paul Ehrlich and Julian Simon had bet on were vigorously pursued by industrial ecologists. Essentially, industrial ecologists began to study how economies function through the lens of physics – just as Georgescu-Roegen had encouraged. Economists had never really taken up the challenge of tracking the world's resources, because it was easier for them to just follow financial transactions.

Through the creation of frameworks for analyzing resource use – and the collection of data – industrial ecology began to emerge as a policy paradigm. Into the new millennium, with the formation of the society, industrial ecology grew in influence and impact. The United Nations developed the International Resource Panel – paralleling the Intergovernmental Panel on Climate Change (IPCC) – but with a focus on studying the efficient use of natural resources. Industrial ecologists formed the core of the new International Resource Panel. Meanwhile, China and the European Union both adopted principles of industrial ecology in establishing policies on the circular economy. Businesses, too, were encouraged to apply methods of industrial ecology by British long-distance solo yachtswoman Ellen MacArthur, through her foundation dedicated to the circular economy. In industrial ecology, global society had developed a new strategy and a new science in its Malthusian struggle. Before describing why it is so promising, let's start with a brief backstory on its origins.

History of Industrial Ecology

Prior to its emergence in the 1990s, industrial ecology had developed with a checkered history. The term *industrial ecology* dates at least as far back as the 1940s, when it was mentioned in the relationship between geography and industrial location.[14] Later, in 1963, it was used in the context of social and behavioural aspects of industrial management.[15] While neither of these usages relates strongly to the modern conception, the idea of developing industry with "no waste" or "closed loops" was employed by Henry Ford, as well as Chicago meat packers of the 1930s. Still, such practices were clearly only rarely achieved by twentieth-century industry. In 1971, the president of the American Association for the Advancement of Science, Dr. Athelstan Spilhaus, published a paper, "The Next Industrial Revolution," in which he proposed the sharing and reuse of material and energy. Spilhaus wrote:

The object of the next industrial revolution is to ensure that there will be no such thing as waste, on the basis that waste is simply some substance that we do not yet have the wit to use.[16]

Around this time, there were attempts to develop industry based on ecological principles in a handful of countries, including Poland, Russia, China, Japan, and, later, Belgium.[17] The greatest practical success was arguably achieved in Japan, although the work in Belgium was particularly advanced.

In the early 1980s, an interdisciplinary group of six Belgian scientists undertook a study of the industrial ecology of Belgium's economy. The group had a well-developed understanding of industrial ecology, noting that "you have to define industrial society as an ecosystem made up of the whole of its means of production, and distribution and consumption networks, as well as the reserves of raw material and energy that it uses and the waste it produces."[18] They described six streams within the Belgian economy – iron, glass, plastic, lead, paper and wood, and food production – using industrial production statistics on energy and material flows. One of the group's principal findings was that the Belgian industrial system was very open, with many areas of specialization, but little linking between sectors. For example, Belgium had a well-developed steel industry with large exports to the rest of Europe, but no connections to the metal construction sector in Belgium. The scientists summarized that there were three dysfunctional characteristics of the Belgian economy:[19]

1. Consumption residues that could potentially be used as a resource were increasingly wasted.

2. Primary energy consumption was increasing only in part due to increasing end consumption, but more due to the inefficient organization of the energy sector and, more broadly, the industrial system.

3. The structure of material circulation in the industrial system inherently generated pollution.

The group's findings were published in 1983, under the title "L'écosystème Belgique : Essai d'écologie industrielle," but were unfortunately soon forgotten. The leader of the project, Francine Toussaint, commented that "we were a voice preaching in the desert."[20] Belgium failed to grasp the economic and environmental opportunities of their initiative.

Malthus Enigma

The principles of industrial ecology had, however, already been adopted with systemic economic benefits about a decade earlier in Japan. In the late 1960s, Japan's Industrial Structure Council, comprising approximately fifty experts from various fields, was commissioned by the Ministry of International Trade and Industry to address the high environmental costs of industrialization. In exploring ways for the Japanese economy to develop with less consumption of materials and greater use of knowledge and information, the council arrived at the idea of examining the economy in an ecological context. Following the release of the council's final report, *A Vision for the 1970s*, in May 1971, the ministry created a working group to further study Japanese industry from an ecological perspective.[21] The group was led by Chihiro Watanabe, who was then with the ministry's Environmental Conservation Bureau and later became a professor at the Tokyo Institute of Technology. The working group conducted comprehensive research of the appropriate scientific literature, consulting with several experts. During a tour of the United States in 1973, Watanabe met with Eugene Odum – one of the brothers of systems ecology, whom we discussed in Chapter 6 – although Odum evidently did not take much interest in the Japanese work.[22] Nonetheless, after publishing a "stimulating" but "philosophical"[23] report to the ministry in May 1972, Watanabe and his group published a practical report with case studies in spring 1973. The work inspired the ministry to develop new policies based on ecological principles. It proved to be timely. In August 1973, the ministry proposed the Sunshine Project, aimed at producing new renewable energy technology. This was two months before the first Oil Crisis. Moreover, a second project, known as the Moonlight Project – focusing on energy efficiency – was started in 1978, just before the second Oil Crisis.[24]

Compared with Japan, notions of industrial ecology had not yet penetrated mainstream industrial strategy in North America and Europe, but this began to change in the early 1990s. A key factor that catalyzed the rise of industrial ecology in the West was an article by Robert Frosch and Nicholas Gallopoulos, both with General Motors, in a special issue of *Scientific American* in September 1989. In their paper, entitled "Strategies for Manufacturing," Frosch and Gallopoulos described how industrial systems could be developed with much lower environmental impacts if they were designed analogous to natural ecosystems. The authors were practical, recognizing that perfect closed-looped industrial ecosystems were difficult to achieve – and they provided extended discussion of progress and challenges with recycling of steel, plastics, and platinum. They also saw the big picture, pointing to a need for industrialized nations to make major changes to their current practices, and for developing countries to leapfrog to more ecologically sound approaches to industry. In reviewing the early history of industrial ecology, Swiss industrial ecologist Suren Erkman observed that the Frosch and Gallopoulos article sparked a strong interest from industry.[25] For a single scientific paper to be so influential was unusual, but it perhaps served as a rallying call because of the status of the first author.

Robert Frosch was a research leader whose career took him to the highest levels within industry, government administration, and academia. Born in New York City in 1928, Frosch grew up in the Bronx and studied undergraduate and graduate degrees in physics at Columbia University. He rose to the position of director of Hudson Laboratories at Columbia, which was contracted to the Office of Naval Research. In 1963, Frosch moved to Washington, DC, taking several roles with the Department of Defense, before becoming assistant secretary for research and development with the Navy. In this role, he was responsible for the Navy's $2.5 billion research and development program.[26] In January 1973, Frosch joined the newly formed United Nations Environment Programme, moving first to Geneva and then to Nairobi, taking one of two assistant executive director positions, under Maurice Strong and Deputy Executive Director Mostafa Tolba.[27] Returning to the United States, Frosch served as the fifth administrator of NASA from 1977 to 1981, during a period that saw continued development of the Space Shuttle program. Following his leadership of NASA, Frosch became vice-president of research at General Motors, where he encouraged the development of electric and hybrid electric vehicles. In 1993, he became a senior research fellow at Harvard University's John F. Kennedy School of Government.

Frosch continued to play a role as the National Academies of Engineering and Sciences both convened key meetings that launched industrial ecology in the early 1990s. Following the Brundtland Report of 1987, the National Academy of Engineering initiated a program in technology and environment. This involved a workshop at Woods Hole, Massachusetts, in August 1988, chaired by Frosch, followed by a symposium the next month at the annual meeting in Washington, DC.[28] The proceedings from the symposium were edited by Jesse Ausubel and Hedy Sladovich, from the National Academy of Engineering.[29] As well as a paper by Frosch, these notably included an influential paper on industrial metabolism by physicist Robert Ayres, discussed further below. Then, on May 20–21, 1991, the National Academy of Sciences hosted a colloquium on industrial ecology in Washington, DC, chaired by

Malthus Enigma

Kumar Patel of AT&T Bell Laboratories.[30] Twenty-three papers from the symposium were published in the *Proceedings of the National Academy of Sciences of the United States of America*, including contributions from Ausubel, Ayres, Faye Duchin, Frosch, Graedel, and Speth. In reflecting on why industrial ecology was useful and timely, as he summarized his thoughts on the colloquium, Ausubel concluded:

> *It is about deepening appreciation of technology and extending what is valued in economics, broadening the domain of what must be considered in engineering design and practice, and ending the isolation of ecology from the man-made world.*[31]

From the meetings at the National Academies, further conferences on industrial ecology followed. Among these was a 1992 conference on industrial ecology and global change held in Snowmass, Colorado, with support from the Office for Interdisciplinary Earth Studies, which was also important for the development of the discipline.[32]

Concurrent with the promotion of industrial ecology by the National Academies was an uptake by industry. This was significantly aided by a consultant named Hardin Tibbs, based in Boston and later San Francisco, who had himself been inspired by Frosch and Gallopoulos's article in *Scientific American*.[33] Tibbs took the central ideas of industrial ecology from Frosch and Gallopoulos and translated them into a business context. He produced a relatively short document, fourteen pages,[34] that was so popular with industry that it sold out quickly, was published again, and spread further through hundreds of Xeroxed copies.[35] Tibbs began with the point that industrial ecology "takes the pattern of the natural environment as a model for solving environmental problems."[36] With a strongly pro-technology perspective, he then explained how, under this paradigm, industry can be developed in sustainable ways, improving social and environmental outcomes through appropriate design. Tibbs then went into further details on key industrial ecology strategies, including creating industrial ecosystems, exemplified by Kalundborg; encouraging dematerialization; improving the metabolic pathways of industry; systematically using green energy sources; and aligning policy with the long-term evolution of industry.

Something else that had been strongly encouraged at the National Academies meetings, as well as by Frosch, was the establishment of educational programs in industrial ecology. Progress was also made in this regard in the early 1990s, at the Norwegian University of Science and Technology (NTNU).[37] The impetus for the program came from industry, from a subsidiary of Norsk Hydro concerned about improving the full environmental performance of aluminum over the life cycle of its use in society. The aluminum company approached the president of NTNU (then the University of Trondheim) and encouraged the university to introduce a course in industrial ecology.[38] NTNU conducted a series of seminars, receiving help from John Ehrenfeld – who had begun teaching industrial ecology to graduate students at MIT – as well as collaborating with Georgia Tech, the Delft University of Technology, and the Technical University of Denmark. Although NTNU is a technical university, it designed the teaching in industrial ecology to be accessible to students with both technical and non-technical backgrounds – achieving the interdisciplinarity that so many universities aspire to. Over the following few years, NTNU gradually piloted various industrial ecology courses and then, in 1998, launched the world's first comprehensive graduate degree program in industrial ecology. By 2014, industrial ecology was being taught at over 400 universities around the world, with over twenty full degree programs established.[39]

Robert Ayres and Industrial Metabolism

Students of industrial ecology soon become familiar with the work of American-born physicist and self-taught economist Robert Ayres, who established the scientific basis for the field. Drawing upon knowledge of physics, chemistry, and biology, Ayres made many contributions to the theoretical foundations of industrial ecology, including the concept of industrial metabolism. Following degrees from the University of Chicago, the University of Maryland, and King's College London, Ayres worked for the Hudson Institute (1962–1967), Resources for the Future (1968), and the International Research and Technology Corporation (1969–1976). He was a professor of engineering and public policy at Carnegie Mellon University from 1979 to 1992, before moving to the prestigious French business school INSEAD. Ayres also had a long association with the International Institute for Applied Systems Analysis (IIASA), in Laxenburg, Austria, spending many summers working there and becoming an Institute Scholar in 2004. He made numerous contributions on a range of topics, such as technological change, the thermodynamic basis of economic growth, energy systems, and dematerialization. Most notable for the field of industrial ecology was his pioneering work on industrial metabolism – the energy and material flows of industrial society.

One of Ayres' earliest and still most impactful research papers was an article co-authored with environmental economist Allen Kneese,

Malthus Enigma

"Production, Consumption, and Externalities."[40] This was published in the *American Economic Review* in 1969, the same year as the first Apollo Moon landings, at the height of the environmental revolution. In the paper, Ayres and Kneese argued, counter to the view of many economists, that so-called "environmental externalities" are not unusual or "exceptional cases" in economies, but are in fact prevalent. Environmental externalities – as discussed in Chapter 7 – are pollutants or wastes that are produced without the polluter bearing the environmental costs. Ayres and Kneese applied the simple fundamental principle of conservation of mass to the US economy, adding up the weights of materials produced in economic sectors such as agriculture and mining, plus net imports. For the years of study – in the mid-1960s – only ten to fifteen percent of the material input to the economy went "into stock" – primarily in the form of buildings and infrastructure – with the rest being released, after consumption, as waste solids, gases, and liquids. Ayres and Kneese proceeded to develop an economic model that could be used to represent the material flows between various producing and consuming sectors of the economy – and the associated wastes. They also discussed, without calculations, how radically different the suite of goods produced would be in a spaceship economy, which would include the full costs of reprocessing residual wastes. Ayres was already developing a sense of the challenges and opportunities of developing circular economies.[41]

In the late 1980s, Ayres' work on material and energy balances was formalized through his development of industrial metabolism.[42] The term was coined in response to early discussions around the scope of the United Nations' emerging International Human Dimensions Programme on Global Environmental Change, which Ayres perceived to have a "major gap," at least in its early form.[43] He developed the industrial metabolism idea through several workshops and meetings in 1987 and 1988, including the National Academy of Engineering meeting at Woods Hole. Ayres defined industrial metabolism to be:

the whole integrated collection of physical processes that convert raw materials and energy, plus labor, into finished products and wastes, in a more or less, steady-state condition.[44]

He noted that industrial metabolism can be seen and described at multiple scales, from manufacturing firms, to regional and national levels, to the whole global economy. The concept was clearly commensurate with the ideas of industrial ecology, so much so that there was early debate as to

Industrial Ecology

whether the emerging discipline might alternatively be called industrial metabolism.

Ayres recognized that the metabolism of industry could be understood in analogous ways to that of an organism, from a thermodynamic perspective. Organisms grow, reproduce, and maintain themselves by ingesting energy-rich, low-entropy material – that is, food – while invariably excreting higher-entropy waste. Industrial facilities similarly take in raw materials and energy, producing useful products for the maintenance of human society, as well as wastes. Moreover, the totality of human industrial systems forms an economic system, which not only is driven by a flow of energy, but is also a self-organizing dissipative system far from equilibrium – again similar to organisms. Thus, industrial metabolism can be understood using the same thermodynamic principles of living systems developed by Schrödinger and Prigogine, as discussed in Chapter 5.

The great challenge of the industrial metabolism, as Ayres explained, is that it is largely composed of open industrial cycles – which produce wastes – unlike the natural cycles of the Earth, which have evolved to become closed. Today, the natural cycles – such as water, carbon/oxygen, nitrogen, and sulphur are, in the absence of human intervention, essentially at steady state. They have relatively constant mass flows recycling through each part of the Earth system – all maintained by a continuous flux of solar energy. Biological processes, such as decomposition by microorganisms, play a substantial role in closing some of these natural cycles. The industrial system, however, is an open one, by which inputs are transformed into wastes, with often limited opportunities for recycling. Ayres provided a variety of material mass balances for the United States, demonstrating examples where recycling ranged from difficult to impossible. The latter category included approximately two million tons of explosives used by the US mining industry circa 1988.[45] Other mass calculations are given for chemicals used in the mining of non-ferrous metals, as well as wastes from the petrochemical feedstock and organic chemicals sector.[46] In another example of the chlorine industry in Europe, Ayres showed that less than half of total output went into final products; approximately one-quarter of the chlorine flow was wasted, with the balance recycled.[47]

Relating to Lovelock's Gaia Hypothesis, discussed in Chapter 6, Ayres noted that the biosphere has not always been composed of closed natural cycles, but that these had evolved over long time scales through the biological evolution of organisms that could stabilize the system.[48] One example is the evolution of blue-green algae – about three billion

Malthus Enigma

years ago – that are thought to be the first organisms capable of recycling CO_2 into sugars, thereby closing the carbon cycle and producing oxygen in the process. The oxygen cycle was later closed by the evolution of aerobic respiration and aerobic photosynthesis. Similarly, the biological processes of nitrification and denitrification evolved to bring balance to the nitrogen cycle. Without the balancing process of denitrification that converts nitrate into stable gaseous nitrogen, the ocean would turn into nitric acid. Ayres recognized that the self-organizing capabilities of natural systems enabled them to evolve and close loops like Gaia – but, of course, these were slow processes that took billions of years to evolve.

Most of the industrial metabolism has been developed in the past 200 years; it contains a larger number of elements than the Earth's natural metabolism – and many of them are ultimately toxic to the Earth system. Ayres argued that for the industrial economy to become sustainable in the long run, it would need to achieve almost complete recycling of the intrinsically toxic and hazardous materials – such as heavy metals – as well as high rates for other materials, such as plastics and paper, that cause environmental problems. He proposed three categories of materials based on their recycling potential:[49]

1. Those that could be recycled economically and technologically under prevalent prices and regulations.
2. Those that could not be economically recycled, although technologically it was feasible.
3. Cases where recycling was technically infeasible with contemporary technologies.

Ayres noted that most structural materials and industrial catalysts were in the first category, while packaging and most refrigerants and solvents were in the second category (as of the late twentieth century). The third category, however, contains the largest number of chemical products, including "coatings, pigments, pesticides, herbicides, germicides, preservatives, flocculants, anti-freezes, explosives, propellants, fire retardants, reagents, detergents, fertilizers, fuels and lubricants."[50] These are where the greatest challenges lay. Ayres' work on industrial metabolism reminds us that beyond global environmental challenges of biodiversity loss and climate change are continual, ongoing challenges of managing the wastes and local pollutants created by industrial society.

In one of his later books, *Industrial Ecology: Towards Closing the Materials Cycle*, Ayres examines a dozen key groupings of metals, compounds, and important waste flows, detailing their predominant use

Industrial Ecology

by industry, environmental impacts, and strategic opportunities for waste reduction. Moving effortlessly between the perspectives of chemist, toxicologist, economist, and engineer, he describes the technological significance of related metal groups, such as aluminum and gallium or zinc and cadmium, discussing current and potential strategies by which an industrialist can reconfigure processes to minimize or reduce environmental impacts or alleviate scarcity concerns. Later comes the key nutrients of phosphorus (addressed alongside fluorine and gypsum) and nitrogen compounds, followed by the chloro-alkali sector and silicon for semiconductors. Opportunities for using or minimizing three major waste streams – post-consumer packaging, scrap tires, and coal ash – are similarly expounded. He ends with a chapter on the possibilities and challenges of constructing industrial ecosystems with wastes from industry providing feedstock for another, and a summary of strategic opportunities and policy levers. Reflective of Ayres' wider work on industrial metabolism, his text provides a brilliant scientific synthesis of the use of resources by industrial society.

Leontief's Legacy

Another important influence on industrial ecology was the technique of input–output analysis developed by the economist Wassily Leontief. Born in Munich, to Russian parents, Leontief was one of the economists that Georgescu-Roegen worked with at Harvard. Around the middle of the twentieth century, Leontief developed techniques for studying the structure of economies in terms of the financial transactions, or physical exchanges of goods, between different sectors. He conducted detailed, practical work piecing together data on the inputs to each economic sector and outputs to others, all comprehensively assembled in a large matrix. Leontief produced the first input–output tables for the United States – for 1919, 1929, and 1939[51] – and received a Nobel Prize for his work. His methods were subsequently adapted by the United States Bureau of Economic Analysis in developing a standard approach for recording input–output data.[52] Leontief also extended his methods to show how the environmental impacts of producing goods and services for final consumption could be traced through an economy, which was particularly useful for industrial ecologists.[53]

The link between Leontief and industrial ecology was fostered by his long-term collaborator, Faye Duchin. In 1975, two years after receiving his Nobel Prize, Leontief moved to New York University to start the Institute for Economic Analysis. Duchin, who completed a PhD in computer science at the University of California, Berkeley, joined Leontief's institute as a faculty member in 1977 and succeeded Leontief as director in 1985. She collaborated closely with Leontief for almost twenty years, before leaving to become dean of humanities and social sciences at Rensselaer Polytechnic Institute in 1996. Duchin's work with Leontief included economic analysis of military spending, the impacts of automation on workers, and an article published in *Science* describing how advancements in computational power were enabling the development of increasingly detailed economic models.[54] Duchin engaged in industrial ecology from its earliest years, presenting a paper on applications of input–output analysis at the National Academy of Sciences colloquium in 1991[55] and attending the formative meeting of the society at the New York Academy of Sciences in 2000.

In one of Duchin's later studies, she used a global input–output model to wrestle with the question of how to feed the world's future population in 2050, while conserving land and water.[56] This, recall, was the challenge we looked at in Chapter 3 – including the worrying analysis by David Tilman. Duchin used data on potentially available rain-fed and irrigated croplands for ten broad regions of the world, as well as data on renewable water supplies. She took a projected global population of 9.7 billion in 2050 and asked how everyone would be fed. To answer, an optimization algorithm was applied to trade between the ten economic regions of the global input–output model. The model was used to find the minimum global use of resources to provide for food consumption in 2050, under different scenarios. In the first of the sustainable scenarios, clearing of forested lands was restricted, and the use of renewable water constrained. The model, however, produced no solution. In other words, the world could not be fed in that scenario. To find a tractable solution, Duchin showed that one of two possible changes was necessary – either food demand in developed countries had to be substantially reduced, or agricultural technology and crop-water management practices in developing countries had to be employed. The input–output model also showed that most of the additional agricultural production would need to occur in Africa or Latin America. Thus, Duchin's analysis echoed our conclusion from Chapter 3 about this century's great Malthusian struggle in Africa.

Duchin's work is just one demonstration of industrial ecologists adapting Leontief's input–output methods to address Malthusian challenges. Industrial ecologists have employed Leontief's input–output technique in the assessment of a wide range of environmental challenges. For several decades, researchers at Carnegie Mellon University in

Industrial Ecology

Pittsburgh produced environmentally extended input–output models of the United States economy. These were used by faculty members at Carnegie Mellon's Green Design Institute to assess the environmental impacts of biofuels, food systems, energy systems, electric vehicles, and other technologies.[57] In Europe, researchers at the Norwegian University of Science and Technology, Leiden University, and other universities have used input–output models to assess the environmental impacts of consumption. This has included, for example, some of the earliest studies of consumption-based carbon footprints of nations and analysis of the environmental impacts of household consumption.[58] Advancements in environmentally extended input–output models linking multiple world regions have been made by researchers in Europe and Australia, for assessing global environmental impacts.[59] These advancements include the development of the Industrial Ecology Virtual Laboratory in Australia by Manfred Lenzen and colleagues. Many Japanese industrial ecologists have also contributed to the advancements in input–output modelling, including the award-winning work of Shinichiro Nakamura, from Waseda University, on the analysis of materials and wastes.[60]

Decoupling

Underlying the great global environmental challenges of biodiversity loss and climate change is a broader issue of how global society obtains and uses natural resources. Indeed, the choice and exploitation of materials, energy, water, and land gives rise to a multitude of environmental problems – as reflected by Ayres' work on industrial metabolism. In 2007, the United Nations Environment Programme (UNEP) recognized that it needed more scientific rigour in studying the use and environmental consequences of resources. The International Resource Panel was thus established as a parallel, but smaller, body to the IPCC and the Intergovernmental Science-Policy Platform on Biodiversity and Ecosystem Services.

The International Resource Panel was launched by UNEP at the World Science Forum in Budapest on November 9, 2007. Notice of the launch described the mandate for the panel and its role in helping the global economy decouple from environmental degradation:

The Resource Panel's objective is to strengthen the global scientific base and to increase international exchange of knowledge and best practices on sustainable resource management. Leading experts in this field, under the chairmanship of Ernst Ulrich Von Weizsaecker and Ismail

Malthus Enigma

> *Serageldin, will focus on assessments and status reports on the environmental impacts of the use of renewable and non-renewable resources over their life cycle, taking into account economic development and supply security issues. The Panel aims to contribute to the global challenge of decoupling economic growth and environmental degradation. Capacity building on data collection and methodologies for assessment as well as prioritisation of resources and products in developing countries and countries with economies in transition is an important working area.*[61]

The panel's membership includes a mix of experienced policy advisors and scientists. The first of the co-chairs, Ernst von Weizsäcker, was a highly influential scientist and politician. Having been a professor of ecology and having worked for the United Nations in New York, von Weizsäcker headed the Institute for European Environmental Policy in Bonn from 1984 to 1991. He sat as an elected member of the German Bundestag from 1998 to 2005, notably chairing the Committee of the Environment from 2002 to 2005; it was during this time that Germany first introduced feed-in tariffs, which spurred the development of the solar, wind, and biomass energy sectors. Among his many other notable exploits are his book *Factor Four: Doubling Wealth, Halving Resource Use* – published in 1995 with Amory and Hunter Lovins – and his founding of the Wuppertal Institute. He also became co-chair of the Club of Rome in 2012. Von Weizsäcker's co-chair on the panel, Ismail Serageldin, had a similarly impressive career working on science, sustainability, and development issues. He spent twenty-eight years with the World Bank, rising to vice-president, before returning to his native Egypt, where he became a television presenter – among his many other activities. Serageldin was later replaced by Ashok Khosla, an Indian environmentalist who held senior positions with UNEP and later became president of the International Union for Conservation of Nature and a member of the Club of Rome. While the International Resource Panel also included many other eminent international statesmen and women, the majority of the technical members of the panel – the scientists who conduct the technical work – have been industrial ecologists. These include Graedel and Bringezu, who attended the New York meeting; Yuichi Moriguchi from the National Institute for Environmental Studies in Japan; and Marina Fischer-Kowalski, a former president of the International Society for Industrial Ecology. More than ten industrial ecologists have been members of the panel.

Industrial Ecology

Among the scientific tools industrial ecologists brought to the International Resource Panel were methods of material flow and stock accounting that had evolved from Robert Ayres' work. Several of the industrial ecologists on the panel – including Bringezu, Fischer-Kowalski, and Moriguchi – had been heavily involved in extending and formalizing national material accounting, working with various international organizations. The World Resources Institute conducted a comparative study of material accounts for Austria, Germany, Japan, the Netherlands, and the US in 1997,[62] followed up by a study, *The Weight of Nations*, in 2000.[63] Eurostat incorporated material accounts into the standard environmental accounts of EU member states.[64] The accounts include the production, extraction, import, and export of materials, classified under the four broad groupings of biomass, fossil fuel carriers, metals, and non-metallic minerals. A similar accounting system was adopted by the Organisation for Economic Co-operation and Development (OECD) in 2008[65] – and within several years, global data sets were established.[66] Through this work with the international community, the industrial ecologists essentially achieved a system of national resources accounting, parallelling the work that was done by Simon Kuznets and others for national economic accounting in the 1930s. The material accounts provided a strong empirical data set for several of the International Resource Panel's publications – including one of its earliest reports, entitled *Decoupling: Natural Resource Use and Environmental Impacts from Economic Growth*.[67]

The panel's first report on decoupling, led by Fischer-Kowalski, describes how system-wide innovation in an era of rapid urbanization offers the possibility to reduce environmental burdens while sustaining economic growth. The *Decoupling* report is one of the most highly viewed reports ever published by UNEP, with over 250,000 downloads.[68] The report begins with laying out the huge challenge of decoupling environmental impacts from economic growth. Over the twentieth century, economic development involved stark increases in the use of natural resources: the extraction of construction materials went up by a factor of thirty-four over the century, ores and minerals by twenty-seven, fossil fuels by twelve, and biomass by 3.6.[69] This was due partly to population growth, but also to changes in technology and lifestyles. The average global material use per person approximately doubled over the century, from 4.6 tonnes to eight to nine tonnes.[70] Taking an industrial ecology perspective, the panel recognized that the environmental impacts associated with these resources could occur over all phases of their life cycle, from extraction and manufacturing through to consumption and

Malthus Enigma

post-use.[71] The panel argued that sustainable economic development must entail an absolute reduction in global resource use, with expanding economic activity and diminishing environmental impacts. Hence, they defined *decoupling* to mean "using less resources per unit of economic output and reducing the environmental impact of any resources that are used or economic activities that are undertaken."[72] The size of this challenge is apparent from the panel's indication that, under business as usual, global resource extraction is expected to triple by 2050.[73]

The report proceeds to outline broad strategies for decoupling. In broadest terms, the panel suggested that the environmental impacts of resource use could be reduced either by substituting, where possible, with less-impacting alternatives, or using existing resources in more benign or efficient ways.[74] With respect to the latter, however, they recognized that it was unrealistic for developing countries to use less resources — and the onus was on developed countries to dramatically reduce resource use through eco-innovation. Indeed, what the International Resource Panel concluded was necessary is what humanity has always had to do under Malthusian struggle — that is, innovate. To be more specific, they called for "unprecedented levels" of "sustainability-oriented innovations," resulting in "radical technological and system change."[75] The International Resource Panel describes more details about specific technologies in later reports but stresses the need for innovation in cities. Echoing Frosch and Gallopoulos, the panel also calls for "as much circularity in economies as is technically feasible."[76] To broadly demonstrate the massive extent of change that is required to attain a sustainable path, a couple scenarios are modelled, beyond business as usual. The most aggressive of these — though not actually reducing global environmental impacts — considers the case where total global resource use is capped at the same value as 2000. This scenario, on equity grounds, would involve a huge contraction in absolute resource use in industrialized countries — with reductions by a factor of between three and five.[77]

Several aspects of the panel's decoupling work are noteworthy in the context of ideas covered in earlier chapters of this book. First is the pro-growth perspective of the International Resource Panel, which is philosophically different from the steady-state economics espoused by Daly and many of the ecological economists that followed him, and, more so, the de-growth movement. The panel goes to some length to distinguish forms of development with positive increases in economic activity from those with steady or declining GDP.[78] They also stress that the decoupling approach entails increasing GDP and increasing quality of life, with a declining use of material resources. In its framework, the panel contrasts the decoupling philosophy with that of some ecological economists who reject the potential for either relative or absolute decoupling. The panel's work embraces the development imperative – to improve the economies of developing nations – which necessarily requires a dose of technological optimism. Herein, perhaps, lies a difference between ecological economists and industrial ecologists.

A second striking difference is that between the type of analysis conducted under the panel's decoupling work and the systems modelling work completed decades earlier by Forrester, Meadows, and the *Limits to Growth* team, discussed in Chapter 6. Forty years on, the data available to the International Resource Panel is substantially richer and deeper than previously available – in part because of the efforts of the industrial ecologists in establishing national material flow accounts. With better data, the science has progressed, too. In the decoupling work – and International Resource Panel studies more generally – the mathematical modelling techniques employed are far more cautious and reserved than the ambitious World2 and World3 models. The scenarios in the *Decoupling* report use relatively simple mathematics. Other International Resource Panel studies have employed environmentally extended economic input–output models – another area that has been expanded and improved by industrial ecologists, as discussed above. These models are far more data driven and more rigorous than the systems models. The industrial ecologists like to employ physical laws – thermodynamics and conservation of mass – as much as possible. There is still some progress to be made, though, in their mathematical models. Not all scenario models used in the decoupling work included physical constraints.[79] Similar concerns have also been raised about the modern-day integrated assessment models that descended from those in the early years of the Club of Rome.[80] So there is still some work to be done in properly applying conservation laws in forward-looking environment–economy models.

Following its report on decoupling, the International Resource Panel has produced a variety of other studies covering topics such as food systems, water use, biofuels, cities, and international trade.[81] Another of its major work streams has been a synthesis of knowledge on the status of the stocks and flows of the world's metal resources. The rigorous detail taken in this work shows how far the science has come since the 1980s, when Paul Ehrlich jumped blindfolded into the bet with Julian Simon on the future price of metals. Led by Tom Graedel of Yale University, the panel's working group on global metal flows has produced a series of six

Malthus Enigma

reports. Starting with an appraisal of metal stocks in society, these also cover recycling rates, the environmental impacts of metals, geological stocks, future demands, and a special report on critical metals. The first of the reports noted the clear importance of metals to human society – which uses over sixty metal elements of the periodic table. Modern society is highly dependent on five of these: iron and manganese, which are the key metals in steel; aluminum, which is used in vehicles, packaging, buildings, and infrastructure; lead, used in batteries; and copper, used in the transmission of electric power.[82] We also make high use of four other metals: chromium and nickel, used in stainless steel; zinc, which inhibits corrosion; and tin, which is used extensively in modern electronics. The panel noted that strong data on the stocks of metals currently used in society exist only for aluminum, copper, iron, lead, and zinc. Nonetheless, drawing upon fifty-four other studies, the panel produced estimates of per-capita stocks for developed and developing countries, as well as global values, for twenty-four metals.[83] Given the high value and unique properties of metals, many are recycled at relatively high rates. Thus, industrial ecologists recognize metals as exemplary materials among those that can be practical for the emerging circular economy. This is the topic we turn to next.

The Circular Economy

On November 27, 2004, intrepid British yachtswoman Ellen MacArthur set sail on a heroic record-breaking circumnavigation of the globe.[84] At 4 p.m., a flotilla of thirty small boats, a cormorant, and a pod of dolphins accompanied her out of Falmouth Harbor into the English Channel. The last of her support team exited on a small boat before she approached the official starting line at 8:10 the next morning. With a goodbye wave from a helicopter, the nervous MacArthur was left alone for her exhilarating and exhausting journey of two and a half months. Her route would take her down through the Atlantic Ocean, then eastward through the Southern Ocean, circumnavigating Antarctica as she passed below the Cape of Good Hope, Australia, New Zealand, and Cape Horn, before returning home back through the Atlantic. To break the record of Frenchman Francis Joyon, she needed to average some 369 miles per day in her seventy-five-foot trimaran, officially called *B&Q*, but affectionately known to her as *Mobi*.

MacArthur was a highly experienced sailor who had made round-the-world and transatlantic record attempts before, but she was pushed to her limits by the journey.[85] A large storm off the coast of Portugal kept her

Industrial Ecology

focused early on – and helped her set a new record solo time for reaching the equator. In the South Atlantic, MacArthur had to switch out her onboard diesel generator, which was required to power essentially all the systems in her cabin, including navigation! Use of her backup generator then led to challenges with her water supply overheating and exhaust fumes in her tiny two-by-two-metre cabin, which she had to turn mechanic to fix. She sailed as close to Antarctica as was safe to avoid icebergs but was shocked to discover several to the north of her as she sped through the southern waters. On Christmas Day, *Mobi* was battered by a massive storm. In fact, much of the trip was a yin-yang between brutal storms and frustrating calms, the latter requiring MacArthur to make multiple exhausting changes of sail – as many as eleven per day – to make progress. She was hunted down by a massive storm that walloped *Mobi* as she approached Cape Horn. Heading back through the Atlantic, she had to climb the main mast to fix some damage. Then she was marooned in the still waters of the Azores, almost causing her to lose her attempt at the record. But on February 7, 2005, MacArthur crossed the finish line, having covered 27,354 miles in seventy-one days, fourteen hours, and eighteen minutes – breaking Joyon's record by more than a day.[86] Averaging only five hours' sleep per day, and no more than three hours at any one time, she was shattered. The British people took to her, though. MacArthur was named BBC Sports Personality of the Year, was made a Dame of the British Empire, and received the Legion of Honour of France.

Although encouraged by emails and satellite phone calls, MacArthur was essentially alone on her journey in one of the most remote, yet inspiring, parts of the world. One evening, after being battered by a storm, she observed a stunning sunset over the Southern Ocean and recorded in her log:

> *I am completely in awe of this place. The beauty of those immense rolling waves is endless and there is a kind of eternal feeling about their majestic rolling that will live on forever.*[87]

MacArthur experienced the Earth in a way that parallels that of Frank Borman and the Apollo 8 crew looking back at the Earthrise, perhaps with a dose of solitude similar to a young Norman Borlaug alone in his isolated Idaho forest watchtower. Her log continued:

> *[T]here is some kind of mesmerizing feeling, some kind of completeness about being here ... I feel this is not so far from the end of the Earth; we are isolated, isolated but on the other hand completely free ... I am glad we have come down here and seen*

Malthus Enigma

> *this storm ... It's a reminder of how small and insignificant we are on this planet – but at the same time what a responsibility we have towards its protection.*[88]

MacArthur's concerns about sustainability and the future of the planet began to grow and take shape in the years after her record circumnavigation of the Earth. There were further experiences that influenced her, including a trip back to the Southern Ocean to visit South Georgia, where she witnessed abandoned communities that had once thrived on the production of whale oil, provoking reflections back on her grandfather's life as a coal miner. MacArthur began to study further herself – learning about sustainable energy, buildings, agriculture, transportation, economics, and industry. In 2007, she took a trip to Barcelona for the start of a large yacht race, in which she decided not to participate. She came to the realization that "sailing round the world again would not achieve anything."[89] She understood that sailing had given her a voice, and she needed to find the best way to use it.[90] During this period, MacArthur was frequently touring and making public presentations. She would tell audiences about her sailing adventures, including her knowledge of what it meant to conserve resources. From months away at sea, sailing the world in a tiny capsule, she had first-hand experience of managing food, water, fuel, and other resources under strict constraints. But she realized that there was a need for a more positive message and that just telling people to "use less" had limited traction.

Eventually, MacArthur found her way to the same ideas of industrial ecology as Frosch and Gallopoulos – indeed of Boulding's Spaceship Earth. In 2009, she founded the Ellen MacArthur Foundation, dedicated to the development of the circular economy. MacArthur realized that there was a need to change the human technological and economic system. But she saw that a far more optimistic message was that we can redesign the future to recycle valuable materials and eliminate pollution. The philosophy was the same as industrial ecology, with Nature providing the guiding principles. "Nature has nowhere to throw things away; it doesn't need to as the waste of one species is food for another,"[91] she wrote. "If waste equals food, then everything changes."[92] Under the banner of the circular economy, the Ellen MacArthur Foundation began to engage with industry and the public with principles of industrial ecology. MacArthur used her extensive corporate network, established through years of sailing sponsorship, to encourage industry in the ways of the circular economy – doubling up on the efforts of Harbin Tibbs and other industrial ecologists since the early 1990s.

MacArthur's adoption of the term *circular economy* was timely, because the concept had already made progress in political arenas. The origins of the concept of the circular economy are complex[93] and can be traced back at least as far as Boulding[94] and Commoner,[95] with industrial ecologists such as Frosch[96] and environmental economists such as David W. Pearce and R. Kerry Turner[97] pushing the idea further in the early 1990s. Elements of the circular economy were already included in policies for waste management, recycling, and extended producer responsibility in European countries, Japan, the US, and elsewhere – some dating from the 1970s.[98] In August 2008, the Chinese government took a more holistic view in adopting the *Circular Economy Promotion Law of the People's Republic of China*. Article 2 of the Chinese law defined the circular economy to be a "general term for the activities of reducing, reusing and recycling in production, circulation and consumption."[99] At the time of the founding of the Ellen MacArthur Foundation, the European Union was beginning to develop policies on the circular economy – and European policy-makers adopted[100] the definition of the foundation:

> *A circular economy is restorative and regenerative by design, and aims to keep products, components, and materials at their highest utility and value at all times. The concept ... is a continuous positive development cycle that preserves and enhances natural capital, optimises resource yields, and minimises system risks by managing finite stocks and renewable flows.*[101]

In recent years, however, industrial ecologists have been learning more about the challenges of establishing circular economies. One insightful study by Willi Haas and colleagues from Alpen Adria University in Vienna used data from the national material flow accounts to address the question: *How Circular is the Global Economy?*[102] The study found that, for 2005, out of the total processed material inflow to human activities of sixty-two gigatonnes, only four gigatonnes was recycled waste materials.[103] The simple answer, then, is that the global economy is only six percent circular – when measured by inflows. A couple reasons can explain this somewhat disappointing finding. First, because the global economy and population are growing, the mass of material inflow is higher than the mass of material outflow. Much of the material inflow – especially construction materials – goes into building up the stock of materials in use by humans. The total amount of waste material coming out of stock, or flowing straight through and exiting as pollution, was forty-five gigatonnes. So the overall recycling rate on the outflow side was about nine percent.

An important second consideration, however, is that the material accounts include all four major components of processed materials: biomass, fossil fuel carriers, metals, and non-metallic minerals. The largest fraction of these is fossil fuel carriers at forty-four percent of processed mass inflows, but most of these are entirely combusted, with little chance of recycling mass. On the other hand, the global recycling rate for metals was found to be seventy-one percent, with construction minerals at thirty-three percent and biomass at fifteen percent.[104] If these recycling rates can be improved further, then the global economy could become more circular. But we also have to stop using fossil fuels and find other ways of reducing our material inflow.

Other studies have noted further challenges with approaching circular economies, among which are the energy requirements of recycling and the potential for rebound effects. The difficulties of achieving circularity have been particularly well highlighted by University of Cambridge industrial ecologists Julian Allwood and Jonathan Cullen. In "Squaring the Circular Economy," Allwood also recognizes the problem of a growing economy, that "if demand is growing, the circle cannot remain closed, and it may be a much more important priority to reduce the rate at which new material is required."[105] Allwood places recycling and other forms of circularity within a broader set of material strategies and stresses the importance of seeking "material efficiency" – in other words, "delivering material services with less input of materials."[106] Allwood also observes that, in some cases, the energy used in recycling materials could exceed that required to make new products from virgin materials. This is particularly the case for products that require "precise and complex mixing of atoms to create high-performance properties."[107] Where the energy is generated from fossil fuels, recycling may make greenhouse gas emissions worse. Allwood reminds us that the goal should be to "meet human needs while minimizing the environmental impact of doing so"[108] and not necessarily to maximize circularity.

Cullen takes the issue of energy required for recycling further in explaining that recycled materials are often of a lower quality than in their first use.[109] For example, when aluminum alloys are melted down, they decrease in quality. The additional energy required to increase the quality of recycled materials should also be considered in assessing the desirability of circularity. Cullen proceeds to develop measures of circularity accounting for the fraction of material recovered and the energy required in doing so.

Industrial Ecology

In cases where circularity does make sense from economic and environmental perspectives, there is potential for the further twist of a rebound effect. As California-based researchers Trevor Zink and Roland Geyer note, it is important to consider the market dynamics.[110] If recycled products or materials compete successfully with production from virgin materials, then the resulting lowering of costs may cause an increase in demand. Rebound effects, discussed in Chapter 4 in the context of energy efficiency, are, however, complicated, and understanding them is still part of our Malthusian struggle.

Industrial Ecology in Technology, Science, and Policy

In wrapping up this chapter, we can ask what industrial ecology has contributed to the wider Malthusian challenge of sustaining the human species within the carrying capacity of the planet. The answer is that it plays a role in all three layers of the Malthus Enigma: as a key direction for technological innovation, as an interdisciplinary science, and as part of the policy regime.

When industrial ecology started in earnest in the 1990s, inspired by Frosch, among others, it was arguably more focused on technological development. Reviewing the early history of industrial ecology, Suren Erkman identified eco-industrial parks and dematerialization of the service sector to be the primary areas of application of the field. Since that time, progress has continued, with the development of industrial symbiosis. There have been some ups and downs, but progress overall. Like all technological development, innovation takes time – it's part of the struggle. The world's economy may be only six percent circular today, and it will never reach one hundred percent – but industrial ecology promises ways to further close material loops, saving materials and avoiding pollution. With further advances in areas such as lightweight design, design for disassembly, remanufacturing, and waste management technologies, industrial ecology will find ways to make more efficient use of our resources. Maybe, if we are smart, we will save some energy, too. This may be subject to rebound effects, but totally decarbonizing our energy supply is necessary regardless.

Addressing the otherwise unforeseen consequences of technological development – such as rebound effects – is where the science of industrial ecology comes in. The scientific dimension of industrial ecology has always been present. In its fullest sense, industrial ecology is the "scientific study of energy and materials flow in industrial-society – and the associated environmental impacts."[111] Industrial ecology is an

213

unusual, interdisciplinary science that bridges from natural sciences to social sciences, picking up knowledge of engineered systems on the way. Industrial ecologists apply the laws of physics to societal systems – from supply chains to cities – to understand the environmental impacts. There is a subtlety here in the contribution of industrial ecology to the second layer of the Malthus Enigma, as I have called it. It's not just about doing the science to learn about the planet that humans inhabit – but also learning in a physical sense how we ourselves grow and impact the world around us. The science of industrial ecology is exemplified by the early work of Ayres tracking material use in economies and conceiving of industrial metabolism. Through the formalization of methods for material and energy flow accounting, industrial ecology links environmental stresses to their underlying human drivers. An example of this is the work of the International Resource Panel on decoupling, as discussed in this chapter. There are further examples, too, including the work by Lenzen and colleagues linking biodiversity loss and trade, and many scientific studies of greenhouse gas emissions on a variety of scales, from households and cities, up to national economies and global infrastructure systems.[112]

The development of laws on the circular economy by the Chinese government and the European Union is a strong statement of industrial ecology moving into public policy. Brad Allenby, who co-authored the leading textbook on industrial ecology, has always envisioned that the discipline would have impacts on policy; he described a framework for this as early as 1999.[113] OECD work on sustainable manufacturing and eco-innovation also recognized that implementation of industrial ecology involved more than just the design of products and processes – also involving the creation of new institutions, organizations, and markets.[114] While industrial ecology has made its way into policies before – such as waste policy, extended producer responsibility, and life-cycle perspectives in California fuel standards[115] – the new laws on the circular economy propel industrial ecology into public policy in a much larger and more comprehensive way. Getting policies to address environmental issues in the right way, however, is the hardest part of the Malthus Enigma. Hopefully, policies on the circular economy will continue to evolve – spurring eco-innovation while adapting to the challenges of adjusting for growth, energy requirements for circularity, and possible rebound effects.

Chapter 9: Conclusions

> *Human societies will continue to stumble. Many will fall. But we have overcome starvation, disease, deprivation, oppression and war. We can overcome ecological crisis.* [1]
>
> Ted Nordhaus and Michael Shellenberger

Malthus was wrong! Malthus was right! Depending on which moment in time one chooses, as well as location on the planet and social standing, it is easy to see how disparate opinions on the Malthusian predicament can arise. If Malthus's thesis is interpreted in its most simplistic terms, then the Cornucopians have it. Malthus fundamentally underestimated the potential of human ingenuity and technological development to provide food for a growing population. From the naive position of a twenty-first-century middle-class resident of a rich, developed country, Malthus may appear to have been spectacularly wrong. After all, the last significant famine due to natural causes – that is, excluding wars – experienced by inhabitants of a Western country was perhaps the Swedish famine of 1867–1869.

But this position can be challenged on two grounds – global inequality and environmental stress. If you put yourself in the sandals of an Ethiopian farmer trying to feed your family during the 1980s famine or consider the predicament of Indian or Pakistani governments facing potential mass starvation before the miracle of Norman Borlaug intervened, then the neo-Malthusian position does not seem so far wrong. Moreover, if one accepts the subtle change in the Malthusian predicament from "feeding all the World's people" to "doing so without destroying the climate and ecosystems upon which we depend," then the Malthusian position becomes even stronger.

Before heading too far in support of either the Malthusian or Cornucopian directions, however, I want to stop – and recast the conclusions in terms of the more complex phenomenon of the Malthus

Malthus Enigma

Enigma that has been explored in this book. The Malthus Enigma is ever present, and there are several layers to it. Regardless of whether you are a wealthy urban resident of the developed world, surrounded by technology that cocoons you from Nature, or a poor rural dweller in the developing world living off the soil, Malthusian struggle is ongoing. In essence, the Malthus Enigma is about continual survival of the human species under a growing population, but we can identify three finer layers, or dimensions, to the predicament.

The first layer of the Malthus Enigma is the necessity to continually innovate. The issue here is that every innovation helping the population survive and increase perpetuates the challenge even further. Moreover, a shift in the nature of this first layer of the Enigma has developed over the past one hundred years – from providing the resources to feed everyone to doing so within the carrying capacity of the biosphere, or planetary boundaries. Arguably, we already have the technology available today to address the great environmental challenge of climate change. But, this is because of a substantial amount of innovation in recent decades in technologies such as renewable energy generation, electric vehicles, and battery storage, among others. Moreover, innovation in technology and management is still necessary to deal with the threat of a sixth mass extinction of species – and to close material loops so that we can live sustainably on Spaceship Earth.

This leads to the second dimension of the Enigma, which is that we are still learning about the nature of the world in which we struggle. Our scientific knowledge of the planet is imperfect – and we must continually strive to understand it. It is akin to humanity playing a long, complex computer game – but still learning the rules as it fumbles along. This statement should not be misconstrued, though. Scientists know a lot about climate change today, and there is a high degree of consensus that humans are contributing to it. We also know enough about the functioning of natural ecosystems to understand that they are under threat, but there are several million species on the planet that have yet to be identified. Our knowledge of the Earth system – and how humans impact it – is still evolving.

The fact that we are still struggling to understand the planet we inhabit clearly leads to contention – which is inherently part of the second layer of the Enigma. Even where the science is robust enough and consensus is reached with reasonable certainty, we still have a third layer of the Enigma, which is that of the policy response. For this third dimension of the Malthus Enigma, we turn away from engineers and scientists – and into the realm of policy-making. Here, the greatest

Conclusions

challenges seem to be if, how, and when markets should be used to address environmental challenges and how to frame these challenges in the context of development.

In these conclusions, I will further explore these three layers of the Malthus Enigma and provide some pointers for the future.

Continuous Technological Innovation

Between his birth in 1766 and the last edition of his *Essay on the Principle of Population* in 1826, Malthus experienced incredible technological change. This included the first steam trains and the first steamboats, as well as hot air balloons, sewing machines, and even the earliest electric battery. The efficiency of the steam engine increased by a factor of about ten – largely due to James Watt – and agricultural output per worker increased by about thirty percent. Still, the significance of these changes had little impact on Malthus's dismal prognosis for humankind.

William Stanley Jevons, writing in 1865, did recognize the importance of technological change, but to a limited extent. He noted how the rise of British manufacturing, combined with opening of trade from repeal of the Corn Laws, moved Britain from corn to coal. Britain was able to break free of its dependence on limited agricultural lands, and hence Malthus's doctrine. Still, Jevons could not see beyond coal as a source of energy – and believed that Britain would regress once the coal ran out. He ended up back at Malthus's position because he did not perceive the potential for alternative forms of energy to develop.

By the middle of the twentieth century, physicists and engineers had developed nuclear power – and there was erroneous talk that it, or even fusion, would provide a limitless source of free energy.[2] If this were true – leaving the issue of nuclear waste disposal aside – the modern characteristics of the Malthus Enigma would still not be diminished. Indeed, easy access to abundant free energy would possibly be one of the worst things that could befall humanity. Yes, it might help address climate change, but the impacts of unconstrained energy use by humans on global ecosystems could be disastrous. Malthus recognized the potential for the human population with unlimited access to food to grow uncontrolled and destroy the surrounding ecosystems. Herman Daly similarly observed that the same potential fate would apply to humans gaining free access to an unlimited energy supply.

These perspectives of Malthus and Daly were expounded further in the *Limits to Growth* study of the early 1970s. The key lesson from

Malthus Enigma

Dennis Meadows and his team was that there are multiple constraints on growth – solving the challenge of one constraint just brings humanity up against the next. In the scenarios of the World3 model, when resource constraints are removed, collapse occurs due to pollution, and when pollution is overcome, the population increases further, until collapsing due to insufficient food.

History teaches us that human innovation keeps solving the population problem, but perhaps we are deluding ourselves. A pessimist could argue that Malthus's fear that with an unlimited supply of food, the human race would be plunged into "irrecoverable misery"[3] has already occurred. The invention of fertilizers through Fritz Haber's fixation of nitrogen, combined with Borlaug's Green Revolution and massive modern automatization of agriculture, today gives us a supply of food that would arguably seem "unlimited" to Malthus. To a late-eighteenth-century farmer worrying over a year of dismal crop yields, what Haber achieved was pure magic! He learned how to extract active nitrogen from the air around us and produce a powder that could be sprinkled on cornfields to dramatically increase their yield. The consequence of access to massive amounts of cheap food is that the human population has ballooned. Yet 670 million people on the planet are still living in extreme poverty, many without adequate access to clean drinking water and basic sanitation.[4] Human settlements have mushroomed into megacities housing tens of millions of people, drawing in huge quantities of energy and materials.[5] We have destroyed large swaths of the ecosystems around the planet, including services we depend on.

If David Tilman's systems framing is correct, this problem will only get worse. Growth in population and affluence – especially in Africa – could see global food demand doubling by mid-century.[6] Unless the developing world adopts high-intensity Western agricultural techniques, Tilman projects that a billion hectares of natural habitat will be destroyed for food production. Yet high-yielding Western agricultural methods produce high quantities of nitrogen pollution and monocrops, both of which undermine long-term resilience of all land uses. It is highly questionable whether current massive agri-tech solutions are sustainable. Others would argue, too, that regionally appropriate agricultural systems, reflecting local culture and Indigenous knowledge, are important, despite the challenges of scaling up regenerative agriculture.

One broad lesson here is that it is better for humanity to be moderately constrained in its access to food and energy than to have easy access to unlimited quantities. This insight is particularly pertinent to our current struggle to revolutionize energy supply systems to massively

Conclusions

reduce greenhouse gas emissions and avoid destructive climate change. Large-scale deployment of dispersed renewables such as wind and solar is emerging as the likely technological solution for this next phase of our Malthusian struggle. The fact that these technologies have only moderate energy returns on the energy invested (EROIs of five to twenty, though increasing) is arguably desirable. Fossil fuels have historically provided much higher energy returns (EROIs of greater than fifty),[7] and who knows what return some future technology might provide. But if humanity has to work a bit harder to access energy supplies, in an equitable way, this is a good thing – putting our tendency to destroy ecosystems in check. Renewables can supply us with enough energy to meet our basic needs – and achieve a high quality of living without generating such an excessive amount of energy.

The good news about renewables – including water power and biomass, as well as wind and solar – is that they can form the backbone of a sustainable energy supply – meeting all of humanity's needs. Renewable energy is a Promethean III technology, as Georgescu-Roegen put it, or "reinvention of fire," as Amory Lovins similarly explained.[8] Several studies on national and global scales have shown that our future energy needs can be largely met using electricity from carbon-free sources. Indeed, the combination of renewables, electric vehicles, household heat pumps, battery storage, and energy-efficiency measures is already beginning to move us toward a low-carbon electric future.[9] Once our energy systems have been decarbonized, this may also open up greener ways for more circular economies to function. This will not only address a broader set of environmental concerns related to waste and pollution, but also recycle precious metals and other components of renewable energy systems. Viable technological solutions for one of our greatest environmental problems – climate change – do exist, should humanity choose to use them.

So what technologies will serve the future populations of our fragile planet? What kinds of energy systems and food systems will prevail? One type of answer to these questions is provided by Ted Nordhaus and Michael Shellenberger – the authors of the provocatively titled book *Break Through: From the Death of Environmentalism to the Politics of Possibility* and the founders of the Breakthrough Institute.[10] In some respects, Nordhaus and Shellenberger are the modern-day heirs of Julian Simon – Cornucopians who hold strongly technologically oriented views of the future – only Nordhaus and Shellenberger are complex eco-modernists from the political left, rather than the right. Just to add some spice here, Ted Nordhaus is also the nephew of Yale's William Nordhaus,

Malthus Enigma

who argued so vehemently against the *Limits to Growth* work of the 1970s. Personally, I do not subscribe to the views of Nordhaus and Shellenberger or the technologies they favour. I have, though, argued that technology plays a key role in the challenge of sustaining future populations. I also think that Nordhaus and Shellenberger's optimistic outlook, as reflected in the quote at the beginning of this chapter, is an important ingredient for inspiring ingenuity and positive change. My role with this book has been to try to be an honest broker presenting the views of both Cornucopians and neo-Malthusians; hence their inclusion.

Nordhaus and Shellenberger's vision of the future is one of large-scale, centralized technologies dominated by powerful actors in the contemporary capitalist system.[11] Theirs is a message of continued economic growth, supported by nuclear power and industrialized agriculture, as well as intensified urbanization. Alternatives to Nordhaus and Shellenberger's world view are possible, however. Energy can be generated from dispersed renewable sources and supplied by household or community-scale systems, possibly intersecting less dominant larger-scale energy systems for added resilience. Alternatives to the widespread use of industrialized agriculture are harder to foresee. The current techniques of organic agriculture cannot feed the planet, but perhaps advancements in agroecology or engagement with Indigenous knowledge will lead to culturally sensitive and regionally appropriate agricultural systems of sufficient yield. Of course, this still requires some technological innovation.

Continued innovation is also still required to close material loops and design our industrial systems to be like waste-free natural ecosystems, as encouraged by Boulding and Frosch. The field of industrial ecology has made good progress in this regard – learning about industrial symbiosis from the Kalundborg example – and developing it further as a practice. To this, we can add the study and redesign of product supply chains, engineering design for disassembly and remanufacturing, and other strategies for dematerialization. The systems-level understanding of the challenges and opportunities of circular economies, provided by industrial ecologists, shows nonetheless that further innovation in technology and management is required.

There are, of course, still skeptics of the potential for technology to help address our environmental challenges.[12] But developing technological solutions has always been a part of our Malthusian struggle. Moreover, the opinions of the technology skeptics are often made without a good understanding of the underlying science, such as the laws of thermodynamics.

Conclusions

Scientific Investigation of the Global Environment

Malthus's treatise on the population problem was written before the laws of thermodynamics had been established, before understanding of the greenhouse effect and climate change, and before the field of ecology had really begun. While the first layer of the Malthus Enigma requires us to continually innovate, doing the science to understand the global environment around us is really the second layer of the Enigma. Scientific progress has come hand-in-hand with technological progress, although the relationship between them is not straightforward.[13] Watt was able to make impressive improvements to the steam engine before the laws of thermodynamics were fully revealed, although he was influenced by scientists around him, such as Joseph Black. Rudolf Diesel, on the other hand, was clearly able to draw upon the laws of thermodynamics in changing the working fluid from steam to air for his diesel engine. Science clearly can enable technological progress, but there is more to the second layer of the Malthus Enigma than discovering scientific principles that help with technological progress; it is also about learning how the Earth functions – and how humans impact it.

It is intriguing to reflect back on the Apollo missions and the formative years of the environmental movement in the late 1960s. Our technology had progressed to the point where we could send people to the Moon, take pictures of the Earth, and even beam back live television images. Yet there was still much we did not understand about the Earth's environment, in particular how we interacted with it. Ehrlich's *Population Bomb* provided an impassioned but scientifically naive understanding of the human condition on Earth; the *Limits to Growth* work supported by the Club of Rome made some important advances in systems modelling and provided insights into Malthusian struggle, though still largely based on a toy model. Even when the full weight of the US government was brought into play with the *Global 2000* study, its findings turned out to be no better than the rival *Resourceful Earth*, funded for just $30,000 from the Cato Institute. Although many of Julian Simon's views were disturbing for environmentalists – and some of his positions, such as on climate change and fisheries, proved to be blatantly wrong – he was right to question the validity of some of the science in the 1970s. The Earthrise observed by the Apollo astronauts corresponded with a necessary new dawn for the science of Earth's environmental systems.

The limitations of our understanding of the Earth's systems are apparent from some of the perspectives on carrying capacity, for

Malthus Enigma

example. Simon was offside to suggest that "the earth's carrying capacity has been increasing" due to increased human knowledge and, moreover, to such an extent that the term had "no useful meaning."[14] Conversely, perhaps Nicholas Georgescu-Roegen was also incorrect to indicate that the Earth's carrying capacity was *decreasing* – a position he arrived at through misunderstanding the second law of thermodynamics. In more recent years, there have been attempts to quantify carrying capacity, which have demonstrated substantial gaps in our knowledge. The efforts of Johan Rockström and colleagues in 2009 to describe planetary boundaries concluded:

> *We have tentatively quantified seven boundaries, but some of the figures are merely our first best guesses. Furthermore, because many of the boundaries are linked, exceeding one will have implications for others in ways that we do not as yet completely understand.*[15]

Updates on the planetary boundaries were made in 2015 and 2023, but a huge amount of uncertainty still remains.[16] Greater attention to net primary production – the global balance of carbon uptake by all biota – may be important for improving understanding of the links between the planetary boundaries.

Arguably, the only planetary boundary that humans have a good handle on is climate change – and this has taken decades of sustained work. Since Guy Stewart Callendar's revival of the nineteenth-century theory of human-induced climate change in 1938, a host of scientists have helped solidify and corroborate the science. These include Gilbert Plass, who determined that CO_2 absorbs infrared radiation at different frequencies from those of water vapour, and Charles David Keeling, who made precise measurements of CO_2 in the atmosphere, but this is just the tip of the iceberg. The effort to understand the global climate system has involved decades of work by thousands of scientists on a diversity of topics, from the atmosphere and oceans to the forests and mountains.

One component of this work has involved the development, over more than fifty years, of global climate models. From the efforts of post–World War II pioneers such as Carl-Gustaf Rossby, Jule Charney, and Norman Phillips to later climate modellers such as James Hansen, global climate models have grown from simple toy models to physically realistic, sophisticated representations of the entire global carbon system. Through the development of computer models within the context of the broader science, humans have developed an advanced understanding of our role in global climate change. As Paul Edwards noted, "Towards the end of the twentieth century global warming became an established fact."[17]

But the road to our understanding of climate change was long and winding – requiring a complex, messy integration of reductionist science and holistic science, as is necessary when studying the Earth system. Given the complexity of the journey, it seems appropriate to recognize the science of humans impacting the Earth as part of the Malthus Enigma.

And this is just the case for climate change – one of the environmental challenges we understand best. Earth system scientists continue to wrestle with the long-term implications of land-use change, biodiversity loss, alterations to the nitrogen and phosphorus cycles, water stress, ocean acidification, ozone depletion, atmospheric aerosol loadings, and a wide variety of other chemical pollutants.[18]

On a more fundamental level, the field of thermodynamics has the potential to be central to a deeper understanding of the second layer of the Malthus Enigma – and to the science of sustainability. From the genius of Sadi Carnot, Rudolf Clausius, and Lord Kelvin in the nineteenth century, it was not until the mid-twentieth century that Erwin Schrödinger and Ilya Prigogine began to realize how thermodynamics applied to living systems. Thermodynamic principles have also long been included in climate models and Earth system science, but there is potential to use them in socio-economic systems, too. Georgescu-Roegen's great insight that thermodynamics underlies all economic activities inspired a generation of ecological economists, who, together with industrial ecologists, have begun to understand how economies function in terms of energy and material flows.[19]

The growing field of industrial ecology will be important for gaining further understanding of how thermodynamics plays out in socio-economic systems. The physical constraints on such human systems can, in theory, be fully described by the laws of thermodynamics, but we need data and an understanding of engineered systems to do so. At the very least, the continued data and experience amassed by industrial ecologists could be used to properly apply the laws of conservation of mass and energy in environment–economy models. This not only might make integrated assessment models more robust, but could also lead to a whole new generation of Earth system models that better incorporate human manufactured capital. Whether such models could begin to cross the divide from scenario tools to holistic science remains an open question. Certainly, better models based on increasingly better data would help the work of the United Nations International Resource Panel, the

Malthus Enigma

Intergovernmental Panel on Climate Change, and others in informing policy.

Policy on Critical Global Environmental Challenges

With respect to the third layer of the Malthus Enigma, the purpose of this book is not to make specific policy recommendations, but rather to describe broad approaches to developing policy for addressing critical global environmental challenges. Central to such approaches, we have seen that the perspective of ecology has a lot to contribute. There are two aspects to the role of ecology. First, as a natural science, the study of ecosystems can help in their preservation – not necessarily by valuing Nature in monetary terms, but by understanding it and cherishing it. Second, as explained in this book, the ecologically based, systems-oriented disciplines of ecological economics and industrial ecology offer the type of physically grounded, integrated approaches necessary to address complex socio-environmental challenges. Ecological economics is a paradigm for policy-making that tackles problems of biodiversity loss and climate change, while also addressing social equity issues, from meeting basic human needs to avoiding overconsumption by the wealthy. Industrial ecology – the study of energy and material use in industrial society – provides analytical power and data to address complex environmental challenges; it is grounded in physics and engineering but includes methods such as input–output analysis adapted from mainstream economics. So the tools of ecology – applied to natural systems and societal systems – are key to developing policies for sustaining humanity.

One of the greatest unknowns affecting policy-makers of the future is, of course, the size of the world's future population – and this very much depends on women's fertility rates. The United Nations projects that global population could be between 9.4 billion and 12.7 billion by the end of the century.[20] Others, such as Stein Vollset and colleagues, suggest global population could peak around 2060 and fall to as low as 6.3 billion by century's end.[21]

I am not going to be bold enough to back either of the predictions. The key factor that will determine the future population is fertility rates, which settle at around 1.9 children per woman in United Nations projections but fall to as low as 1.5 children per woman in the Vollset scenarios. If the United Nations is successful in achieving the Sustainable Development Goals of giving women increased access to education and contraception, then a lower fertility rate might be expected. This all hinges on the broader issue of women's rights internationally. We saw

Conclusions

from the nineteenth-century experiences of Eunice Foote – the woman who discovered the greenhouse effect – how women have been disenfranchised from science. But this is just a small example of a wider systemic problem of women's rights being restricted – a challenge that remains in many countries today.

Irrespective of how the future population grows, human society will need to develop more sustainable forms of consumption. We noted back in Chapter 3 the challenge of environmental impacts stemming from consumption being strongly linked to growing affluence. It is wealth as much as population that drives the demand for food calories that threatens biodiversity globally – and generates increasing amounts of solid waste. Reducing excessive human consumption through lifestyle changes provides a counterpoint to reducing environmental burdens through technology. Some impacts of over-consumption might potentially be averted with the development of more circular economies – but these have some physical constraints. Industrial ecologists have developed the analytical tools and methods to assess consumption-based environmental footprints – and the wider study of sustainable consumption is an important component of ecological economics. Ultimately, though, it would take massive cultural and institutional change to reduce the burden of our consumption on the environment while also addressing widening social inequities.

At the highest political levels, the world knows it faces some stark environmental challenges. These are reflected, for example, in the Organisation for Economic Co-operation and Development's *OECD Environmental Outlook to 2050*,[22] which draws upon the findings of the scientific community in presenting a bleak prognosis of a world toiling under water stress, air pollution, climate change, and loss of biodiversity. Substantial political efforts have been made to at least address the great challenge of climate change – for example, with the COP 21 Paris Agreement. Yet even this may be a fragile agreement subject to the whims of political cycles and aggressive posturing of competitive nation-states.

Finding appropriate policy responses to our environmental crisis in a bottom-up world replete with multiple self-interests is arguably the most challenging layer of our Malthus Enigma. Recall how the work of the Carter Administration on *Global 2000* was thrown out the window when the Reagan era began. But these short-term political challenges are possibly only symptomatic of deeper, longer-term challenges. The very nature of the Malthus Enigma – reflected through its technology and scientific dimensions (the first and second layers, respectively), which are evolving and uncertain – makes the political response very difficult. Given that our scientific framing of the Earth's systems is incomplete, then many of our policies and policy frameworks likely are too.

This difficulty is further apparent when we reflect back on some of the tenuous conclusions that Malthus himself arrived at. To be a church minister yet hold the conviction that the English Poor Laws should be abolished must have been internally wrenching for Malthus. To suggest that the poor should not be subsidized, or fed, as a way of preventing excessive population growth is clearly a difficult position to take.

Malthus was not alone in reaching such a conclusion. Jay Forrester systematically wrestled with the human predicament described by the Club of Rome; he formulated his knowledge in the form of the World2 model and secured the funding for his protégé Meadows to develop the World3 model and the *Limits to Growth* study. Testifying to a US congressional committee in the 1970s, Forrester came to a similar conclusion as Malthus that population growth should be held in check through reduced investments in industry and food production.

As I will explain further below, I think Malthus's position on the Poor Laws was wrong – completely wrong. But first, to bring more light to the predicament, we need to wrestle more deeply with the nature of the competition that humans experience. Is our Malthusian struggle the same as Darwinian struggle – subject to the laws of the jungle? Or do we compete in ways that are beneficial to us all, as espoused in the liberal philosophy of Adam Smith? The connections between Smith, Malthus, and Charles Darwin are worth a brief exploration here, as they have lessons for policy.

In Darwin's famous theory of evolution, the continuance of a plant or animal species comes down to survival of the fittest. Darwinian struggle involves competition both between species and within species. The constant pressure for any species to persist in the jungle intensifies the competition between animals of the same species.[23] Those organisms better able to adapt to their environment and compete tend to survive and reproduce more often, in the process of natural selection.

An academic debate has focused on the question of how much Darwin learned from Malthus in developing his concept of struggle. Darwin had read the sixth edition of Malthus's *Essay on the Principle of Population* before penning *On the Origin of Species* in 1859.[24] The consensus is that Darwin was influenced by Malthus's recognition that Nature is governed by a law of struggle,[25] but that he took the idea further and in different directions. Although Malthus noted that savages will war with each other when subject to shortage of food, he was primarily

Conclusions

concerned with the struggle experienced by the human race as a whole.[26] Moreover, Malthus did not view struggle within a species as resulting in progress or evolution.

The distinction between Darwinian and Malthusian struggle becomes further apparent when we interject the philosophy of Adam Smith. Malthus was, to a point, an adherent to Smith's principle of *laissez-faire* economics.[27] To a large extent, he believed that in a just and moral society, all members of society benefited from competition in free markets. In theory, under *laissez-faire* economics, everyone in society supposedly benefits, whereas under Darwinian struggle, only the fittest survive.

The global reality of our collective human existence is that we experience both Darwinian struggle and Smith's invisible hand of the market. We rely on free markets every day – for small things like buying a coffee or a basket of groceries, to larger purchases such as a car or a house. Although there are many ways in which markets are imperfect, they generally work well for most of us. But there is a segment of society that misses out – that cannot afford housing, a bag of groceries, or even a coffee. Those the market does not provide for are left to Darwinian struggle. And it's not just individuals or families that markets fail to provide for; it can be communities and sometimes entire nations.

Malthus's belief in *laissez-faire* economics, but recognition that the population problem undermined it, is important for the policy layer of the Malthus Enigma. This tension is really at the heart of the matter. Our reliance on the liberty of bottom-up approaches – on the free market – can only go so far. Poorly designed, left unchecked, or inappropriately used, the impacts of free markets on the Earth's essential ecosystems can be disastrous for humanity. This was the reality that Hermann von Wissmann, the British, the Germans, and other Europeans first seriously wrestled with at their convention at the UK Foreign Office in 1900, after decades of game hunters slaying African wildlife. Fear of the destructive force of *laissez-faire* is also what drives the ecological economists to oppose the creation of markets for biodiversity. Somehow, we have to come together for the sake of our civilization and create new laws of the jungle.

This is perhaps where the field of ecological economics could be most important. While I am skeptical about the use of markets for birth permits, as suggested in Daly's *Steady-State Economics*, his ideas on enhanced public control on resource markets may be worth exploring. Rather than continuing with society's experiments on markets for pollution, controlling the exploration and exploitation of fossil fuels and

Malthus Enigma

other resources may be a better way to achieve similar goals. Continuing to work on understanding the value of ecosystem services – and knowing when natural capital should not be bought and sold – are also important contributions of ecological economics.

Use of market solutions is always complex. One of the most notable political leaders who experienced this tension between *laissez-faire* and top-down control was former British Prime Minister Margaret Thatcher. Over a period of eighteen years, Thatcher and her like-minded successor, John Major, disassembled the UK's system of heavily subsidized and state-owned industries. She destroyed the Keynesian, high-tax, government-controlled economy and unleashed the power of the free market. In doing so, she fought and defeated the British coal miners. Closing the coal mines and moving Britain's energy system further toward natural gas and oil, Jevons' "coal question" was firmly confined to history. Many have said that Thatcher did more to support the scientific study of climate change than any other leader of her generation.[28] She raised environmental issues on the global stage, supported the establishment of the Intergovernmental Panel on Climate Change, and opened the UK's Hadley Centre for Climate Change Research.[29] Had she stayed longer in power, her approach to addressing climate change would, no doubt, have been growth focused and market based.[30]

A further reflection on the need to go beyond *laissez-faire* was given by Alexander King, the OECD's director of scientific affairs and co-founder of the Club of Rome. King does not mince his words; toward the end of his memoirs, he remarks: "Realization that our current exploitation of the planet is a time bomb for our eventual destruction has grown considerably in recent years."[31] With decades of experience of delicate international diplomacy, King's essentially parting words are:

The only hope is to begin the construction of a World Solidarity based on the collective will and solidarity of humanity – no easy task to persuade the many societies, ideologies, cultures and religions of our world to work together. This has to be seen as a matter of enlightened self-interest, a form of collective self-interest. It could only happen if people were convinced of the fatal consequences of laissez faire.[32]

Perhaps it is in the words of King, and his co-conspirator Aurelio Peccei, that those looking for answers to our environmental crises can perhaps find most hope. The underlying philosophy of the Club of Rome has always been to address the *global problématique*. The environmental challenges we face – great as they may be – cannot be separated from the mighty challenges of human development – addressing hunger, shelter, education, and security. All of the issues come together in the collective self-interest. This is where Malthus went wrong in his view that the English Poor Laws should be abolished. To remove assistance to the poor as a way of holding population growth in check is just the wrong way to approach the challenge. We know today that providing sustenance, shelter, sanitation, education, health care, and security are far more likely to create conditions that will lower fertility rates. It is in the poorly resourced, low-income countries where many people are facing a Darwinian struggle for food that we see high fertility rates. Therefore, withholding food does not equate to reduced population growth – quite the opposite. The development challenge the world faces today is magnitudes greater than that of nineteenth-century Britain, but it is only within the context of assisting the developing world that the environmental crisis can be approached.

In *One Hundred Pages for the Future*, Peccei once again stresses the importance of having a "sense of the global whole" to "make a reliable diagnosis of the ills of humanity and find appropriate remedies."[33] After reviewing the many challenges faced by humanity, he reiterates: "To approach fundamental questions ... we must see them in the dynamic context of the evolution of the world; we must consider them as individual parts of a broader problématique."[34] The challenge of global development – with ties to global security, international trade, and environmental concerns – is again prominent. "Only the realization that each society has an interest in the progress and welfare of all other societies will open the way to a better future."[35]

In *Break Through*, Ted Nordhaus and Michael Shellenberger add the important ingredient of a positive vision to an integrated approach for addressing global challenges. One of their early chapters on saving the Amazon rainforest exemplifies Peccei's point about placing environmental concerns in a broader context. They discuss the underlying challenges faced by Brazilian society – extreme inequality, massive debt, violence, and insecurity – which help explain why attempts to save the Amazon rainforest cannot be made through a narrow environmental lens. Moreover, through various examples, Nordhaus and Shellenberger demonstrate how the focus on limits and doomsday scenarios fails to inspire change or resonate with hopeful, progressive, social aspirations. As they put it: "Environmental tales of tragedy begin with Nature in harmony and almost always end in a quasi-authoritarian politics."[36] Emphasis on environmental limits in the absence of a fuller compelling message of human development perpetuates inequality, pessimism, and fear, which can regress toward authoritarian oppression.

In closing, then, we can summarize the key ingredients for humanity to flourish under the Malthus Enigma. Policy leaders need to create positive visions for the future and use the new tools of the *Age of Ecology* to inspire the kinds of human innovation and scientific inquiry that are commensurate with living on Spaceship Earth. To do so, they need to draw upon the *Gift of Apollo* – recognition of the unity and fragility of the Earth. They must become super-ecologists, marrying a sense of the global whole with intimate knowledge of the details, always with an eye on the *global problématique*. This may require a new type of economics, with a re-examination of when and how to use markets sustainably. To avoid a quasi-authoritarian state, we somehow need to maintain the helpful elements of *laissez-faire* while avoiding its fatal consequences. Of course, none of this will be easy.

The Malthus Enigma is ongoing, but each generation produces its engineers, scientists, and policy influencers that rise to the challenge. Never has there been a greater need for the next generation to step up and find an inclusive, visionary approach to address our daunting contemporary struggles.

Notes

Chapter 1: The Gift of Apollo

[1] Dimensions of the Saturn V rocket are given in Appendix A of Bilstein, *Stages to Saturn*, 405.
[2] Bilstein, *Stages to Saturn*, xi.
[3] Bilstein, *Stages to Saturn*, 5.
[4] Converted from 28,200 kilometres/hour, given by Bilstein, *Stages to Saturn*, 5.
[5] Borman, *Countdown*, 203; Bilstein, *Stages to Saturn*, 5, gives 39,400 kilometres/hour, or 24,482 miles/hour.
[6] The mass of the third-stage engine with fuel was 120,800 kilograms; empty it was 10,000 kilograms.
[7] Borman, *Countdown*, 211.
[8] Borman, *Countdown*, 212.
[9] Poole, *Earthrise*, 8; from MacLeish, "A Reflection."
[10] Poole, *Earthrise*, 9; quotes from several sources.
[11] Poole, *Earthrise*, 152 and 171.
[12] Poole, *Earthrise*, 171.
[13] Poole, *Earthrise*, 37–40.
[14] Poole, *Earthrise*, 43.
[15] Poole, *Earthrise*, 3; quotes from Norman Cousins.
[16] Quoted in Poole, *Earthrise*, 182.
[17] Environmentalism, of course, existed long before the 1960s, as traced, for example, in Bashford, *Global Population*, and Robertson, *The Malthusian Moment*, among others.
[18] McCormick, *The Global Environmental Movement*, 65–67.
[19] McCormick, *The Global Environmental Movement*, 68.
[20] Poole, *Earthrise*, 152–153.
[21] The founding of UNESCO and UNEP and the pivotal Stockholm conference are detailed in Macekura, *Of Limits and Growth*.
[22] McCormick, *The Global Environmental Movement*, 80.
[23] Egan, *Barry Commoner and the Science of Survival*, provides a biography of Barry Commoner and his role in influencing the US environmental movement.
[24] McCormick, *The Global Environmental Movement*, 84.
[25] McCormick, *The Global Environmental Movement*, 80.
[26] Ehrlich, *The Population Bomb*, 38, draws particular attention to a drop in the

Malthus Enigma

world's food production in 1966, due to agricultural disasters; the one-year drop is shown in per capita terms in Wik et al., *Global Agricultural Performance*, 3.

[27] Ehrlich, *The Population Bomb*, 22–23.
[28] Ehrlich, *The Population Bomb*, 36.
[29] Ehrlich, *The Population Bomb*, 66.
[30] Mayhew (*Malthus*, 150 and 195) explains how the term *Malthusian* changed from meaning adherent of Malthus in the early 1800s to advocating for birth control in the 1870s and became tied to ecology in the mid-1900s.
[31] Ehrlich, *The Population Bomb*, 74.
[32] Ehrlich, *The Population Bomb*, 78.
[33] *People's Daily Online*, "400 Million Births Prevented by One-Child Policy."
[34] Greenhalgh, "Missile Science, Population Science." See also Ridley, "China's One-Child Policy Was Inspired by Western Greens."
[35] These values are from the draft Technical Summary of the Working Group I Contribution to the *Sixth Assessment Report* (IPCC, *Climate Change 2021*, 44); similar values are reported in the IPCC's *Fifth Assessment Report* (Church et al., "Sea Level Change," 1140).
[36] Hallegatte et al., "Future Flood Losses in Major Coastal Cities."
[37] The 21st Conference of the Parties to the United Nations Framework Convention on Climate Change.
[38] See Feldmann and Levermann, "Collapse of the West Antarctic Ice Sheet After Local Destabilization of the Amundsen Basin."
[39] See Dasgupta, *The Economics of Biodiversity*, for a recent comprehensive review of global biodiversity challenges.
[40] Estimates are from the abridged version of Dasgupta, *The Economics of Biodiversity*, 27, Box 5, based on International Union for Conservation of Nature (IUCN), *IUCN Red List of Threatened Species*; Pimm and Raven, "The State of the World's Biodiversity"; and Ceballos et al., "Vertebrates on the Brink."
[41] Barnosky et al., "Has the Earth's Sixth Mass Extinction Already Arrived?"
[42] Corfee-Morlot et al., "Global Warming in the Public Sphere," provides an excellent description of the complex, circular relationship between rational science and socially constructed policy in the context of climate change.
[43] Malthus's first *Essay on the Principle of Population* was published in 1798.
[44] Boulding, "The Economics of the Coming Spaceship Earth."
[45] Rockström et al., "A Safe Operating Space for Humanity"; Steffen et al., "Planetary Boundaries."
[46] Chenoweth and Feitelson, "Neo-Malthusians and Cornucopians Put to the Test."
[47] Sabin, *The Bet*.
[48] Meadows et al., *Dynamics of Growth in a Finite World*; Nordhaus, "World Dynamics"; Nordhaus et al., "Lethal Model 2."

Notes

[49] Daly, "Georgescu-Roegen Versus Solow/Stiglitz"; Daly, "Reply to Solow/Stiglitz."

[50] Chenoweth and Feitelson, "Neo-Malthusians and Cornucopians Put to the Test."

[51] Apostolides et al., "English Agricultural Output and Labour Productivity, 1250–1850"; Kennedy, "A Biophysical Model of the Industrial Revolution."

[52] Erisman et al., "How a Century of Ammonia Synthesis Changed the World"; Stoltzenberg, *Fritz Haber*.

[53] Vietmeyer, *Our Daily Bread*.

[54] McCormick, *The Global Environmental Movement*; Chertow, "The IPAT Equation and Its Variants."

[55] Tilman and Clark, "Food, Agriculture and the Environment."

[56] National Renewable Energy Laboratory, *Renewable Electricity Futures Study*; Energy Research Institute of the National Development and Reform Commission, *China 2050*; International Energy Agency, *Energy Technology Perspectives*.

[57] Lovins, "Energy Strategy."

[58] Murphy and Hall, "Year in Review."

[59] Diesendorf and Wiedmann, "Implications of Trends in Energy Return on Energy Invested."

[60] Carnot, N. L. S., "Reflections on the Motive Power of Heat"; Cardwell, *From Watt to Clausius*.

[61] Schrödinger, *What Is Life?*; Prigogine et al., "Thermodynamics of Evolution."

[62] Rockström et al., "A Safe Operating Space for Humanity"; Steffen et al., "Planetary Boundaries."

[63] Millennium Ecosystem Assessment, *Ecosystems and Human Well-Being*, gives nine million species as an uncertain estimate; Dasgupta, *The Economics of Biodiversity*, 14, suggests there might be eight to twenty million species or more.

[64] Ortiz and Jackson, "Understanding Eunice Foote's 1856 Experiments."

[65] American Institute for Physics, "The Discovery of Global Warming."

[66] Callendar, "Infra-Red Absorption by Carbon Dioxide."

[67] Keeling, "The Concentration and Isotropic Abundance of Carbon Dioxide."

[68] Edwards, *A Vast Machine*.

[69] Fitter and Scott, *The Penitent Butchers*.

[70] Jevons, *The Coal Question*.

[71] Swingland, *Capturing Carbon and Conserving Biodiversity*.

[72] Odum, E. P., *Fundamentals of Ecology*.

[73] Odum, H. T., *Environment, Power and Society*, 1.

[74] Dickinson and Murphy, *Ecosystems*.

[75] In his review of the social and political context to the post-war Malthusian movement in the United States, Robertson (*The Malthusian Moment*, 15 and

Malthus Enigma

23–25), summarizes Leopold's work and its influence on Osborn and Vogt.
[76] von Bertalanffy and Rapoport, *General Systems*.
[77] Forrester, "The Beginning of Systems Dynamics."
[78] Meadows et al., *The Limits to Growth*.
[79] Pauli, *Crusader for the Future*; King, *Let the Cat Turn Round*.
[80] Edwards, *A Vast Machine*.
[81] Georgescu-Roegen, *The Entropy Law and the Economic Process*.
[82] Daly, *Steady-State Economics*.
[83] Røpke, "The Early History of Modern Ecological Economics" and "Trends in the Development of Ecological Economics," tell the history of ecological economics.
[84] Gorman and Solomon, "The Origins and Practice of Emissions Trading."
[85] Drury et al., "Pollution Trading and Environmental Injustice."
[86] Martínez-Alier and Muradian, "Taking Stock."
[87] Erkman, "Industrial Ecology: An Historical View," gives the early history of industrial ecology.
[88] Ayres, "Industrial Metabolism"; Frosch and Gallopoulos, "Strategies for Manufacturing."
[89] Allwood, "Squaring the Circular Economy"; Cullen, "Circular Economy."

Chapter 2: Malthusians and Cornucopians

[1] Daly, *Steady-State Economics*, 43.
[2] A description of Malthus's life is given in Winch, *Malthus*, 11–15.
[3] Ordway, "Possible Limits of Raw Material Consumption," 991, describes Cornucopians as scientists who believe that new technology and discovery can overcome the population challenge posed by Malthus.
[4] Winch, *Malthus*, 8.
[5] Malthus, *An Essay on the Principle of Population*, 108; Winch, *Malthus*, 24.
[6] These are moving average values for wheat yields per acre, gross of tithe and seeds, from Apostolides et al., "English Agricultural Output and Labour Productivity, 1250–1850," Figure 4.
[7] Author's estimate based on Apostolides et al., "English Agricultural Output and Labour Productivity, 1250–1850," Figure 14.
[8] Mayhew, *Malthus*, describes how Malthus riled many groups in English society, from socialist philosophers to romantic poets.
[9] See Boyer, "English Poor Laws," Table 1.
[10] Winch, *Malthus*, 100; originally from Malthus, *An Essay on the Principle of Population*, 2nd ed., 531.
[11] Winch, *Malthus*, 44; originally from Malthus, *An Essay on the Principle of Population*, 2nd ed., 63.
[12] Winch, *Malthus*, 66–67.
[13] Mayhew, *Malthus*, 202–203, identifies several others who opposed the neo-

Notes

Malthusian perspective in the mid-1900s, including John Maddox, Ben Wattenberg, Earl Parker Hanson, and Ester Boserup.

[14] Smith, *Manual of Political Economy*, 35.

[15] See Lowenthal's introduction in Marsh, *Man and Nature*, x–xii.

[16] Warde, *The Invention of Sustainability*, 351.

[17] Marsh, *Man and Nature*, 3; see also comments by Lowenthal in the introduction to Marsh, *Man and Nature*, xx, and the introduction to Thomas, *Man's Role in Changing the Face of the Earth*, xxix.

[18] See Marsh, *Man and Nature*, Chapter 6, and Lowenthal's introduction to Marsh, *Man and Nature*, xx–xvii.

[19] Thomas, *Man's Role in Changing the Face of the Earth*.

[20] This brief discussion of Osborn and Vogt is based on Robertson, *The Malthusian Moment*; the life of Vogt is covered more fully in Mann, *The Wizard and the Prophet*.

[21] *Time*, "Eat Hearty," 27.

[22] Robertson, *The Malthusian Moment*, 37.

[23] Background on Julian Simon is given in Regis, "The Doomslayer."

[24] Regis, "The Doomslayer."

[25] Data from the Food and Agriculture Organization is given in Simon, "Resources, Population, Environment," Table 2.

[26] Simon, "Resources, Population, Environment," 1432.

[27] Simon, "Resources, Population, Environment," 1435.

[28] Ehrlich, "Environmental Disruption," 13.

[29] Sabin, *The Bet*, further describes the wager between Paul Ehrlich and Julian Simon.

[30] Simon, "Environmental Disruption or Environmental Improvement," 39.

[31] Ehrlich, "An Economist in Wonderland," 46.

[32] Regis, "The Doomslayer."

[33] Barney, *Global 2000*, Preface.

[34] Barney, *Improving the Government's Capacity*; referenced in Simon and Kahn, *The Resourceful Earth*, 44.

[35] Barney, *Global 2000*, 1.

[36] Barney, *Global 2000*, i.

[37] Simon and Kahn, *The Resourceful Earth*, 44.

[38] Simon and Kahn, *The Resourceful Earth*, 7.

[39] Simon and Kahn, *The Resourceful Earth*, 2–3.

[40] Simon and Kahn, *The Resourceful Earth*, 6 and 29.

[41] Simon and Kahn, *The Resourceful Earth*, 42.

[42] Simon and Kahn, *The Resourceful Earth*, 45.

[43] The study by Rockström et al., "A Safe Operating Space for Humanity," was updated in Steffen et al., "Planetary Boundaries," and Richardson et al., "Earth Beyond Six of Nine Planetary Boundaries."

[44] Rockström et al., "A Safe Operating Space for Humanity," 31.

Malthus Enigma

[45] Intergovernmental Panel on Climate Change (IPCC), *Climate Change 2014: Synthesis Report*, gives a warming of 0.85°C [0.65°C to 1.06°C] over the period of 1880 to 2012.
[46] Assumes emissions of non-CO_2 greenhouse gases do not significantly increase.
[47] IPCC, *Climate Change 2014: Synthesis Report*, 44.
[48] Amos, "Carbon Dioxide Passes Symbolic Mark."
[49] Steffen et al., "Planetary Boundaries."
[50] Running, "A Measurable Planetary Boundary for the Biosphere," 1458.
[51] Lithotrophic organisms also account for a tiny fraction of primary production.
[52] Several studies have estimated how much of net primary production is appropriated by humans. See, for example, Vitousek et al., "Human Appropriation of the Products of Photosynthesis"; Rojstaczer et al., "Human Appropriation of Photosynthesis Products"; Imhoff et al., "Global Patterns in Human Consumption"; and Haberl et al., "Quantifying and Mapping the Human Appropriation of Net Primary Production."
[53] Petroski, *To Engineer Is Human*.
[54] Ehrlich's role, with Holdren, in developing the IPAT equation is described in Chertow, "The IPAT Equation and Its Variants."
[55] Chenoweth and Feitelson, "Neo-Malthusians and Cornucopians Put to the Test," 54.
[56] Rockström et al., "A Safe Operating Space for Humanity," 474.

Chapter 3: Feeding the World While Saving the Planet

[1] Vietmeyer, *Our Daily Bread*, 59.
[2] Livingstone, *Missionary Travels and Researches in South Africa*, 525.
[3] This part of Livingstone's journey is described in Livingstone, *Missionary Travels and Researches in South Africa*, 526–559.
[4] Livingstone, *Missionary Travels and Researches in South Africa*, 560.
[5] The encounter with the two elephants is described in Livingstone, *Missionary Travels and Researches in South Africa*, 561–563.
[6] Livingstone, *Missionary Travels and Researches in South Africa*, 561.
[7] MacKenzie, *The Empire of Nature*, 94–95.
[8] Livingstone, *Missionary Travels and Researches in South Africa*, 96–100.
[9] Livingstone, *Missionary Travels and Researches in South Africa*, 50–51.
[10] Livingstone, *Missionary Travels and Researches in South Africa*, 148.
[11] Livingstone, *Missionary Travels and Researches in South Africa*, 116
[12] See South African Press Association, "Cape Is World's Extinction Capital."
[13] MacKenzie, *The Empire of Nature*, 202–203.
[14] MacKenzie, *The Empire of Nature*, 202 and 207.
[15] Fitter and Scott, *The Penitent Butchers*, 7.

Notes

[16] By then, the Congo Free State was an independent nation under King Leopold II of Belgium.

[17] MacKenzie, *The Empire of Nature*, 208.

[18] The United Nations General Assembly adopted the Declaration on the Rights of Indigenous Peoples on September 13, 2007. See United Nations Department of Economic and Social Affairs (UN DESA), Social Inclusion Division, "United Nations Declaration on the Rights of Indigenous Peoples."

[19] See, for example, Liboiron, *Pollution Is Colonialism*.

[20] Sexton, "Humans Should Remember They Are Part of Nature, Says Prince Charles."

[21] Lee, "Prince Charles: We Must Learn from Indigenous People on Climate Change."

[22] Fitter and Scott, *The Penitent Butchers*, 8.

[23] Fitter and Scott, *The Penitent Butchers*, 8.

[24] Fitter and Scott, *The Penitent Butchers*, 16.

[25] MacKenzie, *The Empire of Nature*, 162.

[26] MacKenzie, *The Empire of Nature*, 125.

[27] MacKenzie, *The Empire of Nature*, 140.

[28] Humans have begun to transform the Earth system so much that we are now in an era that some scientists call the "Anthropocene." See, for example, Crutzen, "The 'Anthropocene.'"

[29] The International Union for Conservation of Nature was initially called the International Union for the Protection of Nature and also formerly the World Conservation Union. See *Britannica*, "International Union for Conservation of Nature."

[30] Summary statistics of the IUCN's *Red List* are available online. See IUCN, "Summary Statistics."

[31] UN DESA, Population Division, *World Population Prospects 2024: Summary of Results*, 1.

[32] Millennium Ecosystem Assessment, *Ecosystems and Human Well-Being*, iii, Figure A.

[33] UN DESA, Population Division, *World Population Prospects 2019: Highlights*, 1.

[34] UN DESA, Population Division, *World Population Prospects: The 2010 Revision: Highlights and Advance Tables*, 15, Figure 3.

[35] UN DESA, Population Division, *World Population Prospects: The 2010 Revision: Highlights and Advance Tables*, vi.

[36] UN DESA, Population Division, *World Population Prospects: The 2010 Revision: Highlights and Advance Tables*, 11.

[37] UN DESA, Population Division, *World Population Prospects 2019: Highlights*, 9.

[38] UN DESA, Population Division, *World Population Prospects: The 2010 Revision: Highlights and Advance Tables*, xvii.

[39] UN DESA, Population Division, *World Population Prospects: The 2010 Revision: Highlights and Advance Tables*, 2, Table 1.1.
[40] This is the ninety-five percent confidence interval from UN DESA, Population Division, *World Population Prospects 2019: Highlights*, 5.
[41] See Ritchie, "The UN Has Made Population Projections for More Than 50 Years."
[42] Vollset et al., "Fertility, Mortality, Migration, and Population Scenarios," 1285.
[43] UN DESA, Population Division, *World Population Prospects 2019: Highlights*, 9.
[44] Hoornweg et al., "Environment: Waste Production Must Peak This Century."
[45] The scenarios were based on shared socio-economic pathways. For further details, see Hoornweg et al., "Peak Waste."
[46] Wiedmann et al., "Scientists' Warning on Affluence."
[47] That is, unless most people became vegetarian.
[48] Erisman et al., "How a Century of Ammonia Synthesis Changed the World," 637.
[49] Fritz Haber's childhood and years of education are described in Stoltzenberg, *Fritz Haber*, 11–34.
[50] Stoltzenberg, *Fritz Haber*, 77–103.
[51] Stoltzenberg, *Fritz Haber*, 121–153.
[52] Stoltzenberg, *Fritz Haber*, 138.
[53] Stoltzenberg, *Fritz Haber*, 150 and 215.
[54] Smil, *Enriching the Earth*, provides a fuller history of Fritz Haber, Carl Bosch, and the impacts of the industrial synthesis of ammonia in transforming global food supplies.
[55] Erisman et al., "How a Century of Ammonia Synthesis Changed the World," 637.
[56] Erisman et al., "How a Century of Ammonia Synthesis Changed the World," 638.
[57] For Borlaug's early years, see Vietmeyer, *Our Daily Bread*, 1–28.
[58] Bryan, *Rouge*, 79.
[59] Vietmeyer, *Our Daily Bread*, 18.
[60] Vietmeyer, *Our Daily Bread*, 60.
[61] Vietmeyer, *Our Daily Bread*, 154.
[62] Vietmeyer, *Our Daily Bread*, 160.
[63] Vietmeyer, *Our Daily Bread*, 177.
[64] Vietmeyer, *Our Daily Bread*, 171.
[65] Author's calculation based on yields given in Vietmeyer, *Our Daily Bread*, 181 and 182.
[66] Vietmeyer, *Our Daily Bread*, 234.
[67] Ehrlich, *The Population Bomb*.
[68] Vietmeyer, *Our Daily Bread*, 241.

Notes

[69] Goldewijk and Ramankutty, "Land Cover Change Over the Last Three Centuries."

[70] This is a rough estimate based on Goldewijk and Ramankutty, "Land Cover Change Over the Last Three Centuries," Figure 1.

[71] See Millennium Ecosystem Assessment, *Ecosystems and Human Well-Being*, 19, Box 1.1.

[72] Leopold, *A Sand County Almanac*, 47.

[73] Millennium Ecosystem Assessment, *Ecosystems and Human Well-Being*, 18.

[74] Mora et al., "How Many Species Are There?," estimates there are 8.7 million species on land and sea (+/− 1.3 million). This is an estimate of eukaryotes, which includes all animals, plants, fungi, protozoa, and chromista. Dasgupta, *The Economics of Biodiversity*, 14, suggests there might be eight to twenty million species or more with cells containing a distinct nucleus housing genetic material.

[75] Barnosky et al., "Has the Earth's Sixth Mass Extinction Already Arrived?," 52, indicates that approximately 1.9 million species have been named.

[76] Simon and Wildavsky, "Species Loss Revisited."

[77] Millennium Ecosystem Assessment, *Ecosystems and Human Well-Being*, iv.

[78] Millennium Ecosystem Assessment, *Ecosystems and Human Well-Being*, 2.

[79] Millennium Ecosystem Assessment, *Ecosystems and Human Well-Being*, 3.

[80] Millennium Ecosystem Assessment, *Ecosystems and Human Well-Being*, 4.

[81] Millennium Ecosystem Assessment, *Ecosystems and Human Well-Being*, 45.

[82] Millennium Ecosystem Assessment, *Ecosystems and Human Well-Being*, 19, Box 1.1.

[83] Millennium Ecosystem Assessment, *Ecosystems and Human Well-Being*, 31–37.

[84] Lenzen et al., "International Trade Drives Biodiversity Threats."

[85] Lenzen et al., "International Trade Drives Biodiversity Threats," 110.

[86] Lenzen et al., "International Trade Drives Biodiversity Threats," Table S3.3.

[87] Novacek, *The Biodiversity Crisis*; quoted in Barnosky et al., "Has the Earth's Sixth Mass Extinction Already Arrived?"

[88] Barnosky et al., "Has the Earth's Sixth Mass Extinction Already Arrived?," 51.

[89] Barnosky et al., "Has the Earth's Sixth Mass Extinction Already Arrived?," 51.

[90] Barnosky et al., "Has the Earth's Sixth Mass Extinction Already Arrived?," 56.

[91] Tilman et al., "Global Food Demand."

[92] Tilman et al., "Global Food Demand," also predicted a 110 percent increase in the demand for crop protein.

[93] Note that Tilman et al., "Global Food Demand," indicates that population growth to 2050 would be thirty percent. The discrepancy may be attributable to changes in the United Nations forecasted population in 2050, which has been

Malthus Enigma

rising, and the population in the base year. Tilman et al.'s absolute estimates of food demand and impacts in 2050 would be greater if the calculations were repeated today.

[94] Tilman and Clark, "Food, Agriculture and the Environment," 10.
[95] Tilman and Clark, "Food, Agriculture and the Environment," 11, Figure 1.
[96] Calculations are from Tilman and Clark, "Food, Agriculture and the Environment," 10.
[97] Tilman and Clark, "Food, Agriculture and the Environment," 12.
[98] Tilman and Clark, "Food, Agriculture and the Environment," 12.
[99] These are values for 2010 from IPCC, *Climate Change 2014: Synthesis Report*.
[100] Tilman and Clark, "Food, Agriculture and the Environment," 13.
[101] The *OECD-FAO Agricultural Outlook 2019–2028* projected that the area of global agricultural land would remain stable to 2028 due to intensification of agriculture.
[102] Badgley et al., "Organic Agriculture and the Global Food Supply."
[103] Connor, "Organic Agriculture Cannot Feed the World."
[104] Seufert et al., "Comparing the Yields of Organic and Conventional Agriculture."
[105] Smil, *Feeding the World*; Smil, *Enriching the Earth*.
[106] There are modelling studies that address the struggle of feeding the world within planetary boundaries, such as Gerten et al., "Feeding Ten Billion People Is Possible," and Springer and Duchin, "Feeding Nine Billion People Sustainably," discussed in Chapter 8.
[107] UN DESA, Population Division, *World Population Prospects: The 2017 Revision: Key Findings and Advanced Tables*, 32, Table S5, gives the following fertility rates (births per mother) for 2015–2020: Africa, 4.43; Europe, 1.62; North America, 1.86.
[108] UN DESA, Population Division, *World Population Prospects: The 2017 Revision: Key Findings and Advanced Tables*, 1.
[109] UN DESA, Population Division, *World Population Prospects: The 2017 Revision: Key Findings and Advanced Tables*, 1.
[110] Auerbach, *Organic Food Systems*, 11.
[111] Muthee et al., "The Role of Indigenous Knowledge Systems."
[112] Sambo, "Endangered, Neglected, Indigenous Resilient Crops."
[113] Ellis-Jones and Tengberg, "The Impact of Indigenous Soil and Water Conservation Practices."
[114] Richards, *Indigenous Agricultural Revolution*.
[115] Rajasekaran et al., "Indigenous Natural-Resource Management Systems."
[116] DeWalt, "Using Indigenous Knowledge to Improve Agriculture and Natural Resource Management."
[117] Brookfield and Padoch, "Appreciating Agrodiversity," 7.
[118] Petrini, *Slow Food Revolution*.

Notes

[119] Seufert et al., "Comparing the Yields of Organic and Conventional Agriculture." See also Biello, "Will Organic Food Fail to Feed the World?"
[120] Borlaug, "Dr. Norman Borlaug – Organic Farming."
[121] Altieri, *Agroecology*.
[122] Kleppel, *The Emergent Agriculture*, xxv.
[123] The Intergovernmental Science-Policy Platform on Biodiversity and Ecosystem Services published its first *Global Assessment Report* in 2019.
[124] Pereira et al., "Scenarios for Global Biodiversity in the 21st Century."

Chapter 4: One Big Greenhouse

[1] *Time*, "One Big Greenhouse," 61.
[2] Foote, "Circumstances Affecting the Heat of the Sun's Rays"; see also Ortiz and Jackson, "Understanding Eunice Foote's 1856 Experiments."
[3] Foote also reported results with the cylinders placed in the shade.
[4] See Ortiz and Jackson, "Understanding Eunice Foote's 1856 Experiments," endnote 23.
[5] Foote, "Circumstances Affecting the Heat of the Sun's Rays," 383.
[6] Foote, "Circumstances Affecting the Heat of the Sun's Rays," 383.
[7] Shapiro, "Eunice Newton Foote's Early Forgotten Discovery," provides background on Foote's life and work.
[8] Sorenson, "Eunice Foote's Pioneering Work on CO_2 and Climate Warming."
[9] Sorenson, "Eunice Foote's Pioneering Work on CO_2 and Climate Warming."
[10] McNeill, "This Suffrage-Supporting Scientist Defined the Greenhouse Effect"; Shapiro, "Eunice Newton Foote's Early Forgotten Discovery."
[11] *New-York Daily Tribune*, August 26, 1856, p. 7; quoted in Shapiro, "Eunice Newton Foote's Early Forgotten Discovery."
[12] See Ortiz and Jackson, "Understanding Eunice Foote's 1856 Experiments," endnote 9.
[13] Foote, "Circumstances Affecting the Heat of the Sun's Rays."
[14] *Scientific American*, "Scientific Ladies.--Experiments with Condensed Gases."
[15] Quoted in Shapiro, "Eunice Newton Foote's Early Forgotten Discovery."
[16] Foote, "On a New Source of Electrical Excitation."
[17] See Ortiz and Jackson, "Understanding Eunice Foote's 1856 Experiments," endnote 15.
[18] See Ortiz and Jackson, "Understanding Eunice Foote's 1856 Experiments," endnote 5.
[19] Tyndall, "The Bakerian Lecture."
[20] Shapiro, "Eunice Newton Foote's Early Forgotten Discovery."
[21] Wells, *Annual of Scientific Discovery*.
[22] Ortiz and Jackson, "Understanding Eunice Foote's 1856 Experiments."
[23] Shapiro, "Eunice Newton Foote's Early Forgotten Discovery."

Malthus Enigma

[24] *Time*, "One Big Greenhouse," 61.
[25] For the transformation of energy supplies during the Industrial Revolution, see Wrigley, *Energy and the English Industrial Revolution*, and Kennedy, "The Energy Embodied in the First and Second Industrial Revolutions."
[26] Dickinson, *James Watt*, 32.
[27] Watt formalized the measurement of horsepower as a means of charging for his rotative engines. See Dickinson, *James Watt*, 144.
[28] Jevons, *The Coal Question*, 145.
[29] Dickinson, *James Watt*, 90.
[30] The quotation is from the *Bath Chronicle*, March 21, 1776, which gives the source as the *Saturday Post*, Country News, Birmingham, March 11, 1776. See Styles, *Eighteenth Century English Provincial Newspapers*.
[31] Sieferle, *The Subterranean Forest*, 127.
[32] Sieferle, *The Subterranean Forest*, 133.
[33] This virtuous cycle was brilliantly described by Rolf Peter Sieferle in his paper "Europe's Special Course," presented at the 2004 Gordon Research Conference on Industrial Ecology in Oxford, UK.
[34] Iron production also became independent of water power, as coal-powered steam engines could be used.
[35] Sieferle, *The Subterranean Forest*, 118.
[36] Sieferle, *The Subterranean Forest*, 131–132.
[37] Murphy and Hall, "Year in Review," 102.
[38] Murphy and Hall, "Year in Review," 109, Table 2.
[39] Murphy and Hall, "Year in Review," 109, Table 2.
[40] Hall and Cleveland, "Petroleum Drilling and Production in the United States."
[41] Hall and Cleveland, "Petroleum Drilling and Production in the United States," notes that a further 0.4 barrel equivalents per foot were used in the refining process and gives further energy costs related to non-drilling activities.
[42] Murphy and Hall, "Year in Review," 101.
[43] Murphy and Hall, "Year in Review," 108 and 109.
[44] Murphy and Hall, "Year in Review," 114, estimates that the minimum viable EROI for an energy source is about three, because additional energy is required to maintain the metabolism of society beyond that reflected in the EROI calculation.
[45] Murphy and Hall, "Year in Review," 109, Table 2.
[46] Hubbert, "Nuclear Energy and the Fossil Fuels."
[47] Domm, "US Oil Production Tops 10 Million Barrels a Day."
[48] Jevons, *The Coal Question*, 33.
[49] Jevons, *The Coal Question*, 320.
[50] Jevons, *The Coal Question*, 34–35.
[51] Conybeare and Phillips, quoted in Jevons, *The Coal Question*, 37.
[52] Jevons, *The Coal Question*, 207.

Notes

[53] Jevons, *The Coal Question*, 9.
[54] Jevons, *The Coal Question*, 194.
[55] Jevons, *The Coal Question*, 194.
[56] Jevons, *The Coal Question*, 199–200.
[57] Jevons, *The Coal Question*, 145.
[58] Sir William Armstrong, quoted in Jevons, *The Coal Question*, 198.
[59] Jevons, *The Coal Question*, 8.
[60] Jevons, *The Coal Question*, 201.
[61] Jevons, quoted in Alcott, "Jevons' Paradox," 13.
[62] Jevons, quoted in Alcott, "Jevons' Paradox," 13.
[63] Khazzoom, "Economic Implications of Mandated Efficiency."
[64] Brookes, "Energy Policy, the Energy Price Fallacy and the Role of Nuclear Energy"; Brookes, "Energy Efficiency Fallacies Revisited."
[65] See Madlener and Alcott, "Energy Rebound and Economic Growth," and Sorrel, "Jevons' Paradox Revisited."
[66] Sorrel, "Jevons' Paradox Revisited," 1457.
[67] Sorrel, "Jevons' Paradox Revisited," 1456.
[68] Sorrel, "Jevons' Paradox Revisited," 1456.
[69] Sorrel, "Jevons' Paradox Revisited," 1466.
[70] Gillingham et al., "Energy Policy: The Rebound Effect is Overplayed."
[71] Weart, *The Discovery of Global Warming*, provides a comprehensive review of the discovery of global climate change; Corfee-Morlot et al., "Global Warming in the Public Sphere," provides a useful summary.
[72] *Time*, "Eat Hearty," 61.
[73] International Energy Agency, *CO_2 Emissions from Fuel Combustion*, 8.
[74] The American Institute for Physics (AIP) cites several studies indicating that about ninety-seven percent of climate scientists accepted that climate change was anthropogenically induced. See AIP, "The Discovery of Global Warming," note 58.
[75] Arrhenius, *Worlds in the Making*, 53.
[76] This brief description of Arrhenius's life and work draws from Coffey, *Cathedrals of Science*.
[77] Coffey, *Cathedrals of Science*, 3.
[78] Coffey, *Cathedrals of Science*, 28.
[79] Coffey, *Cathedrals of Science*, 26.
[80] Coffey, *Cathedrals of Science*, 30.
[81] Coffey, *Cathedrals of Science*, 31.
[82] AIP, "The Discovery of Global Warming," 3.
[83] Arrhenius, *Worlds in the Making*, 54; AIP, "The Discovery of Global Warming," 4.
[84] The infrared spectrum of a gas shows the frequencies of light waves that are absorbed by it; the specific frequencies depend on the molecular structure of the gas.

[85] Plass, "Carbon Dioxide and Climate," 58.
[86] Callendar's early life is described in Fleming, *The Callendar Effect*, 5–15.
[87] Ekholm, "On the Variations of the Climate."
[88] Fleming, *The Callendar Effect*, 65.
[89] Fleming, *The Callendar Effect*, 71–72.
[90] Callendar, "Infra-Red Absorption by Carbon Dioxide."
[91] Sutherland and Callendar, "The Infra-Red Spectra of Atmospheric Gases."
[92] Fleming, *The Callendar Effect*, 72.
[93] Callendar's work on the FIDO system is described in Fleming, *The Callendar Effect*, 49–60.
[94] Fleming, *The Callendar Effect*, 59.
[95] Keeling's life is briefly described in Harris, "Charles David Keeling"; see also Keeling, "Rewards and Penalties of Monitoring the Earth."
[96] Harris, "Charles David Keeling," 7868.
[97] Keeling, "The Concentration and Isotropic Abundance of Carbon Dioxide."
[98] Charles F. Kennel, quoted in Harris, "Charles David Keeling," 7865.
[99] This section is based on a short biography of Strong at mauricestrong.net. See Strong Foundation, "Short Biography."
[100] Foster, "The Man Who Shaped the Climate Agenda in Paris."
[101] The 21st Session of the Conference of the Parties of the United Nations Framework Convention on Climate Change.
[102] Author's calculations based on data from International Energy Agency statistics.
[103] Clarke et al., "Energy Systems."
[104] Lovins, *The Essential Amory Lovins*, 41.
[105] Lovins, *The Essential Amory Lovins*, 66.
[106] National Renewable Energy Laboratory, *Renewable Electricity Futures Study*, iii.
[107] Energy Research Institute of the National Development and Reform Commission, *China 2050*.
[108] Schroeter, "The Sun is Shining on PV in Saxony."
[109] Schroeter, "The Sun is Shining on PV in Saxony," 59.
[110] Battisti and Corrado, "Evaluation of Technical Improvements."
[111] Kubiszewski et al., "Meta-Analysis of Net Energy Return."
[112] Diesendorf and Wiedmann, "Implications of Trends in Energy Return on Energy Invested."
[113] Bronski et al., *The Economics of Grid Defection*.
[114] Kennedy et al., "Keeping Global Climate Change Within 1.5 C."
[115] Williams et al., *Pathways to Deep Decarbonization in the United States*.
[116] Georgescu-Roegen, "Energetic Dogma, Energetic Economics, and Viable Technology."

Notes

Chapter 5: The Science of Sustainability

[1] Carnot, N. L. S., "Reflections on the Motive Power of Heat," 37.
[2] Carnot, H., "The Life of N. L. Sadi Carnot," 20.
[3] Carnot, H., "The Life of N. L. Sadi Carnot," 22.
[4] Thurston, R. H. "The Work of N. L. Sadi Carnot."
[5] Carnot, N. L. S., "Reflections on the Motive Power of Heat," 37.
[6] Carnot, N. L. S., "Reflections on the Motive Power of Heat," 38.
[7] Carnot, N. L. S., "Reflections on the Motive Power of Heat," 68.
[8] Goldstein and Goldstein, *The Refrigerator and the Universe*, 29.
[9] Goldstein and Goldstein, *The Refrigerator and the Universe*, 32.
[10] Goldstein and Goldstein, *The Refrigerator and the Universe*, 29.
[11] Goldstein and Goldstein, *The Refrigerator and the Universe*, 30.
[12] Cardwell, *From Watt to Clausius*, 27.
[13] Cardwell, *From Watt to Clausius*, 40.
[14] Cardwell, *From Watt to Clausius*, 40.
[15] Cardwell, *From Watt to Clausius*, 45.
[16] The letters are reproduced in Watt and Black, *Partners in Science*.
[17] Brown, *Count Rumford, Physicist Extraordinary*, 17–20.
[18] Brown, *Count Rumford, Physicist Extraordinary*, 24.
[19] Brown, *Count Rumford, Physicist Extraordinary*, 35.
[20] Brown, *Count Rumford, Physicist Extraordinary*, 39.
[21] Brown, *Count Rumford, Physicist Extraordinary*, 60.
[22] Brown, *Count Rumford, Physicist Extraordinary*, 52.
[23] Thompson, "An Inquiry Concerning the Source of the Heat Which Is Excited by Friction."
[24] Lenard, *Great Men of Science*, 172.
[25] Joule measured the work done per unit of heat to be 772 foot-pounds per British thermal unit – which is close to the modern value. See Goldstein and Goldstein, *The Refrigerator and the Universe*, 46–47.
[26] Goldstein and Goldstein, *The Refrigerator and the Universe*, 55.
[27] Cardwell, *From Watt to Clausius*, 243.
[28] A Watt's indicator diagram shows the relationship between pressure and volume during a cycle of a steam engine. Use of the diagram was a trade secret until published in the *Quarterly Journal of Science* in 1822. See Miller, "The Mysterious Case of James Watt's '1785 Steam Indicator.'"
[29] Thomson, "On the Dissipation of Energy," 315.
[30] Thompson, *The Life of William Thomson*, 133.
[31] Cardwell, *From Watt to Clausius*, 241.
[32] Cardwell, *From Watt to Clausius*, 244.
[33] Rankine also independently arrived at conclusions similar to those of Clausius. See Cardwell, *From Watt to Clausius*, 254.
[34] Cardwell, *From Watt to Clausius*, 249.
[35] Cardwell, *From Watt to Clausius*, 259.

Malthus Enigma

[36] Cardwell, *From Watt to Clausius*, 270.
[37] Cardwell, *From Watt to Clausius*, 267–268.
[38] Cardwell, *From Watt to Clausius*, 273.
[39] Cardwell, *From Watt to Clausius*, 207.
[40] Cardwell, *From Watt to Clausius*, 281.
[41] Hubbert, "Nuclear Energy and the Fossil Fuels."
[42] Carnahan et al., "Efficient Use of Energy," iii.
[43] Kleidon, "Empowering the Earth System by Technology."
[44] Kleidon, "A Basic Introduction to the Thermodynamics of the Earth System," 1303.
[45] For more on Schrödinger's life, see Moore, *Schrödinger: Life and Thought*.
[46] Schrödinger, *What Is Life?*, 47.
[47] Schrödinger, *What Is Life?*, 48.
[48] Prigogine, *Étude thermodynamique des phénomènes irréversibles*.
[49] Prigogine et al., "Thermodynamics of Evolution."
[50] Watson, "Gaia."
[51] Watson, "Gaia."
[52] Kleidon, "A Basic Introduction to the Thermodynamics of the Earth System"; Kleidon, "How Does the Earth System Generate and Maintain Thermodynamic Disequilibrium?"
[53] Running, "A Measurable Planetary Boundary for the Biosphere," 1458.
[54] Kleidon, "Beyond Gaia."
[55] Kleidon, "Testing the Effect of Life on Earth's Functioning," quoted in Kleidon, "Beyond Gaia," 272.
[56] Kleidon, "How Does the Earth System Generate and Maintain Thermodynamic Disequilibrium?," 1029, Figure 4.
[57] Kleidon, "How Does the Earth System Generate and Maintain Thermodynamic Disequilibrium?," 1030.
[58] Running, "A Measurable Planetary Boundary for the Biosphere."
[59] Further concerns about changes to net primary production (NPP) are expressed in Doughty et al., "Changing NPP Consumption Patterns in the Holocene," and Running, "Global Aridification and the Decline of NPP."
[60] Rockström et al., "A Safe Operating Space for Humanity."
[61] See Kleidon, "Empowering the Earth System by Technology," on framing a sustainable future in the context of the thermodynamics of the Earth system.

Chapter 6: The Global Problématique

[1] Murawiec and di Paoli, "Club of Rome Founder Alexander King," 21. Note that the *Executive Intelligence Review* Special Report on the Club of Rome is generally a suspect source, as it contains errors and potentially malicious rumours about the club. The report does, however, include what appears to be a genuine interview with co-founder Alexander King – the quotes from the

Notes

interview that are used in this chapter are consistent with other writings on the history of the Club of Rome. King, *Let the Cat Turn Round*, 392, describes how the *Executive Intelligence Review* made malicious attacks on the Club of Rome.

[2] Murawiec and di Paoli, "Club of Rome Founder Alexander King," 21.

[3] The six were Peccei, King, Jantsch, Thiemann, Jean Saint-Geours, and Max Kohnstamm. See King, *Let the Cat Turn Round*, 298.

[4] Meadows et al., *The Limits to Growth*.

[5] "The Predicament of Mankind" was the name of the Club of Rome's project. See Club of Rome, *The Predicament of Mankind*, or the title page of Meadows et al., *The Limits to Growth*. Summaries of the debate include Hecox, "Limits to Growth Revisited"; Bardi, *The Limits to Growth Revisited*; and Hayes, "Computation and the Human Predicament."

[6] Masini, *The Legacy of Aurelio Peccei*, 2.

[7] Pauli, *Crusader for the Future*, 17.

[8] Pauli, *Crusader for the Future*, 27.

[9] Pauli, *Crusader for the Future*, 55.

[10] A fuller list of Peccei's responsibilities is given in Pauli, *Crusader for the Future*, 87.

[11] The Salzburg meeting was attended by the heads of state of Senegal, Mexico, Canada, the Netherlands, Sweden, Ireland, Algeria, Switzerland, and Austria. See Pauli, *Crusader for the Future*, 83.

[12] The paper is reproduced in Pauli, *Crusader for the Future*, Annex 2, and in Malasak and Vapaavuori, *The Club of Rome*.

[13] Pauli, *Crusader for the Future*, 115.

[14] Pauli, *Crusader for the Future*, 105.

[15] Pauli, *Crusader for the Future*, 117.

[16] Brabyn, "Cool Catalyst, Profile of Alexander King," 390.

[17] Murawiec and di Paoli, "Club of Rome Founder Alexander King," 19.

[18] The United Nations meetings were held at UNESCO and the Commission on Science and Technology for Development. See Murawiec and di Paoli, "Club of Rome Founder Alexander King," 20.

[19] See Murawiec and di Paoli, "Club of Rome Founder Alexander King," 21.

[20] King, "The Launch of a Club," 53.

[21] Murawiec and di Paoli, "Club of Rome Founder Alexander King," 21; King, "The Launch of a Club," 53.

[22] For current members, see www.clubofrome.org/members.

[23] Wikipedia, "Club of Rome."

[24] Murawiec and di Paoli, "Club of Rome Founder Alexander King," 22.

[25] Club of Rome, *The Predicament of Mankind*.

[26] Pauli, *Crusader for the Future*, 75.

[27] The project team is listed in Meadows et al., *The Limits to Growth*, 6.

[28] Meadows et al., *The Limits to Growth*; Hayes, "Computation and the Human Predicament."

[29] Meadows et al., *The Limits to Growth*, 24.
[30] Meadows et al., *The Limits to Growth*, 24.
[31] The term comes from a critique of *The Limits to Growth* in Freeman, "Malthus with a Computer."
[32] Pauli, *Crusader for the Future*, 76.
[33] Pauli, *Crusader for the Future*, 78.
[34] Masini, *The Legacy of Aurelio Peccei*, 9.
[35] Pauli, *Crusader for the Future*, 78.
[36] Pauli, *Crusader for the Future*, 79.
[37] Lewis, "To Grow and To Die," 29.
[38] Reinhold, "Mankind Warned of Perils in Growth," 1.
[39] Reinhold, "Mankind Warned of Perils in Growth," 40.
[40] Hecox, "Limits to Growth Revisited," 72–73.
[41] *Nature*, "Another Whiff of Doomsday," 47.
[42] *Nature*, "Another Whiff of Doomsday," 47.
[43] *Nature*, "Another Whiff of Doomsday," 49.
[44] Passell et al., "The Limits to Growth," 1.
[45] Forrester, "The Beginning of Systems Dynamics," 1–4.
[46] Forrester, "The Beginning of Systems Dynamics," 5.
[47] Forrester, "The Beginning of Systems Dynamics," 5.
[48] The beer game is described in Dizikes, "The Many Careers of Jay Forrester," 5–6.
[49] Forrester, "The Beginning of Systems Dynamics," 7.
[50] The sketch is reproduced in Lane and Sterman, "Jay Wright Forrester."
[51] Forrester, "The Beginning of Systems Dynamics," 12.
[52] Hayes, "Computation and the Human Predicament," 190.
[53] King, *Let the Cat Turn Round*, 332; Pauli, *Crusader for the Future*, 75.
[54] Club of Rome, *The Predicament of Mankind*, 31.
[55] King, *Let the Cat Turn Round*, 333.
[56] Murawiec and di Paoli, "Club of Rome Founder Alexander King," 22; also see King, *Let the Cat Turn Round*, 335–336.
[57] Murawiec and di Paoli, "Club of Rome Founder Alexander King," 22.
[58] Murawiec and di Paoli, "Club of Rome Founder Alexander King," 25.
[59] Bardi, *The Limits to Growth Revisited*, 51–54.
[60] King, *Let the Cat Turn Round*, 335.
[61] Mesarović and Pestel, *Mankind at the Turning Point*.
[62] Meadows et al., *The Limits to Growth*, 56–59. Values given here are for years of supply of known reserves at then current usage rates. Lower values are given by Meadows et al. for reserves at growing usage rates.
[63] Meadows et al., *The Limits to Growth*, 72.
[64] King, *Let the Cat Turn Round*, 339.
[65] Bardi, *The Limits to Growth Revisited*, 54.
[66] Meadows et al., *Dynamics of Growth in a Finite World*.

Notes

[67] Nordhaus et al., "Lethal Model 2," 15.
[68] Simmons, "Revisiting *The Limits to Growth*."
[69] Meadows et al., *Limits to Growth: The 30-Year Update*; Turner, "A Comparison of *The Limits to Growth* with 30 Years of Reality"; Turner, "On the Cusp of Global Collapse?"; Bardi, *The Limits to Growth Revisited*; Lang, "Quantitatively Assessing the Role of Higher Education"; Dixson-Declève et al., *Earth for All*.
[70] Lang, "Quantitatively Assessing the Role of Higher Education," Table 4.1, provides a recent comparison based on data up to 2010.
[71] For more on the history of the D-Day landings, see Beevor, *D-Day*.
[72] Edwards, *A Vast Machine*, 117.
[73] Smagorinsky, "The Beginnings of Numerical Weather Prediction," 6.
[74] Edwards, *A Vast Machine*, 112–113.
[75] Edwards, *A Vast Machine*, 118.
[76] Edwards, *A Vast Machine*, 129.
[77] Edwards, *A Vast Machine*, 141.
[78] Phillips, "The General Circulation of the Atmosphere."
[79] Edwards, *A Vast Machine*, 143.
[80] Phillips, "The General Circulation of the Atmosphere."
[81] Edwards, *A Vast Machine*, 146.
[82] Edwards, "A Brief History of Atmospheric General Circulation Modeling."
[83] Edwards, *A Vast Machine*, 172.
[84] Peterson et al., "The Myth of the 1970s Global Cooling Scientific Consensus."
[85] National Research Council, *Carbon Dioxide and Climate*.
[86] Edwards, *A Vast Machine*, 376.
[87] Shabecoff, "Global Warming Has Begun," 1.
[88] Besel, "Accommodating Climate Change Science"; Weart, *The Discovery of Global Warming*; Corfee-Morlot et al., "Global Warming in the Public Sphere"; O'Donnell, "Of Loaded Dice and Heated Arguments"; Pielke, "Policy History of the US Global Change Research Program."
[89] Edwards, *A Vast Machine*, xvi.
[90] Edwards, *A Vast Machine*, 140.
[91] Odum, E. P. "The Emergence of Ecology," 1291.
[92] Barrett, "Eugene Pleasants Odum," 7.
[93] Barrett, "Eugene Pleasants Odum," 4.
[94] Odum, H. T., and E. P. Odum, "Trophic Structure and Productivity of a Windward Coral Reef Community."
[95] Odum, E. P., "The Strategy of Ecosystem Development."
[96] Odum, E. P., and Smalley, "Comparison of Population Energy Flow."
[97] Odum, E. P., *Ecology and our Endangered Life-Support Systems*.
[98] Odum, E. P., *Ecological Vignettes*.
[99] Hall, *Maximum Power*, x.

249

[100] Brown, "Prof. Howard T. Odum 1924–2002."
[101] Odum, H. T., "Energetics of Food Production."
[102] Hagen, *An Entangled Bank*; quoted in Hall, *Maximum Power*, 2.
[103] Odum, H. T., and E. P. Odum, "Trophic Structure and Productivity of a Windward Coral Reef Community."
[104] Odum, H. T., "Trophic Structure and Productivity of Silver Springs, Florida."
[105] Brown, "Prof. Howard T. Odum 1924–2002," 294.
[106] See von Bertalanffy and Rapoport, *General Systems*, and Boulding, "General Systems Theory."
[107] IIASA's integrated assessment framework is described in IIASA, "Integrated Assessment."
[108] King, *Let the Cat Turn Round*, 303.
[109] King, *Let the Cat Turn Round*, 348–349.

Chapter 7: Ecological Economics

[1] Boulding, "The Economics of the Coming Spaceship Earth," 2.
[2] Iglesias, "The Miscommunications and Misunderstandings of Nicholas Georgescu-Roegen," 21.
[3] Pearson formalized techniques of statistical inference that are used to fit curves to data, introducing concepts such as standard deviation, the regression coefficient, and the chi-square test.
[4] Creative destruction describes the tendency for new wealth to grow from the annihilation of old wealth under capitalism.
[5] Iglesias, "The Miscommunications and Misunderstandings of Nicholas Georgescu-Roegen," 24.
[6] Iglesias, "The Miscommunications and Misunderstandings of Nicholas Georgescu-Roegen," 11.
[7] Georgescu-Roegen, *The Entropy Law and the Economic Process*, 281.
[8] Georgescu-Roegen, *The Entropy Law and the Economic Process*, 232.
[9] Georgescu-Roegen, *The Entropy Law and the Economic Process*, 242 and 255.
[10] Levallois, "Can De-Growth Be Considered a Policy Option?"
[11] Georgescu-Roegen, *The Entropy Law and the Economic Process*, 20.
[12] Georgescu-Roegen, *The Entropy Law and the Economic Process*, 20.
[13] Georgescu-Roegen, *The Entropy Law and the Economic Process*, 21.
[14] Ayres, "Sustainability Economics"; Ayres, "The Second Law, the Fourth Law."
[15] Georgescu-Roegen, "Energy and Economic Myths."
[16] Solow, "The Economics of Resources," 11; also see Daly, "Georgescu-Roegen Versus Solow/Stiglitz," 261.
[17] Land and resources were added in some later variants of the model. Useful

Notes

energy was added to the model in Ayres and Voudouris, "The Economic Growth Enigma."

[18] Daly, "Georgescu-Roegen Versus Solow/Stiglitz," 261.
[19] Daly, "Georgescu-Roegen Versus Solow/Stiglitz," 265; Daly, "Reply to Solow/Stiglitz," 273.
[20] Solow, "Georgescu-Roegen Versus Solow/Stiglitz," 268.
[21] Bristow and Kennedy, "Why Do Cities Grow?"
[22] Kennedy, "Biophysical Economic Interpretation of the Great Depression."
[23] Kennedy, "Energy and Capital."
[24] Iglesias, "The Miscommunications and Misunderstandings of Nicholas Georgescu-Roegen," 10.
[25] Syll, "Nicholas Georgescu-Roegen and the Nobel Prize"; also see Røpke, "Trends in the Development of Ecological Economics."
[26] Cleveland and Ruth, "When, Where, and by How Much?," 204.
[27] Iglesias, "The Miscommunications and Misunderstandings of Nicholas Georgescu-Roegen," 19.
[28] Iglesias, "The Miscommunications and Misunderstandings of Nicholas Georgescu-Roegen," 25.
[29] Iglesias, "The Miscommunications and Misunderstandings of Nicholas Georgescu-Roegen," 25–26.
[30] Iglesias, "The Miscommunications and Misunderstandings of Nicholas Georgescu-Roegen," 26.
[31] Cleveland and Ruth, "When, Where, and by How Much?," 205.
[32] Cleveland and Ruth, "When, Where, and by How Much?"
[33] For further influences on the field, including forerunners, see Martínez-Alier and Muradian, "Taking Stock," 2–8.
[34] See René Passet's 1979 image of the human economy and society embodied in the biosphere – for example, in Martínez-Alier and Muradian, "Taking Stock," 2.
[35] Mayumi, "Thermodynamics," 95.
[36] Spash, "The Content, Direction and Philosophy of Ecological Economics," 43.
[37] Spash, "Social Ecological Economics."
[38] Røpke, "Sustainable Consumption"; Spash, "The Content, Direction and Philosophy of Ecological Economics"; Spash and Dobernig, "Theories of (Un)sustainable Consumption"; Wiedmann et al., "Scientists' Warning on Affluence"; Oswald et al., "Large Inequality in International and Intra-National Energy Footprints."
[39] Daly, *Steady-State Economics*, 153–157.
[40] Daly, *Steady-State Economics*, 12.
[41] Daly, *Steady-State Economics*, 11.
[42] Malthus, *Principles of Political Economy*, 227; quoted in Daly, *Steady-State Economics*, 11.

[43] Daly, *Steady-State Economics*, 17.
[44] Daly, *Steady-State Economics*, 16–17.
[45] Daly, *Steady-State Economics*, 56.
[46] Daly, *Steady-State Economics*, 59.
[47] Daly, *Steady-State Economics*, 60.
[48] Daly, *Steady-State Economics*, 69.
[49] Daly, *Steady-State Economics*, 54.
[50] See, for example, Victor, *Managing Without Growth*, and Jackson, *Prosperity Without Growth*.
[51] Røpke, "The Early History of Modern Ecological Economics"; Røpke, "Trends in the Development of Ecological Economics." For a briefer history of ecological economics, see Martínez-Alier and Muradian, "Taking Stock."
[52] Røpke, "The Early History of Modern Ecological Economics," 302.
[53] In particular, self-organizing, dissipative structures. See Prigogine, "Time, Structure and Fluctuations."
[54] They shared related writings by the Ukrainian philosopher Sergei Podolinsky.
[55] Hardin, "The Tragedy of the Commons."
[56] Kennedy, *The Evolution of Great World Cities*.
[57] Kennedy, *The Evolution of Great World Cities*, Chapter 2.
[58] Georgescu-Roegen, *The Entropy Law and the Economic Process*, 282.
[59] Georgescu-Roegen, *The Entropy Law and the Economic Process*, 282.
[60] Daly and Townsend, *Valuing the Earth*.
[61] Costanza et al., "Twenty Years of Ecosystem Services," 2.
[62] Costanza et al., "The Value of the World's Ecosystem Services."
[63] Costanza et al., "Twenty Years of Ecosystem Services," 3.
[64] Costanza et al., "Twenty Years of Ecosystem Services," 3.
[65] Costanza et al., "Twenty Years of Ecosystem Services," 10–14.
[66] See Nabuurs et al., "Agriculture, Forestry and Other Land Uses (AFOLU)." "REDD" stands for "Reducing emissions from deforestation and forest degradation in developing countries."
[67] Costanza et al., "Twenty Years of Ecosystem Services," 13.
[68] Bintliff, "Going to Market in Antiquity."
[69] Daly, *Beyond Growth*, 222–224; noted in Drury et al., "Pollution Trading and Environmental Injustice," 233, note 4.
[70] See, for example, Spash, "The Content, Direction and Philosophy of Ecological Economics."
[71] Coase, "The Federal Communications Commission"; Coase, "The Problem of Social Cost."
[72] Coase, "The Federal Communications Commission," 29; see also Calel, "Carbon Markets," 108.
[73] Crocker, "The Structuring of Atmospheric Pollution Control Systems"; Dales, *Pollution, Property and Prices*.

Notes

[74] Gorman and Solomon, "The Origins and Practice of Emissions Trading," 293.

[75] Gorman and Solomon, "The Origins and Practice of Emissions Trading," 299.

[76] An airshed is a bit like a watershed, though not as rigorously defined. It is a part of the atmosphere, influenced by topographical land features (hills and valleys) through which air pollution spreads in a coherent way. Hence, it is a geographic management unit for air-quality standards.

[77] Gorman and Solomon, "The Origins and Practice of Emissions Trading," 306.

[78] Tietenberg, *Emissions Trading, An Exercise in Reforming Pollution Policy*.

[79] Stavins, *Project 88*.

[80] Gorman and Solomon, "The Origins and Practice of Emissions Trading," 308; see also Shabecoff, "Bush Tells Environmentalists He'll Listen to Them."

[81] Davidson, "Photochemical Oxidant Air Pollution."

[82] Shprentz, *Breath-Taking*; with calculations from Drury et al., "Pollution Trading and Environmental Injustice," 243.

[83] The South Coast Air Quality Management District's audit program is referenced in Drury et al., "Pollution Trading and Environmental Injustice," 265.

[84] Drury et al., "Pollution Trading and Environmental Injustice," 236 and note 54.

[85] The South Coast Air Quality Management District was established in 1976 as the agency responsible for regulating air pollution from stationary sources within the South Coast Air Basin.

[86] Drury et al., "Pollution Trading and Environmental Injustice," 247.

[87] Drury et al., "Pollution Trading and Environmental Injustice," 248.

[88] Drury et al., "Pollution Trading and Environmental Injustice," 251.

[89] Drury et al., "Pollution Trading and Environmental Injustice," 259.

[90] Drury et al., "Pollution Trading and Environmental Injustice," 259.

[91] Schmalensee and Stavins, "Lessons Learned from Three Decades of Experience," 7.

[92] Drury et al., "Pollution Trading and Environmental Injustice," 268.

[93] Drury et al., "Pollution Trading and Environmental Injustice," 269.

[94] Drury et al., "Pollution Trading and Environmental Injustice," 279.

[95] Schmalensee and Stavins, "Lessons Learned from Three Decades of Experience."

[96] Fowlie et al., "What Do Emissions Markets Deliver and to Whom?"

[97] Fowlie et al., "What Do Emissions Markets Deliver and to Whom?"

[98] Schmalensee and Stavins, "Lessons Learned from Three Decades of Experience."

[99] Calel, "Carbon Markets."

[100] Calel, "Carbon Markets," 110.

[101] Intergovernmental Panel on Climate Change, *Climate Change 1995*, 401; from Calel, "Carbon Markets," 110.
[102] Smith and Swierzbinski, "Assessing the Performance of the UK Emissions Trading Scheme."
[103] An archived list of the Chicago Climate Exchange membership is available at https://web.archive.org/web/20100202014432/http://www.chicagoclimatex.com/content.jsf?id=64.
[104] Cameron, "Richard Sandor."
[105] Muûls et al., "Evaluating the EU Emissions Trading System," 1.
[106] Muûls et al., "Evaluating the EU Emissions Trading System," 1.
[107] Calel, "Carbon Markets," 112.
[108] Schmalensee and Stavins, "Lessons Learned from Three Decades of Experience," 14.
[109] Muûls et al., "Evaluating the EU Emissions Trading System," 5.
[110] Schmalensee and Stavins, "Lessons Learned from Three Decades of Experience," 14.
[111] Schmalensee and Stavins, "Lessons Learned from Three Decades of Experience," 14.
[112] Erbach, "Post-2020 Reform of the EU Emissions Trading System."
[113] Corporate Europe Observatory, "EU Emissions Trading."
[114] Erbach, "Post-2020 Reform of the EU Emissions Trading System," 4.
[115] Muûls et al., "Evaluating the EU Emissions Trading System."
[116] See, for example, Spash, "The Content, Direction and Philosophy of Ecological Economics," and Kosoy and Corbera, "Payments for Ecosystem Services."
[117] Swingland, *Capturing Carbon and Conserving Biodiversity*.
[118] Swingland, *Capturing Carbon and Conserving Biodiversity*, 2.
[119] UNEP Finance Initiative, *Demystifying Materiality*; Sukhdev et al., *The Economics of Ecosystems and Biodiversity*.
[120] Swingland, *Capturing Carbon and Conserving Biodiversity*, 11.
[121] Calel, "Carbon Markets," 115.
[122] Schmalensee and Stavins, "Lessons Learned from Three Decades of Experience," 19.
[123] Bonnie et al., "Protecting Terrestrial Ecosystems and the Climate," 318.
[124] Martínez-Alier and Muradian, "Taking Stock," 9.
[125] Spash, "Bulldozing Biodiversity."
[126] Spash, "Bulldozing Biodiversity," 542; from Clark, "Profit Maximization and Extinction of Animal Species."
[127] Spash, "Bulldozing Biodiversity," 544.
[128] Spash, "Bulldozing Biodiversity," 548.
[129] Spash, "Bulldozing Biodiversity," 541.
[130] Spash, "Bulldozing Biodiversity," 542.

Notes

Chapter 8: Industrial Ecology

[1] Ehrenfeld and Gertler, "Industrial Ecology in Practice."
[2] Ehrenfeld and Gertler, "Industrial Ecology in Practice," 72, Table 2.
[3] Wastes avoided per year were as follows: 200,000 tonnes of ash, 80,000 tonnes of scrubber sludge, 2,800 tonnes of hydrogen sulphide, one million cubic metres of wastewater sludge, 1,500 to 2,500 tonnes of sulphur dioxide, and 130,000 tonnes of carbon dioxide; from Ehrenfeld and Gertler, "Industrial Ecology in Practice," 72, Table 2.
[4] Chertow, "'Uncovering' Industrial Symbiosis."
[5] International Synergies, "Projects: NISP®."
[6] Dr. Shi Lei, associate professor at the Tsinghua University School of Environment, personal communication with the author; further details are in Kennedy et al., "Infrastructure for China's Ecologically Balanced Civilization." Also see Ghisellini et al., "A Review on Circular Economy," for a further review of circular economy initiatives in China.
[7] See, for example, International Society for Industrial Ecology (ISIE), *Rising to Global Challenges*.
[8] Other prominent academics included Jesse Ausubel, an environmental scholar from The Rockefeller University; Robert Socolow, a physicist from Princeton University; David Allen, an engineer now with the University of Texas at Austin; and Faye Duchin, an economist from Rensselaer Polytechnic Institute. Fitting with the desire to grow a new society, younger professors were also in attendance: Scott Matthews, from Carnegie Mellon University, and Arpad Horvath, from the University of California, Berkeley. Along with René Kleijn, two others made the long trip from Europe: Helge Brattebø, who headed the industrial ecology program at the Norwegian University of Science and Technology, and Stefan Bringezu, from the influential German Wuppertal Institute for Climate, Environment and Energy. Tadatomo Suga of the University of Tokyo travelled all the way from Japan. Representing industry were Brad Allenby of AT&T, Robert Pfahl of Motorola, and James Fava of Five Winds International. Completing the party was an enthusiastic group of policy experts: Makarand "Mak" Dehejia, a former vice-president with the World Bank Group; David Rejeski, from the United States Environmental Protection Agency; Emily Matthews, from the World Resources Institute; and Richard Podolsky, from the New York Academy of Sciences.
[9] Ernest Lowe, a scientist from RPP International, also offered a few written comments in absentia.
[10] ConAccount was a network under the title Coordination of Regional and National Material Flow Accounting for Environmental Sustainability. See Bringezu et al., *Concerted Action: Coordination of Regional and National Material Flow Accounting for Environmental Sustainability (ConAccount)*.
[11] Personal communication from Reid Lifset; a Yale University communication

is given at https://news.yale.edu/2001/02/08/industry-meets-environment-new-international-society-yale (accessed December 12, 2024).
[12] See Fischer-Kowalski, "Society's Metabolism Part I," and Fischer-Kowalski and Hüttler, "Society's Metabolism Part II," for the development of societal metabolism.
[13] Kennedy et al., "The Changing Metabolism of Cities," defines urban metabolism, but the idea comes from Wolman, "The Metabolism of Cities."
[14] Renner, "Geography of Industrial Localization."
[15] Barnard, "Education for Management."
[16] Spilhaus, "The Next Industrial Revolution," 324.
[17] Erkman, "Industrial Ecology: An Historical View"; Salmi and Toppinen, "Embedding Science in Politics"; Zhu et al., "Efforts for a Circular Economy in China."
[18] Erkman, "Industrial Ecology: An Historical View," 3; originally from Billen et al., *L'écosystème Belgique*.
[19] Erkman, "Industrial Ecology: An Historical View," 4; originally from Billen et al., *L'écosystème Belgique*.
[20] Erkman, "Industrial Ecology: An Historical View," 4.
[21] Erkman, "Industrial Ecology: An Historical View," 4.
[22] Erkman, "Industrial Ecology: An Historical View," 4.
[23] Erkman, "Industrial Ecology: An Historical View," 4.
[24] Further discussion on industrial ecology in Japan is given in Richards and Fullerton, *Industrial Ecology*.
[25] Erkman, "Industrial Ecology: An Historical View," 5.
[26] See NASA, "Robert A. Frosch."
[27] Frosch's move to UNEP is discussed in a 1981 interview with David DeVorkin. See Frosch, "Robert Frosch – Session IV."
[28] Ausubel and Sladovich, *Technology and Environment*, vi.
[29] Ausubel and Sladovich, *Technology and Environment*, vi.
[30] Papers from the colloquium are summarized in Jelinski et al., "Industrial Ecology: Concepts and Approaches."
[31] Ausubel, "Industrial Ecology: Reflections on a Colloquium," 880.
[32] Fifty scientists attended the conference in Snowmass. For proceedings, see Socolow et al., *Industrial Ecology and Global Change*.
[33] Erkman, "Industrial Ecology: An Historical View," 6.
[34] Tibbs, "Industrial Ecology: An Environmental Agenda for Industry."
[35] Erkman, "Industrial Ecology: An Historical View," 6.
[36] Tibbs, "Industrial Ecology: An Environmental Agenda for Industry," 167.
[37] Marstrander et al., "Teaching Industrial Ecology to Graduate Students."
[38] Marstrander et al., "Teaching Industrial Ecology to Graduate Students," 119.
[39] Finlayson et al., "Postsecondary Education in Industrial Ecology Across the World."
[40] Ayres and Kneese, "Production, Consumption, and Externalities."

Notes

[41] Another example was Ayres' study retrospectively estimating emissions in the Hudson River Valley over a hundred-year period using data on industrial activity. See Ayres et al., *An Historical Reconstruction of Major Pollutant Levels in the Hudson-Raritan Basin*.

[42] Ayres' work on industrial metabolism is published in several papers. One of the earliest versions is the paper Ayres presented at the Woods Hole workshop (Ayres, "Industrial Metabolism"); here, however, I draw upon a later working paper, Ayres, "Industrial Metabolism: Work in Progress."

[43] Ayres, "Industrial Metabolism: Work in Progress," 2.

[44] Ayres, "Industrial Metabolism: Work in Progress," 3.

[45] Ayres, "Industrial Metabolism: Work in Progress," 6.

[46] Ayres, "Industrial Metabolism: Work in Progress," 6–7.

[47] Ayres, "Industrial Metabolism: Work in Progress," 9.

[48] Ayres, "Industrial Metabolism: Work in Progress," 5.

[49] Ayres, "Industrial Metabolism: Work in Progress," 11.

[50] Ayres, "Industrial Metabolism: Work in Progress," 11.

[51] Leontief, *The Structure of the American Economy*; Leontief, *Studies in the Structure of the American Economy*.

[52] Meade, "The U.S. Benchmark IO Table."

[53] Leontief, "Environmental Repercussions and the Economic Structure."

[54] Leontief and Duchin, *Military Spending: Facts and Figures*; Leontief and Duchin, *The Future Impact of Automation on Workers*; Leontief et al., "New Approaches in Economic Analysis."

[55] Duchin, "Industrial Input-Output Analysis."

[56] Springer and Duchin, "Feeding Nine Billion People Sustainably."

[57] For example publications, see www.cmu.edu/gdi/publications/index.html (accessed May 14, 2022).

[58] See, for example, Peters, "From Production-Based to Consumption-Based National Emission Inventories"; Hertwich and Peters, "Carbon Footprint of Nations"; and Ivanova et al., "Environmental Impact Assessment of Household Consumption."

[59] See Tukker and Dietzenbacher, "Global Multiregional Input–Output Frameworks"; Lenzen et al., "The Global MRIO Lab"; and Lenzen et al., "The Challenges and Opportunities of Constructing Input–Output Frameworks in a Virtual Laboratory."

[60] Nakamura won the ISIE's Society Prize for his work on waste input–output models. See, for example, Nakamura and Kondo, "Input-Output Analysis of Waste Management."

[61] World Science Forum, "Launching the UNEP International Panel for Sustainable Resource Management."

[62] Adriaanse et al., *Resource Flows*.

[63] Matthews et al., *The Weight of Nations*.

[64] The first guide was Eurostat, *Economy-Wide Material Flow Accounts and*

Malthus Enigma

Derived Indicators, published in 2001. More recent guidelines are given in Eurostat, *Economy-Wide Material Flow Accounts: Compilation Guide 2013*.
[65] OECD, *Measuring Material Flows and Resource Productivity*.
[66] See, for example, Krausmann et al., "The Global Sociometabolic Transition"; Schandl and West, "Resource Use and Resource Efficiency in the Asia-Pacific Region"; and Schaffartzik et al., "The Global Metabolic Transition."
[67] UNEP, *Decoupling*.
[68] ISIE, *Rising to Global Challenges*, 25.
[69] UNEP, *Decoupling*, xiii.
[70] UNEP, *Decoupling*, 18.
[71] UNEP, *Decoupling*, 1.
[72] UNEP, *Decoupling*, xiii.
[73] UNEP, *Decoupling*, xv.
[74] UNEP, *Decoupling*, 18.
[75] UNEP, *Decoupling*, 30.
[76] UNEP, *Decoupling*, 32.
[77] UNEP, *Decoupling*, 30.
[78] See UNEP, *Decoupling*, 34, Figure 3.1; based on Gallopín, *A Systems Approach to Sustainability*.
[79] UNEP, *Decoupling*, 28.
[80] Pauliuk et al., "Industrial Ecology in Integrated Assessment Models."
[81] The International Resource Panel reports are available at www.resourcepanel.org/reports (accessed June 18, 2018).
[82] UNEP, *Metal Stocks in Society*, 10 and Appendix 1, 31.
[83] UNEP, *Metal Stocks in Society*, Appendix 2, 38.
[84] The beginning of MacArthur's record-breaking journey is described in Chapter 9 of MacArthur, *Full Circle*, 114–121.
[85] Details of the journey are from MacArthur, *Full Circle*, 122–215.
[86] MacArthur, *Full Circle*, 198.
[87] MacArthur, *Full Circle*, 137.
[88] MacArthur, *Full Circle*, 137.
[89] MacArthur, *Full Circle*, 317.
[90] MacArthur, *Full Circle*, 318.
[91] MacArthur, *Full Circle*, 362.
[92] MacArthur, *Full Circle*, 364.
[93] Bocken et al., "Taking the Circularity to the Next Level."
[94] Boulding, "The Economics of the Coming Spaceship Earth."
[95] Commoner, *The Closing Circle*.
[96] Frosch, "Industrial Ecology: A Philosophical Introduction."
[97] Pearce and Turner, *Economics of Natural Resources and the Environment*.
[98] Ghisellini, et al., "A Review on Circular Economy."
[99] China Council for International Cooperation on Environment and

Notes

Development, "Circular Economy Promotion Law of the People's Republic of China."

[100] Moreau et al., "Coming Full Circle," 498, notes that the Ellen MacArthur Foundation's definition was used in European Commission, *Closing the Loop*.

[101] Ellen MacArthur Foundation, "Concept: What Is a Circular Economy?"; from Moreau et al., "Coming Full Circle," 498.

[102] Haas et al., "How Circular Is the Global Economy?"

[103] Haas et al., "How Circular Is the Global Economy?," 770, Figure 2.

[104] Haas et al., "How Circular Is the Global Economy?," Table S1.

[105] Allwood, "Squaring the Circular Economy," 446.

[106] Allwood, "Squaring the Circular Economy," 446.

[107] Allwood, "Squaring the Circular Economy," 446.

[108] Allwood, "Squaring the Circular Economy," 446.

[109] Cullen, "Circular Economy."

[110] Zink and Geyer, "Circular Economy Rebound."

[111] A generally accepted definition of industrial ecology is given by Robert White, 1994, president of the National Academy of Engineering: "Industrial ecology is the study of the flows of materials and energy in industrial and consumer activities, of the effects of these flows on the environment, and of the influences of economic, political, regulatory, and social factors on the flow, use, and transformation of resources." See White, "Preface."

[112] References for the examples given here include Druckman and Jackson, "The Carbon Footprint of UK Households"; Hertwich et al., "Integrated Life-Cycle Assessment of Electricity-Supply Scenarios"; Kennedy and Corfee-Morlot, "Past Performance and Future Needs for Low-Carbon, Climate-Resilient Infrastructure"; Kennedy et al., "Methodology for Inventorying Greenhouse Gas Emissions from Global Cities"; and Müller et al., "Carbon Emissions of Infrastructure Development." Many industrial ecologists have conducted studies related to climate change, leading and contributing to the work of the IPCC. See ISIE, *Rising to Global Challenges*.

[113] Allenby, *Industrial Ecology: Policy Framework and Implementation*.

[114] See OECD, *Sustainable Manufacturing and Eco-Innovation*, 15, Figure 5, Conceptual relationships between sustainable manufacturing and eco-innovation.

[115] ISIE, *Rising to Global Challenges*.

Chapter 9: Conclusions

[1] Nordhaus and Shellenberger, *Break Through*, 151.

[2] See, for example, Weinberg and Hammond, "Limits to the Use of Energy."

[3] Malthus, *Principles of Political Economy*, 227; quoted in Daly, *Steady-State Economics*, 11.

[4] United Nations, "Goal 1: End Poverty in All Its Forms Everywhere."

259

[5] Kennedy et al., "Energy and Material Flows of Megacities."
[6] Relative to 2005; see Tilman et al., "Global Food Demand."
[7] Murphy and Hall, "Year in Review."
[8] Lovins, *Reinventing Fire*.
[9] See, for example, Stewart et al., "The Electric City as a Solution to Sustainable Urban Development."
[10] The mission of the Breakthrough Institute is available at thebreakthrough.org/about (accessed June 5, 2022). For more on its origins, see Kallis and Bliss, "Post-Environmentalism."
[11] The technological perspectives of Nordhaus and Shellenberger have changed over time, but both were big proponents of nuclear energy. See Kallis and Bliss, "Post-Environmentalism."
[12] See, for example, Huesemann and Huesemann, *Techno-Fix*.
[13] Allen, *The British Industrial Revolution in Global Perspective*, Chapter 10, explores the links between scientists and investors during the Industrial Revolution.
[14] Simon and Kahn, *The Resourceful Earth*, 45.
[15] Rockström et al., "A Safe Operating Space for Humanity," 475.
[16] Steffen et al., "Planetary Boundaries"; Richardson et al., "Earth Beyond Six of Nine Planetary Boundaries."
[17] Edwards, *A Vast Machine*, 8.
[18] Rockström et al., "A Safe Operating Space for Humanity"; Steffen et al., "Planetary Boundaries."
[19] Arguably, more progress has been made applying the first law of thermodynamics than the second. See Hammond and Winnett, "The Influence of Thermodynamic Ideas on Ecological Economics," for a critique of thermodynamics in ecological economics. Examples of progress made are given in Bristow and Kennedy, "Why Do Cities Grow?," 212.
[20] United Nations Department of Economic and Social Affairs, Population Division, *World Population Prospects 2019: Highlights*, 5.
[21] Vollset et al., "Fertility, Mortality, Migration, and Population Scenarios."
[22] Organisation for Economic Co-operation and Development, *OECD Environmental Outlook to 2050*.
[23] Bowler, "Malthus, Darwin, and the Concept of Struggle," 634.
[24] Bowler, "Malthus, Darwin, and the Concept of Struggle," 638.
[25] Bowler, "Malthus, Darwin, and the Concept of Struggle," 635.
[26] Bowler, "Malthus, Darwin, and the Concept of Struggle," 638.
[27] Smith's principle of *laissez-faire* economics was borrowed from the French Physiocrats.
[28] In contrast, US President Lyndon B. Johnson had been briefed about global warming as early as 1965 but took no action. See Nuccitelli, "Scientists Warned the President About Global Warming."
[29] See Margaret Thatcher Foundation, "Speech at 2nd World Climate

Conference," and commentaries on her environmental legacy in Vidal, "Margaret Thatcher: An Unlikely Green Hero?," and Booker, "Was Margaret Thatcher the First Climate Sceptic?"

[30] See Montague, "Who Was Responsible for Thatcher's Climate Change U-Turn?"
[31] King, *Let the Cat Turn Round*, 402.
[32] King, *Let the Cat Turn Round*, 402.
[33] Peccei, *One Hundred Pages for the Future*, 131.
[34] Peccei, *One Hundred Pages for the Future*, 130–131.
[35] Peccei, *One Hundred Pages for the Future*, 14.
[36] Nordhaus and Shellenberger, *Break Through*, 131.

Bibliography

Adriaanse, A., S. Bringezu, A. L. Hammond, Y. Moriguchi, E. Rodenburg, D. Rogich, and H. Schütz. *Resource Flows: The Material Basis of Industrial Economies.* Washington, DC: World Resources Institute, 1997.

American Institute for Physics (AIP). "The Discovery of Global Warming." Accessed August 9, 2016. https://history.aip.org/climate/index.htm.

Alcott, B. "Jevons' Paradox." *Ecological Economics* 54, no. 1 (2005): 9–21.

Allen, R. C. *The British Industrial Revolution in Global Perspective.* Cambridge University Press, 2009.

Allenby, B. R. *Industrial Ecology: Policy Framework and Implementation.* New York: Prentice Hall, 1999.

Allwood, J. M. "Squaring the Circular Economy: The Role of Recycling Within a Hierarchy of Material Management Strategies." In *Handbook of Recycling*, edited by E. Worrell and A. Markus, 445–477. Elsevier, 2014.

Altieri, M. A. *Agroecology: The Science of Sustainable Agriculture.* CRC Press, 2018.

Amos, J. "Carbon Dioxide Passes Symbolic Mark." *BBC News*, May 10, 2013.

Apostolides, A., S. Broadberry, B. Campbell, M. Overton, and B. Van Leeuwen. "English Agricultural Output and Labor Productivity, 1250–1850: Some Preliminary Estimates." November 26, 2008. https://web.archive.org/web/20200325030912/https://warwick.ac.uk/fac/soc/economics/staff/sbroadberry/wp/agriclongrun4.pdf.

Arrhenius, S. *Worlds in the Making: The Evolution of the Universe.* Harper & Brothers, 1908.

Auerbach, R. *Organic Food Systems: Meeting the Needs of Southern Africa.* Wallingford, UK: CABI Publishing, 2020.

Ausubel, J. H. "Industrial Ecology: Reflections on a Colloquium." *Proceedings of the National Academy of Sciences of the United States of America* 89, no. 3 (1992): 879–884.

Ausubel, J. H., and A. E. Sladovich, eds. *Technology and Environment*, National Academy of Engineering. Washington, DC: National Academy Press, 1989.

Ayres, R. U., L. W. Ayres, J. A. Tarr, and R. C. Widgery. *An Historical Reconstruction of Major Pollutant Levels in the Hudson-Raritan Basin, 1880–1980.* Vol. 1, *Summary.* Office of Oceanography and Marine

Assessment, National Ocean Service, National Oceanic and Atmospheric Administration, US Department of Commerce, 1988.
Ayres, R. U. "Sustainability Economics: Where Do We Stand?" *Ecological Economics* 67, no. 2 (2008): 281–310.
Ayres, R. U. "The Second Law, the Fourth Law, Recycling and Limits to Growth." *Ecological Economics* 29, no. 3 (1999): 473–483.
Ayres, R. U. "Industrial Metabolism." In *Technology and Environment*, edited by J. H. Ausubel and H. E. Sladovich, 23–49. Washington, DC: National Academy Press, 1989.
Ayres, R. U. "Industrial Metabolism: Work in Progress." INSEAD Working Paper 97/09/EPS. Institut européen d'administration des affaires (INSEAD), February 1997. https://flora.insead.edu/fichiersti_wp/inseadwp1997/97-09.pdf.
Ayres, R. U., and A. V. Kneese. "Production, Consumption, and Externalities." *The American Economic Review* 59, no. 3 (1969): 282–297.
Ayres, R. U., and L. W. Ayres. *Industrial Ecology: Towards Closing the Materials Cycle*. Cheltenham, UK: Edward Elgar, 1996.
Ayres, R. U., and V. Voudouris. "The Economic Growth Enigma: Capital, Labour and Useful Energy?" *Energy Policy* 64 (January 2014): 16–28.
Badgley, C., J. Moghtader, E. Quintero, E. Zakem, M. J. Johnson-Chappell, K. Avilés-Vázquez, A. Samulon, and I. Perfecto. "Organic Agriculture and the Global Food Supply." *Renewable Agriculture and Food Systems* 22, no. 2 (2007): 86–108.
Bardi, U. *The Limits to Growth Revisited*. Springer Briefs in Energy: Energy Analysis, edited by C. A. S. Hall. Springer Science + Business Media, 2011.
Barnard, F. "Education for Management Conceived as a Study of Industrial Ecology." *Journal of Vocational Education & Training* 15, no. 30 (1963): 22–26.
Barney, G. O. *Improving the Government's Capacity to Analyze and Predict Conditions and Trends of Global Population, Resources, and Environment*. Gerald O. Barney and Associates, 1982.
Barney, G. O. *Global 2000: The Report to the President*. 2nd ed. Washington, DC: Seven Locks Press, 1988.
Barnosky, A. D., N. Matzke, S. Tomiya, G. O. U. Wogan, B. Swartz, T. B. Quental, C. Marshall, et al. "Has the Earth's Sixth Mass Extinction Already Arrived?" *Nature* 471, no. 7336 (2011): 51–57.
Barrett, G. W. "Eugene Pleasants Odum 1913–2002." Chap. 10 in *Biographical Memoirs*, Volume 87, 317–332. Washington, DC: The National Academies Press, 2005.
Bashford, A. *Global Population: History, Geopolitics, and Life on Earth*. Columbia University Press, 2014.
Battisti, R., and A. Corrado. "Evaluation of Technical Improvements of Photovoltaic Systems Through Life Cycle Assessment Methodology." *Energy* 30, no. 7 (2005): 952–967.

Bibliography

Beevor, A. *D-Day: The Battle for Normandy*. New York; Toronto: Viking, 2009.

Besel, R. D. "Accommodating Climate Change Science: James Hansen and the Rhetorical/Political Emergence of Global Warming." *Science in Context* 26, no. 1 (2013): 137–152.

Biello, D. "Will Organic Food Fail to Feed the World?" *Scientific American*, April 25, 2012.

Billen, G., F. Toussaint, P. Peters, M. Sapir, A. Steenhout and J.-P. Vanderborght. *L'écosystème Belgique : Essai d'écologie industrielle*. Brussels: Centre de recherche et d'information socio-politiques, 1983.

Bilstein, R. E. *Stages to Saturn: A Technological History of the Apollo/Saturn Launch Vehicles*. NASA History Series. Washington, DC: National Aeronautics and Space Administration, 1980.

Bintliff, J. "Going to Market in Antiquity." In *Zu Wasser und zu Land: Verkehrswege in der antiken Welt. Stuttgarter Kolloquium zur Historischen Geographie des Altertums 7, 1999*, edited by Eckart Olshausen and Holger Sonnabend, 209–250. Stuttgart: Franz Steiner, 2002.

Bocken, N. M., E. A. Olivetti, J. M. Cullen, J. Potting, and R. Lifset. "Taking the Circularity to the Next Level: A Special Issue on the Circular Economy." *Journal of Industrial Ecology* 21, no. 3 (2017): 476–482.

Bonnie, R., M. Carey, and A. Petsonk. "Protecting Terrestrial Ecosystems and the Climate Through a Global Carbon Market." In *Capturing Carbon and Conserving Biodiversity: The Market Approach*, edited by I. Swingland, 309–331. London, UK: Earthscan, 2003.

Booker, C. "Was Margaret Thatcher the First Climate Sceptic?" *The Telegraph*. June 12, 2010. Accessed September 5, 2018. https://www.telegraph.co.uk/comment/columnists/christopherbooker/7823477/Was-Margaret-Thatcher-the-first-climate-sceptic.html.

Borlaug, Norman. "Dr. Norman Borlaug – Organic Farming." Interview by Paul Underwood. Wessels Living History Farm, January 3, 2023. Accessed May 3, 2020. https://www.youtube.com/watch?v=1uKiNqHedBU.

Borman, F. *Countdown: An Autobiography*. New York: Silver Arrow Books, 1988.

Boulding, K. E. "General Systems Theory—The Skeleton of Science." *Management Science* 2, no. 3 (1956): 197–208.

Boulding, K. E. "The Economics of the Coming Spaceship Earth." In *Environmental Quality in a Growing Economy: Essays from the Sixth RFF Forum*, edited by H. Jarrett, 3–14. Baltimore, MD: Resources for the Future / Johns Hopkins University Press, 1966.

Bowler, P. J. "Malthus, Darwin, and the Concept of Struggle." *Journal of the History of Ideas* 37, no. 4 (1976): 631–650.

Boyer, G. "English Poor Laws." *EH.net Encyclopedia*. Accessed December 31, 2012. https://eh.net/encyclopedia/english-poor-laws/.

Brabyn, H. "Cool Catalyst, Profile of Alexander King." *New Scientist*, August 24, 1972.
Bringezu, S., S. Moll, M. Fischer-Kowalski, R. Kleijn, V. Palm, R. U. Ayres, P. Bourdeau, P. H. Brunner, M. Jänicke, and H. U. de Haes. *Concerted Action: Coordination of Regional and National Material Flow Accounting for Environmental Sustainability (ConAccount): Summary Final Report*. ENV4-CT96-0266. December 1997.
Bristow, D., and C. A. Kennedy. "Why Do Cities Grow? Insights from Nonequilibrium Thermodynamics at the Urban and Global Scales." *Journal of Industrial Ecology* 19, no. 2 (2015): 211–221.
Britannica. "International Union for Conservation of Nature." Accessed May 6, 2018. https://www.britannica.com/topic/International-Union-for-Conservation-of-Nature.
Bronski, P., J. Creyts, L. Guccione, M. Madrazo, J. Mandel, B. Rader, and H. Tocco. *The Economics of Grid Defection: When and Where Distributed Solar Generation Plus Storage Competes with Traditional Utility Service*. Rocky Mountain Institute, 2014.
Brookes, L. "Energy Efficiency Fallacies Revisited." *Energy Policy* 28, no. 6–7 (2000): 355–366.
Brookes, L. "Energy Policy, the Energy Price Fallacy and the Role of Nuclear Energy in the UK." *Energy Policy* 6, no. 2 (1978): 94–106.
Brookfield, H., and C. Padoch. "Appreciating Agrodiversity: A Look at the Dynamism and Diversity of Indigenous Farming Practices." *Environment: Science and Policy for Sustainable Development* 36, no. 5 (1994): 6–45.
Brown, M. T. "Prof. Howard T. Odum 1924–2002." *Energy* 28, no. 4 (2003): 293–301.
Brown, S. C. *Count Rumford, Physicist Extraordinary*. Science Study Series. Garden City, NY: Anchor Books, 1962.
Bryan, F. R. *Rouge: Pictured in Its Prime*. Detroit, MI: Wayne State University Press, 2003.
Calel, R. "Carbon Markets: A Historical Overview." *Wiley Interdisciplinary Reviews: Climate Change* 4, no. 2 (2013): 107–119.
Callendar, G. S. "Infra-Red Absorption by Carbon Dioxide with Special Reference to Atmospheric Radiation." *Quarterly Journal of the Royal Meteorological Society* 67, no. 291 (1941): 263–275.
Callendar, G. S. "The Artificial Production of Carbon Dioxide and its Influence on Temperature." *Quarterly Journal of the Royal Meteorological Society* 64, no. 275 (1938): 223–240.
Cameron, James. "Richard Sandor." *Time*. Heroes of the Environment. October 17, 2007.
Cardwell, D. S. L. *From Watt to Clausius: The Rise of Thermodynamics in the Early Industrial Age*. London: Heinemann, 1971.
Carnahan, W., K. W. Ford, A. Prosperetti, G. I. Rochlin, A. H. Rosenfeld, M. H. Ross, J. E. Rothberg, G. M. Seidel, and R. H. Socolow. "Efficient Use

Bibliography

of Energy: A Physics Perspective." Study Report 399. American Physical Society, January 1975.

Carnot, H. "The Life of N. L. Sadi Carnot." Chap. 2 in *Reflections on the Motive Power of Heat*, edited by R. H. Thurston. 2nd ed. New York: Wiley, 1897.

Carnot, N. L. S. "Reflections on the Motive Power of Heat and on Machines Fitted to Develop that Power." Chap. 3 in *Reflections on the Motive Power of Heat*, edited by R. H. Thurston. 2nd ed. New York: Wiley, 1897.

Ceballos, G., P. R. Ehrlich, and P. H. Raven. "Vertebrates on the Brink as Indicators of Biological Annihilation and the Sixth Mass Extinction." *Proceedings of the National Academy of Sciences of the United States of America* 117, no. 24 (2020): 13596–13602.

Chenoweth, J., and E. Feitelson. "Neo-Malthusians and Cornucopians Put to the Test: Global 2000 and The Resourceful Earth Revisited." *Futures* 37, no. 1 (2005): 51–72.

Chertow, M. R. "'Uncovering' Industrial Symbiosis." *Journal of Industrial Ecology* 11, no. 1 (2007): 11–30.

Chertow, M. R. "The IPAT Equation and Its Variants." *Journal of Industrial Ecology* 4, no. 4 (2000): 13–29.

China Council for International Cooperation on Environment and Development. "Circular Economy Promotion Law of the People's Republic of China." *Beijing Review*, December 4, 2008.

Church, J. A., P. U. Clark, A. Cazenave, J. M. Gregory, S. Jevrejeva, A. Levermann, M. A. Merrifield, et al. "Sea Level Change." Chap. 13 in *Climate Change 2013: The Physical Science Basis: Contribution of Working Group I to the Fifth Assessment Report of the Intergovernmental Panel on Climate Change*. Cambridge and New York: Cambridge University Press, 2013.

Clapeyron, É. "Memoir on the Motive Power of Heat." *Journal de l'École polytechnique* 14 (1834): 153–190.

Clark, C. W. "Profit Maximization and Extinction of Animal Species." *Journal of Political Economy* 81, no. 4 (1973): 950–961.

Clarke, L., Y.-M. Wei, A. de la Vega Navarro, A. Garg, A. N. Hahmann, S. Khennas, et al. "Energy Systems." Chap. 6 in Intergovernmental Panel on Climate Change (IPCC), *Climate Change 2022: Mitigation of Climate Change. Contribution of Working Group III to the Sixth Assessment Report of the Intergovernmental Panel on Climate Change*. Cambridge University Press, 2022.

Cleveland, C. J., and M. Ruth. "When, Where, and by How Much Do Biophysical Limits Constrain the Economic Process?: A Survey of Nicholas Georgescu-Roegen's Contribution to Ecological Economics." *Ecological Economics* 22, no. 3 (1997): 203–223.

Club of Rome. *The Predicament of Mankind: Quest for Structured Responses to Growing World-Wide Complexities and Uncertainties: A Proposal*. Geneva: The Club of Rome, 1970.

Coase R. H. "The Problem of Social Cost." *Journal of Law and Economics* 3 (October 1960): 1–44.
Coase, R. H. "The Federal Communications Commission," *Journal of Law and Economics* 2 (October 1959): 1–40.
Coffey, P. *Cathedrals of Science: The Personalities and Rivalries That Made Modern Chemistry*. Oxford University Press, 2008.
Commoner, B. *The Closing Circle: Nature, Man, and Technology*. New York: Alfred A. Knopf, 1971.
Connor, D. J. "Organic Agriculture Cannot Feed the World." *Field Crops Research* 106, no. 2 (2008): 187.
Corfee-Morlot, J., M. Maslin, and J. Burgess. "Global Warming in the Public Sphere." *Philosophical Transactions of the Royal Society A: Mathematical, Physical and Engineering Sciences* 365, no. 1860 (2007): 2741–2776.
Corporate Europe Observatory. "EU Emissions Trading: 5 Reasons to Scrap the ETS." October 26, 2015. Accessed August 16, 2018. https://corporateeurope.org/en/environment/2015/10/eu-emissions-trading-5-reasons-scrap-ets.
Costanza, R., R. d'Arge, R. de Groot, S. Farber, M. Grasso, B. Hannon, K. Limburg, et al. "The Value of the World's Ecosystem Services and Natural Capital." *Nature* 387, no. 6630 (1997): 253–260.
Costanza, R., R. de Groot, L. Braat, I. Kubiszewski, L. Fioramonti, P. Sutton, S. Farber, and M. Grasso. "Twenty Years of Ecosystem Services: How Far Have We Come and How Far Do We Still Need To Go?" *Ecosystem Services* 28 (December 2017): 1–16.
Crocker T. D. "The Structuring of Atmospheric Pollution Control Systems." Chap. 5 in *The Economics of Air Pollution*, edited by H. Wolozin. New York: W. W. Norton (1966): 61–86.
Crutzen, P. J. "The 'Anthropocene.'" In *Earth System Science in the Anthropocene*, edited by E. Ehlers and T. Krafft, 13–18. Berlin and Heidelberg: Springer, 2006.
Cullen, J. M. "Circular Economy: Theoretical Benchmark or Perpetual Motion Machine?" *Journal of Industrial Ecology* 21, no. 3 (2017): 483–486.
Daily, G., ed. *Nature's Services: Societal Dependence on Natural Ecosystems*. Island Press, 1997.
Dales J. H. *Pollution, Property and Prices: An Essay in Policy-Making and Economics*. Toronto: University of Toronto Press, 1968.
Daly, H. E. "Georgescu-Roegen Versus Solow/Stiglitz." *Ecological Economics* 22, no. 3 (1997): 261–266.
Daly, H. E. "Reply to Solow/Stiglitz." *Ecological Economics* 22, no. 3 (1997): 271–273.
Daly, H. E. *Beyond Growth: The Economics of Sustainable Development*. Beacon Press, 1997.
Daly, H. E. *Steady-State Economics: The Economics of Biophysical Equilibrium and Moral Growth*. San Francisco: Freeman & Co., 1977.

Bibliography

Daly, H. E., and K. N. Townsend. *Valuing the Earth: Economics, Ecology, Ethics.* MIT Press, 1996.

Dasgupta, P. *The Economics of Biodiversity: The Dasgupta Review.* London: HM Treasury, 2021.

Davidson, A. "Photochemical Oxidant Air Pollution: A Historical Perspective." *Studies in Environmental Science* 72 (1998): 393–405.

DeWalt, B. R. "Using Indigenous Knowledge to Improve Agriculture and Natural Resource Management." *Human Organization* 53, no. 2 (1994): 123–131.

Dickinson, G., and K. Murphy. *Ecosystems: A Functional Approach.* London: Routledge, 1998.

Dickinson, H. W. *James Watt: Craftsman and Engineer.* University of Cambridge Press, 1935.

Diesendorf, M., and T. Wiedmann. "Implications of Trends in Energy Return on Energy Invested (EROI) for Transitioning to Renewable Electricity." *Ecological Economics* 176 (October 2020): 106726.

Dixson-Declève, S., O. Gaffney, J. Ghosh, J. Randers, J. Rockström, and P. E. Stoknes. *Earth for All: A Survival Guide for Humanity.* New Society Publishers, 2022.

Dizikes, P. "The Many Careers of Jay Forrester." *MIT Technology Review.* June 23, 2015.

Domm, P. "US Oil Production Tops 10 Million Barrels a Day for First Time Since 1970." *CNBC*, January 31, 2018. Accessed May 26, 2018. https://www.cnbc.com/2018/01/31/us-oil-production-tops-10-million-barrels-a-day-for-first-time-since-1970.html.

Doughty, C. E., S. Faurby, A. Wolf, Y. Malhi, and J. C. Svenning. "Changing NPP Consumption Patterns in the Holocene: From Megafauna-'Liberated' NPP to 'Ecological Bankruptcy.'" *The Anthropocene Review* 3, no. 3 (2016): 174–187.

Druckman, A., and T. Jackson. "The Carbon Footprint of UK Households 1990–2004: A Socio-Economically Disaggregated, Quasi-Multi-Regional Input–Output Model." *Ecological Economics* 68, no. 7 (2009): 2066–2077.

Drury, R. T., M. E. Belliveau, J. S. Kuhn, and S. Bansal. "Pollution Trading and Environmental Injustice: Los Angeles' Failed Experiment in Air Quality Policy." *Duke Environmental Law & Policy Forum* 9, no. 2 (Spring 1999): 231–290.

Duchin, F. "Industrial Input-Output Analysis: Implications for Industrial Ecology." *Proceedings of the National Academy of Sciences of the United States of America* 89, no. 3 (1992): 851–855.

Edwards, P. N. "A Brief History of Atmospheric General Circulation Modeling." *International Geophysics* 70 (July 2000): 67–90.

Edwards, P. N. *A Vast Machine: Computer Models, Climate Data and the Politics of Global Warming.* MIT Press, 2010.

Egan, M. *Barry Commoner and the Science of Survival: The Remaking of American Environmentalism.* MIT Press, 2009.

Ehrenfeld, J., and N. Gertler. "Industrial Ecology in Practice: The Evolution of Interdependence at Kalundborg." *Journal of Industrial Ecology* 1, no. 1 (1997): 67–79.

Ehrlich, P. R. "An Economist in Wonderland." *Social Science Quarterly* 62, no. 1 (1981): 44–49.

Ehrlich, P. R. "Environmental Disruption: Implications for the Social Sciences." *Social Science Quarterly* 62, no. 1 (1981): 7–22.

Ehrlich, P. R., *The Population Bomb*. New York: Ballantine Books, 1968.

Ekholm, N. "On the Variations of the Climate of the Geological and Historical Past and their Causes." *Quarterly Journal of the Royal Meteorological Society* 27, no. 117 (1901): 1–62.

Ellen MacArthur Foundation. "Concept: What Is a Circular Economy? A Framework for an Economy That Is Restorative and Regenerative by Design." Accessed December 5, 2015. https://web.archive.org/web/20151210052202/www.ellenmacarthurfoundation.org/circular-economy/overview/concept.

Ellis-Jones, J., and A. Tengberg. "The Impact of Indigenous Soil and Water Conservation Practices on Soil Productivity: Examples from Kenya, Tanzania and Uganda." *Land Degradation & Development* 11, no. 1 (2000): 19–36.

Energy Research Institute of the National Development and Reform Commission. *China 2050 High Renewable Energy Penetration Scenario and Roadmap Study*. Energy Foundation China, 2015.

Erbach, Gregor. "Post-2020 Reform of the EU Emissions Trading System." PE 621.902. European Parliamentary Research Service, May 2018. Accessed August 15, 2018. https://www.europarl.europa.eu/RegData/etudes/BRIE/2018/621902/EPRS_BRI(2018)621902_EN.pdf.

Erisman, J. W., M. A. Sutton, J. Galloway, Z. Klimont, and W. Winiwarter. "How a Century of Ammonia Synthesis Changed the World." *Nature Geoscience* 1, no. 10 (2008): 636–639.

Erkman, S. "Industrial Ecology: An Historical View." *Journal of Cleaner Production* 5, no. 1–2 (1997): 1–10.

European Commission. *Closing the Loop—An EU Action Plan for the Circular Economy*. COM(2015) 614/2. Brussels: European Commission, 2015.

Eurostat. *Economy-Wide Material Flow Accounts and Derived Indicators: A Methodological Guide*. Luxembourg: European Statistical Office, 2001.

Eurostat. *Economy-Wide Material Flow Accounts: Compilation Guide 2013*. Luxembourg: European Statistical Office, 2013.

Feldmann, J., and A. Levermann. "Collapse of the West Antarctic Ice Sheet After Local Destabilization of the Amundsen Basin." *Proceedings of the National Academy of Sciences of the United States of America* 112, no. 46 (2015): 14191–14196.

Bibliography

Finlayson, A., K. Markewitz, and J.-M. Frayret. "Postsecondary Education in Industrial Ecology Across the World." *Journal of Industrial Ecology* 18, no. 6 (2014): 931–941.

Fischer-Kowalski, M. "Society's Metabolism: The Intellectual History of Materials Flow Analysis, Part I, 1860–1970." *Journal of Industrial Ecology* 2, no. 1 (1998): 61–78.

Fischer-Kowalski, M., and W. Hüttler. "Society's Metabolism: The Intellectual History of Materials Flow Analysis, Part II, 1970–1998." *Journal of Industrial Ecology* 2, no. 4 (1998): 107–136.

Fitter, R., and P. Scott. *The Penitent Butchers*. London: Collins, 1978.

Fleming, J. R. *The Callendar Effect: The Life and Work of Guy Stewart Callendar (1898–1964), The Scientist Who Established the Carbon Dioxide Theory of Climate Change*. Boston, MA: American Meteorological Society, 2007.

Foote, E. "On a New Source of Electrical Excitation." In *Proceedings of the American Association for the Advancement of Science. Eleventh Meeting. Held at Montreal, Canada East, August, 1857*, edited by Joseph Lovering, 123–126. Cambridge: Joseph Lovering, 1858.

Foote, E. "Circumstances Affecting the Heat of the Sun's Rays." *The American Journal of Science and Arts* 22 (September 1856): 382–383.

Forrester, J. W. "The Beginning of Systems Dynamics." Banquet talk at the International Meeting of the System Dynamics Society, Stuttgart, Germany, July 13, 1989.

Foster, P. "The Man Who Shaped the Climate Agenda in Paris, Maurice Strong, Leaves a Complicated Legacy." *National Post*, November 29, 2015.

Fowlie, M., S. P. Holland, and E. T. Mansur. "What Do Emissions Markets Deliver and to Whom? Evidence from Southern California's NOx Trading Program." *American Economic Review* 102, no. 2 (2012): 965–993.

Freeman, C. "Malthus with a Computer." *Futures* 5, no. 1 (1973): 5–13.

Frosch, R. A. "Robert Frosch – Session IV." Interview by David DeVorkin. *American Institute of Physics, Niels Bohr Library & Archives: Oral History Interviews*, September 15, 1981. Accessed June 3, 2018. https://www.aip.org/history-programs/niels-bohr-library/oral-histories/28066-4.

Frosch, R. A. "Industrial Ecology: A Philosophical Introduction." *Proceedings of the National Academy of Sciences of the United States of America* 89, no. 3 (1992): 800–803.

Frosch, R. A., and N. E. Gallopoulos. "Strategies for Manufacturing." *Scientific American* 261, no. 3 (1989): 144–152.

Gallopín, G. *A Systems Approach to Sustainability and Sustainable Development*. Santiago: United Nations Publication, Sustainable Development and Human Settlements Division, Economic Commission for Latin America, 2003.

Georgescu-Roegen, N. "Energetic Dogma, Energetic Economics, and Viable Technology." In *Advances in the Economics of Energy and Resources*, edited by J. R. Moroney. Greenwich: JAI Press, 1982.

Georgescu-Roegen, N. "Energy and Economic Myths." *Southern Economic Journal* 41, no. 3 (1975): 347–381.

Georgescu-Roegen, N. *The Entropy Law and the Economic Process*. Harvard University, 1971.

Gerten, D., V. Heck, J. Jägermeyr, B. L. Bodirsky, I. Fetzer, M. Jalava, M. Kummu, et al. "Feeding Ten Billion People Is Possible Within Four Terrestrial Planetary Boundaries." *Nature Sustainability* 3, no. 3 (2020): 200–208.

Ghisellini, P., C. Cialani, and S. Ulgiati, S. "A Review on Circular Economy: The Expected Transition to a Balanced Interplay of Environmental and Economic Systems." *Journal of Cleaner Production* 114 (2016): 11–32.

Gillingham, K., M. J. Kotchen, D. S. Rapson, and G. Wagner. "Energy Policy: The Rebound Effect is Overplayed." *Nature* 493, no. 7433 (2013): 475–476.

Goldewijk, K. K., and N. Ramankutty. "Land Cover Change over the Last Three Centuries due to Human Activities: The Availability of New Global Data Sets." *GeoJournal* 61, no. 4 (2004): 335–344.

Goldstein, M and I. F. Goldstein. *The Refrigerator and the Universe: Understanding the Laws of Energy*. Cambridge, MA: Harvard University Press, 1995.

Gorman, H., and B. Solomon. "The Origins and Practice of Emissions Trading." *Journal of Policy History* 14, no. 3 (2002): 293–320.

Greenhalgh, S. "Missile Science, Population Science: The Origins of China's One-Child Policy." *The China Quarterly* 182 (June 2005): 253–276.

Haas, W., F. Krausmann, D. Wiedenhofer, and M. Heinz. "How Circular is the Global Economy? An Assessment of Material Flows, Waste Production, and Recycling in the European Union and the World in 2005." *Journal of Industrial Ecology* 19, no. 5 (2015): 765–777.

Haberl, H., K. H. Erb, F. Krausmann, V. Gaube, A. Bondeau, C. Plutzar, M. Fischer-Kowalski, et al. "Quantifying and Mapping the Human Appropriation of Net Primary Production in Earth's Terrestrial Ecosystems." *Proceedings of the National Academy of Sciences of the United States of America* 104, no. 31 (2007): 12942–12947.

Hagen, J. B. *An Entangled Bank: The Origins of Ecosystem Ecology*. Rutgers University Press, 1992.

Hall, C. A. S., and C. J. Cleveland. "Petroleum Drilling and Production in the United States: Yield per Effort and Net Energy Analysis." *Science* 211 (1981): 576–579.

Hall, C. A. S., ed. *Maximum Power: The Ideas and Applications of H. T. Odum*. Colorado University Press, 1995.

Bibliography

Hallegatte, S., C. Green, R. J. Nicholls, and J. Corfee-Morlot. "Future Flood Losses in Major Coastal Cities." *Nature Climate Change* 3, no. 9 (2013): 802–806.

Hammond, G. P., and A. B. Winnett. "The Influence of Thermodynamic Ideas on Ecological Economics: An Interdisciplinary Critique." *Sustainability* 1 no. 4 (2009): 1195–1225.

Hardin, G. "The Tragedy of the Commons." *Science* 162, no. 3859 (1968): 1243–1248.

Harris, D. C. "Charles David Keeling and the Story of Atmospheric CO_2 Measurements." *Analytical Chemistry* 82, no. 19 (2010): 7865–7870.

Hayes, B. "Computation and the Human Predicament." *American Scientist* 100, no. 3 (2012): 186–191.

Hecox, W. E. "Limits to Growth Revisited: Has the World Modeling Debate Made Any Progress?" *Boston College Environmental Affairs Law Review* 5, no. 1 (1976): 65–96.

Hertwich, E. G., and G. P. Peters. "Carbon Footprint of Nations: A Global, Trade-Linked Analysis." *Environmental Science and Technology* 43, no. 16 (2009): 6414–6420.

Hertwich, E. G., T. Gibon, E. A. Bouman, A. Arvesen, S. Suh, G. A. Heath, L. Shi, et al. "Integrated Life-Cycle Assessment of Electricity-Supply Scenarios Confirms Global Environmental Benefit of Low-Carbon Technologies." *Proceedings of the National Academy of Sciences of the United States of America* 112, no. 20 (2015): 6277–6282.

Hoornweg, D., P. Bhada-Tata, and C. A. Kennedy. "Environment: Waste Production Must Peak This Century." *Nature* 502, no. 7473 (2013): 615–617.

Hoornweg, D., P. Bhada-Tata, and C. A. Kennedy. "Peak Waste: When Is It Likely to Occur?" *Journal of Industrial Ecology* 19, no. 1 (2015): 117–128.

Hubbert, M. K. "Nuclear Energy and the Fossil Fuels." Paper presented at the Spring Meeting of the Southern District Division of Production, American Petroleum Institute, Plaza Hotel, San Antonio, Texas, March 7–9, 1956.

Huesemann, M., and J. Huesemann. *Techno-Fix: Why Technology Won't Save Us or the Environment.* New Society Publishers, 2011.

Iglesias, S. L. "The Miscommunications and Misunderstandings of Nicholas Georgescu-Roegen." Honors thesis, Trinity College of Arts and Sciences, Duke University, 2009.

Imhoff, M. L., L. Bounoua, T. Ricketts, C. Loucks, R. Harriss, and W. T. Lawrence. "Global Patterns in Human Consumption of Net Primary Production." *Nature* 429, no. 6994 (2004): 870–873.

Intergovernmental Panel on Climate Change (IPCC). *Climate Change 2021: The Physical Science Basis. Working Group I Contribution to the Sixth Assessment Report of the Intergovernmental Panel on Climate Change.* Cambridge and New York: Cambridge University Press, 2021.

Intergovernmental Panel on Climate Change (IPCC). *Climate Change 2014: Synthesis Report. Contribution of Working Groups I, II and III to the Fifth Assessment Report of the Intergovernmental Panel on Climate Change.* Geneva, Switzerland: IPCC, 2014.

Intergovernmental Panel on Climate Change (IPCC). *Climate Change 1995: Economic and Social Dimensions of Climate Change*: Contribution of Working Group III to the Second Assessment Report of the Intergovernmental Panel on Climate Change. Cambridge and New York: Cambridge University Press, 1996.

Intergovernmental Science-Policy Platform on Biodiversity and Ecosystem Services (IPBES). *Global Assessment Report on Biodiversity and Ecosystem Services of the Intergovernmental Science-Policy Platform on Biodiversity and Ecosystem Services.* Edited by E. S. Brondizio, J. Settele, S. Díaz, and H. T. Ngo (editors). Bonn, Germany: IPBES Secretariat, 2019. https://doi.org/10.5281/zenodo.3831673.

International Energy Agency (IEA). *CO_2 Emissions from Fuel Combustion 2012.* OECD Publishing, 2012.

International Energy Agency (IEA). *Energy Technology Perspectives 2020.* IEA, 2020.

International Institute for Applied Systems Analysis (IIASA). "Integrated Assessment." https://previous.iiasa.ac.at/web/home/research/researchPrograms/Energy/IAMF.en.html.

International Synergies. "Projects: NISP®." Accessed October 2014. https://international-synergies.com/ourprojects/nisp/.

International Union for Conservation of Nature (IUCN). "Summary Statistics." Accessed May 6, 2018. https://www.iucnredlist.org/resources/summary-statistics.

International Union for Conservation of Nature (IUCN). *IUCN Red List of Threatened Species.* Version 2020-2, 2020. https://www.iucnredlist.org.

International Society for Industrial Ecology (ISIE). *Rising to Global Challenges: 25 Years of Industrial Ecology.* ISIE, 2015.

Ivanova, D., K. Stadler, K. Steen-Olsen, R. Wood, G. Vita, A. Tukker, and E. G. Hertwich. "Environmental Impact Assessment of Household Consumption." *Journal of Industrial Ecology* 20, no. 3 (2016): 526–536.

Jackson, T. *Prosperity Without Growth: Economics for a Finite Planet.* Routledge, 2009.

Jelinski, L. W., T. E. Graedel, R. A. Laudise, D. W. McCall, and C. K. Patel. "Industrial Ecology: Concepts and Approaches." *Proceedings of the National Academy of Sciences of the United States of America* 89, no. 3 (1992): 793–797.

Jevons, W. S. *The Coal Question: An Inquiry Concerning the Progress of the Nation, and the Probable Exhaustion of our Coal-Mines.* London: Macmillan, 1866.

Bibliography

Kallis, G., and S. Bliss. "Post-Environmentalism: Origins and Evolution of a Strange Idea." *Journal of Political Ecology* 26, no. 1 (2019): 466–485.

Keeling, C. D. "Rewards and Penalties of Monitoring the Earth." *Annual Review of Energy and the Environment* 23, no. 1 (1998): 25–82.

Keeling, C. D. "The Concentration and Isotropic Abundance of Carbon Dioxide in the Atmosphere." *Tellus A: Dynamic Meteorology and Oceanography* 12 (1960): 200–203.

Kennedy C. A. "Energy and Capital." *Journal of Industrial Ecology* 24, no. 5 (2020): 1047–1058.

Kennedy, C. A. "Biophysical Economic Interpretation of the Great Depression: A Critical Period of an Energy Transition." *Journal of Industrial Ecology* 27, no. 4 (2023): 1197–1211.

Kennedy, C. A., I. Stewart, A. Facchini, I. Cersosimo, R. Mele, B., Chen, C. Dubeux, et al. "Energy and Material Flows of Megacities." *Proceedings of the National Academy of Sciences of the United States of America* 112, no. 19 (2015): 5985–5990.

Kennedy, C. A., J. Steinberger, B. Gasson, Y. Hansen, T. Hillman, M. Havránek, G. V. Mendez, et al. "Methodology for Inventorying Greenhouse Gas Emissions from Global Cities." *Energy Policy* 38, no. 9 (2010): 4828–4837.

Kennedy, C. A. "A Biophysical Model of the Industrial Revolution." *Journal of Industrial Ecology* 25, no. 3 (2021): 663–676.

Kennedy, C. A. "The Energy Embodied in the First and Second Industrial Revolutions." *Journal of Industrial Ecology* 24, no. 4 (2020): 887–898.

Kennedy, C. A. *The Evolution of Great World Cities: Urban Wealth and Economic Growth.* University of Toronto Press, 2011.

Kennedy, C. A., and J. Corfee-Morlot. "Past Performance and Future Needs for Low-Carbon, Climate-Resilient Infrastructure – An Investment Perspective." *Energy Policy* 59 (2013): 773–783.

Kennedy, C. A., I. D. Stewart, M. I. Westphal, A. Facchini, and R. Mele. "Keeping Global Climate Change Within 1.5 C Through Net Negative Electric Cities." *Current Opinion in Environmental Sustainability* 30 (2018): 18–25.

Kennedy, C. A., J. Corfee-Morlot, and M. Zhong. "Infrastructure for China's Ecologically Balanced Civilization." *Engineering* 2, no. 4 (2016): 414–425.

Kennedy, C. A., J. Cuddihy, and J. Engel-Yan. "The Changing Metabolism of Cities." *Journal of Industrial Ecology* 11, no. 2 (2007): 43–59.

Khazzoom, J. D. "Economic Implications of Mandated Efficiency in Standards for Household Appliances." *Energy Journal* 1, no. 4 (1980): 21–40.

King, A. "The Launch of a Club." In *The Club of Rome: The Dossiers: 1965–1984*, edited by P. Malasak and M. Vapaavuori, 55–59. Finnish Association for the Club of Rome, 2005.

King, A. *Let the Cat Turn Round: One Man's Traverse of the Twentieth Century.* London: CPTM, 2006.

Kleidon, A. "A Basic Introduction to the Thermodynamics of the Earth System far from Equilibrium and Maximum Entropy Production." *Philosophical Transactions of the Royal Society of London. Series B: Biological Sciences* 365, no. 1545 (2010): 1303–1315.

Kleidon, A. "Beyond Gaia: Thermodynamics of Life and Earth System Functioning." *Climatic Change* 66, no. 3 (2004): 271–319.

Kleidon, A. "Empowering the Earth System by Technology: Using Thermodynamics of the Earth System to Illustrate a Possible Sustainable Future of the Planet." In *Strategies for Sustainability of the Earth System*, edited by P. A. Wilderer, M. Grambow, M. Molls, and K. Oexle, 433–444. Springer: Cham, 2022.

Kleidon, A. "How Does the Earth System Generate and Maintain Thermodynamic Disequilibrium and What Does It Imply for the Future of the Planet?" *Philosophical Transactions of the Royal Society A* 370, no. 1962 (2012): 1012–1040.

Kleidon, A. "Testing the Effect of Life on Earth's Functioning: How Gaian is the Earth System?" *Climatic Change* 52, no. 4 (2002): 383–389.

Kleppel, G. S. *The Emergent Agriculture: Farming, Sustainability and the Return of the Local Economy*. New Society Publishers, 2014.

Kosoy, N., and E. Corbera. "Payments for Ecosystem Services as Commodity Fetishism." *Ecological Economics* 69, no. 6 (2010): 1228–1236.

Krausmann, F., M. Fischer-Kowalski, H. Schandl, and N. Eisenmenger. "The Global Sociometabolic Transition." *Journal of Industrial Ecology* 12, no. 5–6 (2008): 637–656.

Kubiszewski, I., C. J. Cleveland, and P. K. Endres. "Meta-Analysis of Net Energy Return for Wind Power Systems." *Renewable Energy* 35, no. 1 (2010): 218–225.

Lane, D. C., and J. D. Sterman. "Jay Wright Forrester." Chap. 20 in *Profiles in Operations Research: Pioneers and Innovators*, edited by S. Gass and A. Assad, 363–386. New York: Springer, 2011.

Lang, T. "Quantitatively Assessing the Role of Higher Education in Global Sustainability." PhD thesis, University of Toronto, 2016.

Lee, D. "Prince Charles: We Must Learn from Indigenous People on Climate Change." *BBC News*, May 19, 2022. https://www.bbc.com/news/uk-61517750.

Lenard, P. *Great Men of Science: A History of Scientific Progress*. London: G. Bell & Sons Ltd., 1938.

Lenzen, M., A. Geschke, M. D. A. Rahman, Y. Xiao, J. Fry, R. Reyes, E. Dietzenbacher, et al. "The Global MRIO Lab: Charting the World Economy." *Economic Systems Research* 29, no. 2 (2017): 158–186.

Lenzen, M., D. Moran, K. Kanemoto, B. Foran, L. Lobefaro, and A. Geschke. "International Trade Drives Biodiversity Threats in Developing Nations." *Nature* 486, no. 7401 (2012): 109–112.

Lenzen, M., T. Wiedmann, A. Geschke, J. Lane, P. Daniels, S. Kenway, and J. Boland. "The Challenges and Opportunities of Constructing Input–Output

Bibliography

Frameworks in a Virtual Laboratory: The New NeCTAR Industrial Ecology Lab." Paper presented at the 20th International Congress on Modelling and Simulation (MODSIM2013), Adelaide, South Australia, Australia, 2013.

Leontief, W. W. *Studies in the Structure of the American Economy: Theoretical and Empirical Explorations in Input-Output Analysis.* Oxford University Press, 1953.

Leontief, W. W. *The Structure of the American Economy, 1919–1929.* Cambridge: Harvard University Press, 1941. Reprinted and enlarged. New York: Oxford University Press, 1951.

Leontief, W. W., and F. Duchin. *The Future Impact of Automation on Workers.* New York: Oxford University Press, 1986.

Leontief, W. W. "Environmental Repercussions and the Economic Structure: An Input-Output Approach." *Review of Economics and Statistics* 52, no. 3 (1970): 262–271.

Leontief, W. W., and F. Duchin. *Military Spending: Facts and Figures, Worldwide Implications, and Future Outlook.* New York: Oxford University Press, 1983.

Leontief, W. W., F. Duchin, and D. B. Szyld. "New Approaches in Economic Analysis." *Science* 228, no. 4698 (1985): 419–422.

Leopold, A. *A Sand County Almanac.* New York: Oxford University Press, 1966.

Levallois, C. "Can De-Growth Be Considered a Policy Option? A Historical Note on Nicholas Georgescu-Roegen and the Club of Rome." *Ecological Economics* 69, no. 11 (2010): 2271–2278.

Lewis, A. "To Grow and To Die." *New York Times*, January 29, 1972.

Liboiron, M. *Pollution Is Colonialism.* Duke University Press, 2021.

Livingstone, D. *Missionary Travels and Researches in South Africa.* London: John Murray, 1857.

Lovins, A. B. "Energy Strategy: The Road Not Taken." *Foreign Affairs* 55 (1976): 65–96. Reprinted as Chap. 6 in Lovins, A. B., *The Essential Amory Lovins: Selected Writings.* London: Earthscan, 2011.

Lovins, A. B. *Reinventing Fire: Bold Business Solutions for the New Energy Era.* Chelsea Green Publishing, 2013.

Lovins, A. B. *The Essential Amory Lovins: Selected Writings.* London: Earthscan, 2011.

MacArthur, E. *Full Circle: My Life and Journey.* London, UK: Penguin Michael Joseph, 2010.

Macekura, S. *Of Limits and Growth.* Cambridge University Press, 2015.

MacKenzie, J. M. *The Empire of Nature: Hunting, Conservation and British Imperialism.* Manchester: Manchester University Press, 1988.

MacLeish, A. "A Reflection: Riders on Earth Together, Brothers in Eternal Cold." *New York Times*, December 25, 1968.

Malthus Enigma

Madlener, R., and B. Alcott. "Energy Rebound and Economic Growth: A Review of the Main Issues and Research Needs." *Energy* 34, no. 3 (2009): 370–376.

Malasak, P., and M. Vapaavuori, eds. *The Club of Rome: The Dossiers: 1965–1984*. Finnish Association for the Club of Rome, 2005.

Malthus, T. R. *Principles of Political Economy*. London: W. Pickering, 1820.

Malthus, T. R. *An Essay on the Principle of Population, As It Affects the Future Improvement of Society, with Remarks on the Speculations of Mr. Godwin, M. Condorcet, and Other Writers*. London: J. Johnson, 1798.

Malthus, T. R. *An Essay on the Principle of Population; Or, a View of Its past and Present Effects on Human Happiness; With an Inquiry Into Our Prospects Respecting the Future Removal or Mitigation of the Evils Which It Occasions*. 2nd ed. London: J. Johnson, 1803.

Manabe, S., and R. T. Wetherald. "The Effects of Doubling the CO_2 Concentration on the Climate of a General Circulation Model." *Journal of the Atmospheric Sciences* 32, no. 1 (1975): 3–15.

Mann, C. C. *The Wizard and the Prophet: Two Remarkable Scientists and Their Dueling Visions to Shape Tomorrow's World*. Knopf, 2018.

Margaret Thatcher Foundation. "Speech at 2nd World Climate Conference." November 6, 1990. Accessed September 5, 2018. https://www.margaretthatcher.org/document/108237.

Marsh, G. P. *Man and Nature: Or, Physical Geography as Modified by Human Action*. Harvard University Press, 1965.

Marstrander, R., H. Brattebø, K. Røine, and S. Støren. "Teaching Industrial Ecology to Graduate Students: Experiences at the Norwegian University of Science and Technology." *Journal of Industrial Ecology* 3, no. 4 (1999): 117–130.

Martínez-Alier, J., and R. Muradian. "Taking Stock: The Keystones of Ecological Economics." In *Handbook of Ecological Economics*, edited by J. Martínez-Alier and R. Muradian, 1–25. Edward Elgar Publishing, 2015.

Masini, E. B. *The Legacy of Aurelio Peccei and the Continuing Relevance of his Anticipatory Vision*. The Club of Rome; Fondazione Aurelio Peccei; and European Support Centre of the Club of Rome, Vienna, 2006.

Matthews, E., C. Amann, S. Bringezu, M. Fischer-Kowalski, W. Hüttler, R. Kleijn, Y. Moriguchi, et al. *The Weight of Nations—Material Outflows from Industrial Economies*. Washington, DC: World Resources Institute, 2000.

Mayhew, R. J. *Malthus: The Life and Legacies of an Untimely Prophet*. Harvard University Press, 2014.

Mayumi, K. T. "Thermodynamics: Relevance, Implications, Misuse and Ways Forward." In *Routledge Handbook of Ecological Economics*, edited by C. L. Spash, 89–98. Routledge, 2017.

McCormick, J. *The Global Environmental Movement*. 2nd ed. Chichester: Wiley, 1995.

Bibliography

McNeill, L. "This Suffrage-Supporting Scientist Defined the Greenhouse Effect But Didn't Get the Credit, Because Sexism." *Smithsonian Magazine*, December 5, 2016. Accessed May 20, 2022. https://www.smithsonianmag.com/science-nature/lady-scientist-helped-revolutionize-climate-science-didnt-get-credit-180961291/.

Meade, D. S. "The US Benchmark IO Table History, Methodology, and Myths." Paper presented at the 18th Inforum World Conference, Hikone, Japan, September 5–10, 2010. http://inforumweb.inforumecon.com/papers/conferences/2010/Meade.pdf.

Meadows, D. H., D. L. Meadows, J. Randers, and W. W. Behrens. *The Limits to Growth*. 2nd ed. New York: Universe Books, 1974.

Meadows, D. H., J. Randers, and D. L. Meadows. *Limits to Growth: The 30-Year Update*. White River Junction, VT: Chelsea Green Publishing, 2004.

Meadows, D. L., W. Behrens, D. H. Meadows, R. Naill, J. Randers, and E. Zahn. *Dynamics of Growth in a Finite World*. Cambridge, MA: Wright-Allen Press, 1974.

Mesarović, M. D., and E. Pestel. *Mankind at the Turning Point: The Second Report to the Club of Rome*. New York: E. P. Dutton & Co., 1974.

Millennium Ecosystem Assessment. *Ecosystems and Human Well-Being: Biodiversity Synthesis*. Washington, DC: World Resources Institute, 2005.

Miller, D. P. "The Mysterious Case of James Watt's '1785 Steam Indicator': Forgery or Folklore in the History of an Instrument?" *International Journal for the History of Engineering & Technology* 81 (2011): 129–150.

Montague, Brendan. "Who Was Responsible for Thatcher's Climate Change U-Turn?" *DeSmog*. Updated May 19, 2015. Accessed September 5, 2018. https://www.desmog.com/2015/05/19/who-was-responsible-thatcher-s-climate-change-u-turn/.

Moore, W. *Schrödinger: Life and Thought*. Cambridge University Press, 1989.

Mora, C., D. P. Tittensor, S. Adl, A. G. B. Simpson, and B. Worm. "How Many Species Are There on Earth and in the Ocean?" *PLoS Biology* 9, no. 8 (2011): e1001127

Moreau, V., M. Sahakian, P. Griethuysen, and F. Vuille. "Coming Full Circle: Why Social and Institutional Dimensions Matter for the Circular Economy." *Journal of Industrial Ecology* 21, no. 3 (2017): 497–506.

Müller, D. B., G. Liu, A. N. Løvik, R. Modaresi, S. Pauliuk, F. S. Steinhoff, and H. Brattebø. "Carbon Emissions of Infrastructure Development." *Environmental Science & Technology* 47, no. 20 (2013): 11739–11746.

Murawiec, L., and D. di Paoli. "Club of Rome Founder Alexander King Discusses His Goals and Operations." *Executive Intelligence Review* 8, no. 25 (1981): 18–29.

Murphy, D. J., and C. A. S. Hall. "Year in Review – EROI or Energy Return on (Energy) Invested." *Annals of the New York Academy of Science* 1185, no. 1 (2010): 102–118.

Muthee, D. W., G. K. Gwademba, and J. M. Masinde. "The Role of Indigenous Knowledge Systems in Enhancing Agricultural Productivity in Kenya." *Eastern Africa Journal of Contemporary Research* 1, no. 1 (2019): 34–45.

Muûls, M., J. O. Colmer, R. A. Martin, and U. J. Wagner. "Evaluating the EU Emissions Trading System: Take It or Leave It? An Assessment of the Data After Ten Years." Grantham Institute Briefing Paper No. 21, October 2016.

Nabuurs, G.-J., R. Mrabet, A. Abu Hatab, M. Bustamante, H. Clark, P. Havlík, J. House, et al. "Agriculture, Forestry and Other Land Uses (AFOLU)." Chap. 7 in Intergovernmental Panel on Climate Change (IPCC), *Climate Change 2022: Mitigation of Climate Change. Contribution of Working Group III to the Sixth Assessment Report of the Intergovernmental Panel on Climate Change*. Cambridge University Press, 2022.

Nakamura, S., and Y. Kondo. "Input-Output Analysis of Waste Management." *Journal of Industrial Ecology* 6, no. 1 (2002): 39–63.

National Aeronautics and Space Administration (NASA). "Robert A. Frosch." Accessed June 3, 2018. https://www.nasa.gov/people/robert-a-frosch/.

National Renewable Energy Laboratory. *Renewable Electricity Futures Study*. Edited by M. M. Hand, S. Baldwin, E. DeMeo, J. M. Reilly, T. Mai, D. Arent, G. Porro, M. Meshek, and D. Sandor. 4 vols. National Renewable Energy Laboratory, 2012.

National Research Council. *Carbon Dioxide and Climate: A Scientific Assessment*: Report of an Ad Hoc Study Group on Carbon Dioxide and Climate, Woods Hole, Massachusetts, July 23–27, 1979.

Nature. "Another Whiff of Doomsday." *Nature* 236, no. 5341 (1972): 47–49.

Nordhaus, T., and M. Shellenberger. *Break Through: From the Death of Environmentalism to the Politics of Possibility*. Houghton Mifflin Harcourt, 2007.

Nordhaus, W. D. "World Dynamics: Measurement Without Data." *The Economic Journal* 83, no. 332 (1973): 1156–1183.

Nordhaus, W. D., R. N. Stavins, and M. L. Weitzman. "Lethal Model 2: The Limits to Growth Revisited." *Brookings Papers on Economic Activity*, no. 2 (1992): 1–59.

Novacek, M. J., ed. *The Biodiversity Crisis: Losing What Counts*. The New Press, 2001.

Nuccitelli, D. "Scientists Warned the US President About Global Warming 50 Years Ago Today." *The Guardian*, November 5, 2015. Accessed September 5, 2018. https://www.theguardian.com/environment/climate-consensus-97-per-cent/2015/nov/05/scientists-warned-the-president-about-global-warming-50-years-ago-today.

O'Donnell, T. M. "Of Loaded Dice and Heated Arguments: Putting the Hansen-Michaels Global Warming Debate in Context." *Social Epistemology* 14, no. 2–3 (2000): 109–127.

Odum, E. P. *Fundamentals of Ecology*. Philadelphia, PA, and London, England: W. B. Saunders Co., 1953.

Bibliography

Odum, E. P., and A. E. Smalley. "Comparison of Population Energy Flow of a Herbivorous and a Deposit-Feeding Invertebrate in a Salt Marsh Ecosystem." *Proceedings of the National Academy of Sciences of the United States of America* 45, no. 4 (1959): 617–622.

Odum, E. P. "The Emergence of Ecology as a New Integrative Discipline." *Science* 195, no. 4284 (1977): 1289–1293.

Odum, E. P. "The Strategy of Ecosystem Development." *Science* 164, no. 3877 (1969): 262–270.

Odum, E. P. *Ecological Vignettes: Ecological Approaches to Dealing with Human Predicaments.* Taylor & Francis, 1998.

Odum, E. P. *Ecology and Our Endangered Life-Support Systems.* Sinauer Associates, 1989.

Odum, H. T. "Energetics of Food Production." Chap. 3 in Vol. 3 of *The World Food Problem: A Report of the President's Science Advisory Committee, Report of the Panel on the World Food Supply*, 55–94. Washington, DC: The White House, 1967.

Odum, H. T. "Trophic Structure and Productivity of Silver Springs, Florida." *Ecological Monographs* 27, no. 1 (1957): 55–112.

Odum, H. T. *Environment, Power and Society.* New York: Wiley-Interscience, 1971.

Odum, H. T. *Systems Ecology: An Introduction.* New York: Wiley, 1983.

Odum, H. T., and E. P. Odum. "Trophic Structure and Productivity of a Windward Coral Reef Community on Eniwetok Atoll." *Ecological Monographs* 25, no. 3 (1955): 291–320.

Ordway, S. H., Jr. "Possible Limits of Raw-Material Consumption." In *Man's Role in Changing the Face of the Earth*, edited by L. T. Thomas, 987–1019. Chicago: University of Chicago Press, 1956.

Organisation for Economic Co-operation and Development (OECD). *Measuring Material Flows and Resource Productivity.* Vol. 1, *The OECD Guide.* Paris: OECD Publishing, 2008.

Organisation for Economic Co-operation and Development (OECD). *OECD Environmental Outlook to 2050: The Consequences of Inaction.* Paris: OECD Publishing, 2012.

Organisation for Economic Co-operation and Development (OECD). *Sustainable Manufacturing and Eco-Innovation: Framework, Practices and Measurement: Synthesis Report*, Paris: OECD Publishing, 2009.

Organisation for Economic Co-operation and Development (OECD) and Food and Agriculture Organization of the United Nations (FAO). *OECD-FAO Agricultural Outlook 2019–2028.* Paris: OECD Publishing, 2019.

Ortiz, J. D., and R. Jackson. "Understanding Eunice Foote's 1856 Experiments: Heat Absorption by Atmospheric Gases." *Notes and Records* 76, no. 1 (2022): 67–84.

Oswald, Y., A. Owen, and J. K. Steinberger. "Large Inequality in International and Intra-National Energy Footprints Between Income Groups and Across Consumption Categories." *Nature Energy* 5, no. 3 (2020): 231–239.

Passell, Peter, Marc Roberts, and Leonard Ross. "The Limits to Growth." *New York Times Book Review*, April 2, 1972.

Pauli, G. A. *Crusader for the Future: A Portrait of Aurelio Peccei*. Oxford: Pergamon Press, 1987.

Pauliuk, S., A. Arvesen, K. Stadler, and E. G. Hertwich. "Industrial Ecology in Integrated Assessment Models." *Nature Climate Change* 7, no. 1 (2017): 13–20.

Pearce, D., and R. K. Turner. *Economics of Natural Resources and the Environment*. London: Harvester Wheatsheaf, 1990.

Peccei, A. *One Hundred Pages for the Future: Reflections of the President of the Club of Rome*. Pergamon Press, 1981. US edition. Pergamon/Elsevier, 2016.

People's Daily Online. "400 Million Births Prevented by One-Child Policy." *People's Daily Online*, October 28, 2011. Accessed March 27, 2012. https://web.archive.org/web/20120325025201/http://english.people.com.cn/90882/7629166.html.

Pereira, H. M., P. W. Leadley, V. Proença, R. Alkemade, J. P. Scharlemann, J. F. Fernandez-Manjarrés, L. Chini, et al. "Scenarios for Global Biodiversity in the 21st Century." *Science* 330, no. 6010 (2010): 1496–1501.

Peters, G. P. "From Production-Based to Consumption-Based National Emission Inventories." *Ecological Economics* 65, no. 1 (2008): 13–23.

Peterson, T. C., W. M. Connolley, and J. Fleck. "The Myth of the 1970s Global Cooling Scientific Consensus." *Bulletin of the American Meteorological Society* 89, no. 9 (2008): 1325–1337.

Petrini, C. *Slow Food Revolution: A New Culture for Dining and Living*. Rizzoli International Publications, 2006.

Petroski, H. *To Engineer Is Human: The Role of Failure in Successful Design*. St. Martins Press, 1985.

Phillips, N. A. "The General Circulation of the Atmosphere: A Numerical Experiment." *Quarterly Journal of the Royal Meteorological Society* 82, no. 352 (1956): 123–164.

Pielke, R. A. "Policy History of the US Global Change Research Program: Part I. Administrative Development." *Global Environmental Change* 10, no. 1 (2000): 9–25.

Pimm, S. L., and P. H. Raven. "The State of the World's Biodiversity." In *Biological Extinction: New Perspectives*, edited by P. Dasgupta, P. H. Raven, and A. McIvor. Cambridge: Cambridge University Press, 2019.

Plass, G. N. "Carbon Dioxide and Climate." *Scientific American* 201 (July 1959): 41–47. Abridged reprint with commentary by J. R. Fleming and G. Schmidt in *American Scientist* 98, no. 1 (2010): 58–62.

Poole, R. *Earthrise: How Man First Saw the Earth*. Newhaven: Yale University Press, 2008.

Prigogine, I. *Étude thermodynamique des phénomènes irréversibles*. Liege: Desoer, 1947.

Bibliography

Prigogine, Ilya. "Time, Structure and Fluctuations." Nobel Lecture, December 8, 1977. https://www.nobelprize.org/prizes/chemistry/1977/prigogine/lecture/.

Prigogine, L., G. Nicolis, and A. Babloyantz. "Thermodynamics of Evolution." *Physics Today* 25, no. 12 (1972): 38–44.

Rajasekaran, B., D. M. Warren, and S. C. Babu. "Indigenous Natural-Resource Management Systems for Sustainable Agricultural Development—A Global Perspective." *Journal of International Development* 3, no. 3 (1991): 387–401.

Regis, E. "The Doomslayer." *Wired*, February 1, 1997.

Reinhold, R. "Mankind Warned of Perils in Growth." *New York Times*, February 27, 1972.

Renner, G. T. "Geography of Industrial Localization." *Economic Geography* 23, no. 3 (1947): 167–189.

Richards, D. G., and A. B. Fullerton, eds. *Industrial Ecology: U.S.–Japan Perspectives*. Washington, DC: National Academy Press, 1994.

Richards, P. *Indigenous Agricultural Revolution: Ecology and Food Production in West Africa*. Routledge, 1985.

Richardson, K., W. Steffen, W. Lucht, J. Bendtsen, S. E. Cornell, J. F. Donges, M. Drüke, et al. "Earth Beyond Six of Nine Planetary Boundaries." *Science Advances* 9, no. 37 (2023): eadh2458.

Ridley, M. "China's One-Child Policy Was Inspired by Western Greens." *The Rational Optimist* (blog), January 18, 2014. http://www.rationaloptimist.com/blog/chinas-one-child-policy-was-inspired-by-western-greens/, or https://www.mattridley.co.uk/blog/chinas-one-child-policy-was-inspired-by-western-greens/.

Ritchie, Hannah. "The UN Has Made Population Projections for More Than 50 Years – How Accurate Have They Been?" *Our World in Data*, June 20, 2023. https://ourworldindata.org/population-projections.

Robertson, T. *The Malthusian Moment: Global Population Growth and the Birth of American Environmentalism*. Rutgers University Press, 2012.

Rockström, J., W. Steffen, K. Noone, Å. Persson, F. S. Chapin III, E. F. Lambin, B. Nykvist, et al. "A Safe Operating Space for Humanity." *Nature* 461, no. 7263 (2009): 472–475.

Rojstaczer, S., S. M. Sterling, and N. J. Moore. "Human Appropriation of Photosynthesis Products." *Science* 294, no. 5551 (2001): 2549–2552.

Røpke, I. "Sustainable Consumption: Transitions, Systems and Practices." In *Handbook of Ecological Economics*, edited by J. Martínez-Alier and R. Muradian, 332–359. Edward Elgar Publishing, 2015.

Røpke, I. "The Early History of Modern Ecological Economics." *Ecological Economics* 50, no. 3 (2004): 293–314.

Røpke, I. "Trends in the Development of Ecological Economics from the Late 1980s to the Early 2000s." *Ecological Economics* 55, no. 2 (2005): 262–290.

Running S. W. "A Measurable Planetary Boundary for the Biosphere." *Science* 337, no. 6101 (2012): 1458–1459.

Running, S. W. "Global Aridification and the Decline of NPP: A Commentary on Projected Increases in Global Terrestrial Net Primary Productivity Loss Caused by Drought Under Climate Change by Dan Cao, Jiahua Zhang, Jiaqi Han, Tian Zhang, Shanshan Yang, Jingwen Wang, Foyez Ahmed Prodhan, and Fengmei Yao." *Earth's Future* 10, no. 11 (2022): e2022EF003113.

Sabin, P. *The Bet: Paul Ehrlich, Julian Simon, and Our Gamble over Earth's Future*. Yale University Press, 2013.

Salmi, O., and A. Toppinen, "Embedding Science in Politics: 'Complex Utilization' and Industrial Ecology as Models of Natural Resource Use." *Journal of Industrial Ecology* 11, no. 3 (2007): 93–111.

Sambo, B. E. "Endangered, Neglected, Indigenous Resilient Crops: A Potential Against Climate Change Impact for Sustainable Crop Productivity and Food Security." *IOSR Journal of Agriculture and Veterinary Science* 7, no. 2 (2014): 34–41.

Sandbach, F. "The Rise and Fall of the Limits to Growth Debate." *Social Studies of Science* 8, no. 4 (1978): 495–520.

Schaffartzik, A., A. Mayer, S. Gingrich, N. Eisenmenger, C. Loy, and F. Krausmann. "The Global Metabolic Transition: Regional Patterns and Trends of Global Material Flows, 1950–2010." *Global Environmental Change* 26, no. 1 (2014): 87–97.

Schandl, H., and J. West. "Resource Use and Resource Efficiency in the Asia-Pacific Region." *Global Environmental Change* 20, no. 4 (2010): 636–647.

Schmalensee, R., and R. Stavins. "Lessons Learned from Three Decades of Experience with Cap-and-Trade." NBER Working Paper No. w21742. National Bureau of Economic Research, November 2015.

Schrödinger, E. *What Is Life? The Physical Aspect of the Living Cell*. Cambridge University Press, 1944.

Schroeter, S. "The Sun is Shining on PV in Saxony." *European Energy Review*, January/February 2009.

Scientific American. "Scientific Ladies.--Experiments with Condensed Gases." *Scientific American* 12, no. 1 (September 1856): 5.

Seufert, V., N. Ramankutty, and J. A. Foley. "Comparing the Yields of Organic and Conventional Agriculture." *Nature* 485, no. 7397 (2012): 229–232.

Sexton, C. "Humans Should Remember They Are Part of Nature, Says Prince Charles." *Earth.com*, December 30, 2020. https://www.earth.com/news/humans-should-remember-they-are-part-of-nature-says-prince-charles/.

Shabecoff, P. "Bush Tells Environmentalists He'll Listen to Them." *New York Times*, December 1, 1988.

Shabecoff, P. "Global Warming Has Begun, Expert Tells Senate." *New York Times*, June 23, 1988.

Bibliography

Shapiro, M. "Eunice Newton Foote's Early Forgotten Discovery." *Physics Today*, August 23, 2021. Accessed May 20, 2022. https://physicstoday.scitation.org/do/10.1063/pt.6.4.20210823a/full/.

Shprentz, D. *Breath-Taking: Premature Mortality Due to Particulate Air Pollution in 239 American Cities.* New York: Natural Resources Defense Council, 1996.

Sieferle, R. P. "Europe's Special Course." Paper presented at the 2004 Gordon Research Conference in Industrial Ecology, Oxford, UK. August 1–6, 2004.

Sieferle, R. P. *The Subterranean Forest: Energy Systems and the Industrial Revolution.* White Horse Press, 2001.

Simmons, M. R. "Revisiting *The Limits to Growth*: Could The Club of Rome Have Been Correct, After All? (Part One)" *Resilience*. September 30, 2000. Accessed December 29, 2015. https://web.archive.org/web/20151007062239/http://www.resilience.org/stories/2000-09-30/revisiting-limits-growth-could-club-rome-have-been-correct-after-all-part-one.

Simon, J. L. "Environmental Disruption or Environmental Improvement." *Social Science Quarterly* 62, no. 1 (1981): 30–43.

Simon, J. L. "Resources, Population, Environment: An Oversupply of False Bad News." *Science* 208, no. 4451 (1980): 1431–1437.

Simon, J. L., and A. Wildavsky. "Species Loss Revisited." In *The State of Humanity*, edited by Julian L. Simon, 346–361. Basil Blackwell, 1995.

Simon, J. L., and H. Kahn. "Introduction." In *The Resourceful Earth: A Response to "Global 2000,"* edited by J. L. Simon and H. Kahn, 1–51. Oxford, UK: Blackwell, 1984.

Smagorjnsky, J. "The Beginnings of Numerical Weather Prediction and General Circulation Modeling: Early Recollections." *Advances in Geophysics* 25 (1983): 3–37.

Smil, V. *Enriching the Earth: Fritz Haber, Carl Bosch, and the Transformation of World Food Production.* MIT Press, 2004.

Smil, V. *Feeding the World: A Challenge for the Twenty-First Century.* MIT Press, 2001.

Smith, E. P. *A Manual of Political Economy.* New York: Putnam, 1853.

Smith, S., and J. Swierzbinski. "Assessing the Performance of the UK Emissions Trading Scheme." *Environmental and Resource Economics* 37, no. 1 (2007): 131–158.

Socolow, R., C. Andrews, F. Berkhout, and V. Thomas, eds. *Industrial Ecology and Global Change.* Cambridge and New York: Cambridge University Press, 1994.

Solow, R. M. "Georgescu-Roegen vs. Solow/Stiglitz—Reply." *Ecological Economics* 22, no. 3 (1997): 267–268.

Solow, R. M. "The Economics of Resources or the Resources of Economics." *American Economic Review* 64, no. 2 (1974): 1–14.

Sorenson, R. P. "Eunice Foote's Pioneering Work on CO_2 and Climate Warming." *Search and Discovery* (2011): 70092. January 31, 2011. Accessed 20 May 2022. https://www.searchanddiscovery.com/documents/2011/70092sorenson/ndx_sorenson.pdf.

Sorrell, S. "Jevons' Paradox Revisited: The Evidence for Backfire from Improved Energy Efficiency." *Energy Policy* 37, no. 4 (2009): 1456–1469.

South African Press Association. "Cape Is World's Extinction Capital." *Independent Online*, August 14, 2014.

Spash, C. L. "Bulldozing Biodiversity: The Economics of Offsets and Trading-in Nature." *Biological Conservation* 192 (December 2015): 541–551.

Spash, C. L. "Social Ecological Economics." In *Routledge Handbook of Ecological Economics*, edited by C. L. Spash, 3–16. Routledge, 2017.

Spash, C. L. "The Content, Direction and Philosophy of Ecological Economics." In *Handbook of Ecological Economics*, edited by J. Martínez-Alier and R. Muradian, 26–47. Edward Elgar Publishing, 2015.

Spash, C. L., and K. Dobernig. "Theories of (Un)sustainable Consumption." In *Routledge Handbook of Ecological Economics*, edited by C. L. Spash, 203–213. Routledge, 2017.

Spilhaus, A. "The Next Industrial Revolution." *Proceedings of the American Philosophical Society* 115, no. 4 (1971): 324–327.

Springer, N. P., and F. Duchin. "Feeding Nine Billion People Sustainably: Conserving Land and Water Through Shifting Diets and Changes in Technologies." *Environmental Science & Technology* 48, no. 8 (2014): 4444–4451.

Stavins, R. N. *Project 88 – Harnessing Market Forces to Protect Our Environment: Initiatives for the New President. A Public Policy Study Sponsored by Senator Timothy E. Wirth, Colorado, Senator John Heinz, Pennsylvania*. Washington, DC, 1988.

Steffen, W., K. Richardson, J. Rockström, S. E. Cornell, I. Fetzer, E. M. Bennett, R. Biggs, et al. "Planetary Boundaries: Guiding Human Development on a Changing Planet." *Science* 347, no. 6223 (2015): 736 and 1259855.

Stewart, I. D., C. A. Kennedy, A. Facchini, and R. Mele. "The Electric City as a Solution to Sustainable Urban Development." *Journal of Urban Technology* 25, no 1. (2018): 3–20.

Stoltzenberg, D. *Fritz Haber: Chemist, Nobel Laureate, German, Jew*. Philadelphia: Chemical Heritage Press, 2004.

Strong Foundation. "Short Biography." Accessed May 19, 2018. https://mauricestrong.net/index.php?option=com_content&view=article&id=15&Itemid=24.

Styles, J. A., ed. *Eighteenth Century English Provincial Newspapers*. Brighton, Sussex, England: Harvester Press Microfilm Publications, 1985.

Sukhdev, P., H. Wittmer, C. Schröter-Schlaack, C. Nesshöver, J. Bishop, P. T. Brink, B. Simmons, et al. *The Economics of Ecosystems and Biodiversity:*

Bibliography

Mainstreaming the Economics of Nature: A Synthesis of the Approach, Conclusions and Recommendations of TEEB. Geneva: United Nations Environment Programme, 2010.

Sutherland, G. B. B., and G. S. Callendar. "The Infra-Red Spectra of Atmospheric Gases Other Than Water Vapour." *Reports on Progress in Physics* 9 (1944): 118–128.

Swingland, I. R., ed. *Capturing Carbon and Conserving Biodiversity: The Market Approach.* Earthscan, 2003.

Syll, L. "Nicholas Georgescu-Roegen and the Nobel Prize." *Lars P. Syll* (blog). May 16, 2012. Accessed December 2, 2017. https://larspsyll.wordpress.com/2012/05/16/nicholas-georgescu-roegen-and-the-nobel-prize-in-economics/.

Thomas, W. L., Jr., ed. *Man's Role in Changing the Face of the Earth: An International Symposium under the Co-chairmanship of Carl O. Sauer, Marston Bates, and Lewis Mumford.* University of Chicago Press, 1956.

Thompson, Benjamin. "An Inquiry Concerning the Source of the Heat Which Is Excited by Friction." *Philosophical Transactions of the Royal Society of London* 88 (January 1798): 80–102.

Thompson, S. P. *The Life of William Thomson: Baron Kelvin of Largs.* 2 vols. London: Macmillan, 1910.

Thomson, W. (Lord Kelvin). "On a Universal Tendency in Nature to the Dissipation of Mechanical Energy." *Proceedings of the Royal Society of Edinburgh* 3 (1857): 139–142.

Thomson, W. (Lord Kelvin). "On a Universal Tendency in Nature to the Dissipation of Mechanical Energy." *Philosophical Magazine Series* 4, no. 25 (1852) 304–306.

Thomson, W. (Lord Kelvin). "On the Dissipation of Energy." *The Fortnightly Review*, March 1, 1892.

Thurston, R. H. "The Work of N. L. Sadi Carnot." Chap. 1 in *Reflections on the Motive Power of Heat*, edited by R. H. Thurston. 2nd ed. New York: Wiley, 1897.

Tibbs, H. *Industrial Ecology: An Environmental Agenda for Industry.* Global Business Network, 1993.

Tietenberg, T. H. *Emissions Trading, An Exercise in Reforming Pollution Policy.* Resources for the Future, 1985.

Tilman, D., and M. Clark. "Food, Agriculture and the Environment: Can We Feed the World and Save the Earth?" *Daedalus* 144, no. 4 (2015): 8–23.

Tilman, D., C. Balzer, J. Hill, and B. L. Befort. "Global Food Demand and the Sustainable Intensification of Agriculture." *Proceedings of the National Academy of Sciences of the United States of America* 108, no. 50 (2011): 20260–20264.

Time. "Eat Hearty." *Time.* November 8, 1948, Economics.
Time. "One Big Greenhouse." *Time.* May 28, 1956, Science.

Tukker, A., and E. Dietzenbacher. "Global Multiregional Input–Output Frameworks: An Introduction and Outlook." *Economic Systems Research* 25, no. 1 (2013): 1–19.

Turner, G. M. "A Comparison of *The Limits to Growth* with 30 Years of Reality." *Global Environmental Change* 18, no. 3 (2008): 397–411.

Turner, G. M. "On the Cusp of Global Collapse? Updated Comparison of the Limits to Growth with Historical Data." *GAIA-Ecological Perspectives for Science and Society* 21, no. 2 (2012): 116–124.

Tyndall, J. "The Bakerian Lecture: On the Absorption and Radiation of Heat by Gases and Vapours, and on the Physical Connexion of Radiation, Absorption, and Conduction." *The London, Edinburgh, and Dublin Philosophical Magazine and Journal of Science* 22, no. 146 (1861): 169–194.

United Nations Environment Programme (UNEP). *Decoupling: Natural Resource Use and Environmental Impacts from Economic Growth.* UNEP, 2011.

United Nations Environment Programme (UNEP). *Metal Stocks in Society: Scientific Synthesis.* UNEP, 2010.

United Nations Environment Programme (UNEP) Finance Initiative. *Demystifying Materiality: Hardwiring Biodiversity and Ecosystem Services into Finance.* CEO Briefing. UNEP Finance Initiative, 2010.

United Nations Department of Economic and Social Affairs (UN DESA), Population Division. *World Population Prospects: The 2010 Revision: Highlights and Advance Tables.* ESA/P/WP.220. United Nations, 2011.

United Nations Department of Economic and Social Affairs (UN DESA), Population Division. *World Population Prospects: The 2017 Revision: Key Findings and Advanced Tables.* ESA/P/WP/248. United Nations, 2017.

United Nations Department of Economic and Social Affairs (UN DESA), Population Division. *World Population Prospects 2019: Highlights.* ST/ESA/SER.A/423. United Nations, 2019.

United Nations Department of Economic and Social Affairs (UN DESA), Population Division. *World Population Prospects 2024: Summary of Results.* UN DESA/POP/2024/TR/NO. 9. United Nations, 2024.

United Nations Department of Economic and Social Affairs (UN DESA), Social Inclusion Division. "United Nations Declaration on the Rights of Indigenous Peoples." Accessed May 14, 2022. https://social.desa.un.org/issues/indigenous-peoples/united-nations-declaration-on-the-rights-of-indigenous-peoples.

United Nations. "Goal 1: End Poverty in All Its Forms Everywhere." Accessed December 23, 2024. https://www.un.org/sustainabledevelopment/poverty/.

Victor, P. A. *Managing Without Growth: Slower by Design, Not Disaster.* Edward Elgar Publishing, 2008.

Vidal, J. "Margaret Thatcher: An Unlikely Green Hero?" *Environment Blog, The Guardian.* April 9, 2013. Accessed September 5, 2018.

Bibliography

https://www.theguardian.com/environment/blog/2013/apr/09/margaret-thatcher-green-hero.

Vietmeyer, N. *Our Daily Bread: The Essential Norman Borlaug.* Lorton, VA: Bracing Books, 2011.

Vitousek, P. M., P. R. Ehrlich, A. H. Ehrlich, and P. A. Matson. "Human Appropriation of the Products of Photosynthesis." *BioScience* 36, no. 6 (1986): 368–373.

Vollset, S. E., E. Goren, C.-W. Yuan, J. Cao, A. E. Smith, T. Hsiao, C. Bisignano, et al. "Fertility, Mortality, Migration, and Population Scenarios for 195 Countries and Territories from 2017 to 2100: A Forecasting Analysis for the Global Burden of Disease Study." *Lancet* 396, no. 10258 (2020): 1285–1306.

von Bertalanffy, L., and A. Rapoport. *General Systems: Yearbook of the Society for the Advancement of General System Theory.* Vol 1. The Society, 1956.

Warde, P. *The Invention of Sustainability: Nature and Destiny, c. 1500–1870.* Cambridge University Press, 2018.

Watson, A. "Gaia." *New Scientist Inside Science* 48 (1991): 1–4.

Watt, J., and J. Black. *Partners in Science: Letters of James Watt and Joseph Black.* Constable & Company Limited, 1970.

Weart, S. R. *The Discovery of Global Warming.* Harvard University Press, 2008.

Weinberg, A. M., and R. P. Hammond. "Limits to the Use of Energy: The Limit to Population Set by Energy Is Extremely Large, Provided That the Breeder Reactor Is Developed or That Controlled Fusion Becomes Feasible." *American Scientist* 58, no. 4 (1970): 412–418.

Wells, D. A., ed. "Heat of the Sun's Rays." In *Annual of Scientific Discovery: Or, Year-Book of Facts in Science and Art for 1857*, 159–160. Boston: Gould and Lincoln, 1857.

White, R. M. "Preface." In *The Greening of Industrial Ecosystems*, edited by B. R. Allenby and D. J. Richards, v–vi. Washington, DC: National Academy Press, 1994.

Wiedmann, T., M. Lenzen, L. T. Keyßer, and J. K. Steinberger. "Scientists' Warning on Affluence." *Nature Communications* 11, no. 1 (2020): 1–10.

Wik, M., P. Pingali, and S. Brocai. *Background Paper for the World Development Report 2008: Global Agricultural Performance: Past Trends and Future Prospects.* Washington, DC: World Bank, 2008.

Wikipedia. "Club of Rome." Accessed October 4, 2015. https://en.wikipedia.org/wiki/Club_of_Rome.

Williams, J. H., B. Haley, F. Kahrl, J. Moore, A. D. Jones, M. S. Torn, and H. McJeon. *Pathways to Deep Decarbonization in the United States.* Sustainable Development Solutions Network and Institute for Sustainable Development and International Relations, 2014.

Winch, D. *Malthus*, Oxford: Oxford University Press, 1987.

Wolman, A. "The Metabolism of Cities." *Scientific American* 213, no. 3 (1965): 178–193.
World Science Forum. "Launching the UNEP International Panel for Sustainable Resource Management." Accessed June 8, 2018. http://www.sciforum.hu/previous-fora/2007/programme/special-sessions/launching-the-unep-international-panel-for-sustainable-resource-management.html.
Wrigley, E. A. *Energy and the English Industrial Revolution*. Cambridge: Cambridge University Press, 2010.
Zhu, J., C. Fan, H. Shi, and L. Shi. "Efforts for a Circular Economy in China: A Comprehensive Review of Policies." *Journal of Industrial Ecology* 23, no. 1 (2019): 110–118.
Zink, T., and R. Geyer. "Circular Economy Rebound." *Journal of Industrial Ecology* 21, no. 3 (2017): 593–602.

Index

Afghanistan, 57–58
Africa, 12, 15, 43–44, 46–47, 49, 66, 68–71, 99, 166, 202, 218
agricultural land, 11–12, 27, 69, 136
agricultural production, 48, 90, 202
agricultural systems, 62, 70
agriculture, 4, 11–12, 46–47, 51, 57, 61, 66–67, 69–70, 135–36, 148, 171, 198, 210, 218
agroecology, 72, 220
air pollution, viii, 16, 19, 31, 129, 144, 176, 225
Allenby, Brad, 190, 214
Allwood, Julian, 212
aluminum, 197, 201, 208
Amazon rainforest, 229
ammonium, 53–54
Ampère, André-Marie, 108, 117
amphibians, 61–62, 64
Ångström, Knut, 93–94, 104
animals, 15, 30, 32, 42–43, 46, 62, 65, 84, 151, 226
Apollo, 1–21, 35, 209, 230
Apollo missions, 3–5, 221
Arrhenius, Svante, 14, 91–95, 104
Asia, 61, 69, 183
atmosphere, 13, 33, 35–36, 53, 63, 75–76, 82, 85, 91–92, 94–97, 107, 126–27, 147–48, 222
AT&T, 190
Auerbach, Raymond, 69–70
Australia, 58, 188, 203, 208

Ausubel, Jesse, 190, 195–96
Ayres, Robert, ix, 160, 196–201

Badgley, Catherine, 12, 67–68
Badische Anilin- und Soda-Fabrik (BASF), 53
Barcelona, 170, 210
Bardi, Ugo, 145
BASF (Badische Anilin- und Soda-Fabrik), 53
Battelle Institute, 134–35
Bavaria, 116
Belgium, 87, 193
biodiversity, viii, x, 8–9, 12, 14–16, 20, 37, 40, 46–47, 54, 59–60, 63, 69, 72, 155, 164, 171, 173, 181–85, 203, 225, 227
biodiversity loss, viii, 8, 14, 19, 27, 33, 35, 40, 47, 50, 54, 60, 62–64, 66–67, 69, 72–73, 129, 139, 164, 174, 185, 191, 200, 203, 223–25
biofuels, 86, 203, 207
biogeochemistry, 123, 151
biology, 28, 197
biomass, 80, 101, 123, 205, 212, 219
biomes, 8, 61, 172–73, 181
biophysical, 129, 139, 164
biosphere, 17, 39, 85, 128, 163, 199, 216
biota, 36, 127–29, 222
birds, 32, 46, 60–62, 64
births, 7, 48, 52, 136, 166, 217, 227

Malthus Enigma

Black, Joseph, 81, 110, 115, 221
Boltzmann, Ludwig, 122
Borlaug, Norman, 55–58, 62, 64, 68, 72, 160, 215
Borman, Frank, 1–3, 8, 209
Boston, 116, 135, 141, 196
Boulding, Kenneth, 9, 18, 29, 153, 157, 163, 166, 211, 220
Boulton, Matthew, 81–83, 102
Bringezu, Stefan, 190, 204–5
Britain, 11, 24–25, 44, 82, 84, 87–88, 217
Bucharest, 158
Buenos Aires, 133–34

calcium sulphide, 188
California, ix, 5, 19, 31, 89, 97, 172, 175–76, 178–79, 202
Callendar, Guy Stewart, 14, 79, 94–96, 104, 110, 222
caloric theory, 110, 114, 116–17, 119
Cambridge, 23, 118, 137, 212
Canada, 58, 97–99, 122, 135, 188
Canadian International Development Agency, 99
Cape Horn, 208–9
capital, 109, 136, 160, 184
capitalist system, x
carbon, 36, 75, 97, 105, 127, 129, 164, 180, 182, 184
carbon cycle, 127–29, 148, 200
carbon dioxide, 75, 92, 95
carbon emissions, 9, 16, 19, 66
carbonic acid, 76
carbon sequestration, 62, 182, 184
carbon trading, 179–80, 182
carbon uptake, 127, 222
Cardwell, John, 115, 119–21
Carnegie Mellon University, 197, 202
carnivores, 153
Carnot
    Lazare, 107–8, 119

Sadi, 108–14, 117–21, 125, 129, 148, 160, 223
carrying capacity, 9, 11, 13, 17–18, 24, 29, 34–35, 37–38, 68, 71, 79, 139, 159, 163, 174, 213, 216, 221–22
CFCs, 125
Charney, Jule, 18, 147, 149, 222
chemical energy, 36
chemicals, 53, 83, 199
chemist, 201
chemistry, 52–54, 93, 114–15, 133, 151, 197
Chertow, Marian, 189
Chicago, 29, 147, 174, 197
Chicago Climate Exchange, 19, 179–80
China, 7, 12, 20, 49, 102, 132–33, 188–89, 192–93, 211
chlorofluorocarbons, 125, 176
chromium, 31, 208
circular economy, 20, 51, 192, 208, 210–12, 214, 219–20, 225
cities, viii, 5, 18, 32, 82, 95, 103, 141, 161, 174, 176, 206–7, 214
Clapeyron, Émile, 118
Clausius, Rudolf, 13, 110–11, 118–20, 125, 129, 223
Clean Air Act, 5, 175–76
Clean Development Mechanism, 179–80
Cleveland, Cutler, 85–86, 162, 169
climate, 14, 17, 33, 46, 62, 68, 77, 79, 92–93, 127, 149, 183, 190, 215
climate change, viii, 7–8, 15, 19, 35, 64, 70, 73, 78–80, 86, 92–96, 99–100, 103, 105, 128, 139, 144, 164, 185, 191, 200, 203, 216–17, 219, 221–25, 228
Climate Exchange PLC, 180
Club of Rome, 18, 131–35, 137–38, 141–43, 154, 157, 159, 204, 207, 221, 226, 228

292

Index

CO_2, 6, 14, 36, 91–97, 99, 126, 144, 149, 222
CO_2 concentrations, 94–97, 128, 144, 150
coal, 11, 15, 75, 80, 82–85, 87–89, 91, 102–3, 111, 144, 188, 217
Coal Question, 87–88, 228
Columbia University, 27, 195
Commoner, Barry, 6, 11, 29, 191, 211
commons, tragedy of the, 171
communities, 18, 24, 56, 103, 173, 177, 190, 227
computer models, 14, 33, 138, 148, 222
computers, 91, 94, 133, 136, 140–41, 148
conservation, 14, 18, 28, 43, 45, 115, 159, 166, 172, 191, 198, 207, 223
conservation of energy, 13, 25, 110, 114, 117–20, 129, 153, 155
consumption, 18, 50–51, 63, 164, 182, 193–94, 198, 203, 205, 211, 225
contraception, 25, 49, 224
copper, 31, 208
coral reefs, 61
corn, 26, 88, 217
Corn Laws, 26, 84, 88, 142, 217
Cornucopians, 10, 16, 23–39, 47, 58, 71, 215, 219–20
Costanza, Robert, 19, 169–70, 172–74, 181
cropland, 61, 66–67
crops, 11, 27, 46, 54–55, 57, 62, 65–66, 68, 71, 218
Cullen, Jonathan, 212
cycle, 25, 113
  biogeochemical, x, 129
  hydrological, 125, 127–29

Daly, Herman, ix, 10, 19–20, 23, 29, 161–72, 174, 182, 185, 206, 217, 227

Darwin, Charles, 17, 226
death, 6, 23, 48, 58, 118, 219
death rates, 48–49, 137, 145
decoupling, 164, 203–7, 214
deforestation, 27, 174
degradation, environmental, viii, 18, 203–4
Delft University of Technology, 197
dematerialization, 196–97, 213, 220
diesel, 12, 121
diets, 46, 66
discovery, 76–77, 104, 115, 129
diversity, viii, 20, 46, 60, 62, 70, 128, 222
Drury, Richard, 177–78, 183
Duchin, Faye, 196, 201–2
Duke University, 151, 161

Earth Day, 3
Earthrise, 3–4, 209, 221
Earth Summit, 15, 99, 179
Earth system, ix, 35–36, 40, 46–48, 80, 104, 111, 126–27, 129, 148, 150, 199–200, 216, 223
eco-industrial parks, 189
eco-innovation, 206, 214
École polytechnique, 108–9, 118
ecological economics, 5, 19–20, 27, 29, 39, 104, 152, 155, 157–85, 191, 224–25, 227–28
ecological economists, x, 17, 19–20, 162–64, 167–68, 170–74, 181–82, 184, 190, 206–7, 227
ecologists, 8, 16–18, 29–30, 65, 152, 163, 168, 170, 172, 185
ecology, 3–5, 9, 16–18, 21, 36, 39–40, 59, 72, 105, 111, 150–53, 170, 172, 185, 191, 196, 204, 221, 224, 230
economic growth, 164–65, 197, 204–5, 220
economics, ix, 5, 16, 27, 39, 83, 87, 105, 111, 132, 155, 157–58,

293

161–62, 164, 169, 171–72, 184–85, 196, 210, 227, 230
economists, 9–10, 28, 138, 142, 144, 157, 159, 161–63, 168, 170, 185, 192, 198, 201
ecosystems, 10, 18, 20, 38, 40, 60–62, 64–66, 68, 71, 150–51, 153, 168–73, 187, 193, 215, 218–19, 224, 227
ecosystem services, 40, 47, 59, 62, 72, 128–29, 171–74, 181–82, 184, 203, 228
Edinburgh, 81, 115
EDVAC (Electronic Discrete Variable Automatic Computer), 147
Edwards, Paul, 149–50, 222
efficiency, 81, 89–91, 113, 121–23, 136, 217
Ehrenfeld, John, 187–90, 197
Ehrlich, Paul, ix, 5–8, 11, 29–31, 34, 37, 39, 58, 142, 157, 160, 172, 191–92, 207, 221
electricity, 55, 100, 102, 117–18, 160, 219
electric vehicles, 73, 103, 203, 216, 219
Electronic Discrete Variable Automatic Computer (EDVAC), 147
Electronic Numerical Integrator and Computer (ENIAC), 147
Ellen MacArthur Foundation, 210–11
emissions, 6, 85, 91, 166, 175, 177–81, 183
 anthropogenic, 78–79, 93, 95
 human-produced, 14
 net-zero, 184
 nitrogen oxides, 176–77
 reducing, 176, 179, 183
 sources, 178
 sulphate, 14, 148
emissions trading, 175–76, 183
energy
 mechanical, 112, 120
 solar, 160, 199
 useful, 123, 128
energy and material flows, 63, 153, 168, 191, 193, 223
energy consumption, 15, 90
energy crisis, 165, 169
energy efficiency, 15, 89, 101, 121–22, 169, 176, 194, 213
energy flows, 17, 80, 168, 170
energy sources, 80, 85–86, 217
energy supply, 54, 79–80, 84–85, 88, 213
energy systems, 16–17, 73, 80, 82, 197, 203, 219
 renewable, 219
energy use, 66, 80, 89–90, 101, 103, 122, 169
engine, 13, 81–82, 112–13, 121
engineering, ix, 2, 38, 48, 101, 118, 132, 152, 195, 197, 224
engineers, x, 38, 81, 109–10, 121–22, 125, 190, 201, 216–17, 230
England, 5, 23–26, 81, 84, 87, 90, 94, 125, 143
ENIAC (Electronic Numerical Integrator and Computer), 147
entropy, 120–25, 163, 168
Entropy Law, 19, 159–62, 165, 168, 171
environment, ix, 5, 15, 17, 30–32, 34, 59, 98–99, 124, 150, 152, 165, 169–72, 180, 190, 195, 204, 223, 225–26
environmental agenda, 45
environmental challenges, viii, 6, 12, 15, 19–20, 27, 34, 164, 191, 202, 217, 220, 223, 225, 228
environmental conditions, 127
environmental consciousness, 49, 157
environmental crises, 225, 228–29
environmental damage, 44, 174
environmental disasters, 4
environmental evangelists, 6

Index

environmental impacts, 11, 51, 54, 58, 63, 65–66, 72, 174, 189–91, 201, 203–6, 208, 214
environmental issues, 30–32, 191, 214
environmentalists, 18, 37, 96, 138, 183, 221
environmental policy, 15–16, 20, 151
environmental pollution, 5, 29, 168, 176
environmental problems, 31, 71, 196, 200, 203
Environmental Protection Agency. *See* EPA
environmental revolution, 4, 28, 30, 78, 198
environmental scientists, 13, 35, 190
environmental stresses, 14, 54, 58, 80, 98, 214–15
environmental systems, 14, 34, 110, 221
EPA (Environmental Protection Agency), 175–76
equations, 92, 135–36, 159
EROI (Energy Return on Energy Investment), 85–86, 103, 169, 219
Europe, 116, 122, 133, 189, 193–94, 199, 203
European Climate Exchange, 180
European Union, 20, 179–81, 192, 211, 214
experiments, 19, 59, 75–79, 92, 104, 116–17, 174, 178, 181, 189
exports, 63, 121, 205
externalities, environmental, 198
extinction, 8, 43, 62, 184
extinction rates, 61, 64

famine, 25, 46, 57–58, 215
farm, 11, 59
farmers, 55–57, 62, 69–70, 72
farming, 68, 90
 organic, 67–68, 71
farmland, 11–12, 33, 58, 64, 66–68
fauna, 39, 45–47, 59
feed-in tariffs, 100, 181, 204
fertility rates, 48–50, 69, 166, 224
fertilizers, 8, 25, 51–52, 57–58, 66–67, 79, 111, 125, 188, 200, 218
 chemical, 47, 51–52, 55
FIDO (Fog Investigation and Dispersal Operation), 96
Finland, 188
first law of thermodynamics, 110, 114, 117, 122
Fischer-Kowalski, Marina, 204–5
fish, 15, 30, 33, 44, 46, 63, 153
fisheries, 171, 221
flora, 39, 46–47, 59
Florida, 151–53, 170
Fog Investigation and Dispersal Operation (FIDO), 96
food, 6–7, 9, 12, 24, 26–27, 30, 32–33, 37, 40, 48, 56–57, 59, 62–63, 65–66, 79, 88, 90, 111, 127, 165, 199, 210, 215, 217–18, 226, 229
food chains, 6, 46
food consumption, 27, 202
food crisis, 165
food prices, 24
food production, 6, 11–12, 24–25, 27, 30, 47, 57, 65–66, 136–37, 142, 146, 152, 169, 172, 193, 218, 226
food riots, 6
food scarcity, 27
food source, 39
foodstuffs, 65–66
food supplies, 9–11, 25, 90
food systems, 62, 203, 207, 219
food waste, 71
food webs, global, 36
Foote, Elisha, 76, 78
Foote, Eunice, 76–79, 91, 104, 225
forestry, 46, 55, 66, 189

Malthus Enigma

forests, 6, 28, 32, 173, 184, 222
Forrester, Jay, 18, 131, 135, 139–45, 151–53, 207, 226
fossil fuels, 12, 76, 78, 80, 85, 87, 91, 97, 100, 103, 128, 205, 212, 219, 227
fragmentation, 46, 64
France, 23, 44, 46, 87, 107–9, 118, 209
Frosch, Robert, 187–88, 191, 194–96, 206, 210–11, 213, 220

Gaia, 125–26, 200
Gaia Hypothesis, 3, 125–27, 199
gases, 53, 75–78, 86, 109, 113–15, 198
gasoline, 12, 86, 91, 163
GCMs (general circulation models), 139, 147–50
GDP, 65, 173–74
general circulation models. *See* GCMs
General Electric, 141
General Motors, 194–95
Geneva, 134–35, 195
Georgescu-Roegen, Nicholas, 19, 29, 103–4, 157–63, 165, 169, 171, 184–85, 192, 201, 219, 222–23
Georgia, 3, 151
Germany, 44, 53–54, 92, 100–102, 108–9, 135, 188, 204–5
Geyer, Roland, 213
Gift of Apollo, 1–21, 230
Graedel, Tom, 189–90, 196, 204, 207
grain, 25–26, 57
greenhouse effect, 14, 16, 73, 76–78, 80, 91, 93–94, 98, 104–5, 149, 221, 225
greenhouse gas emissions, 19, 50, 66–67, 87, 103, 129, 149, 154, 179, 181, 212, 214, 219
greenhouse gases, 14, 91, 99, 149, 176

Green Revolution, 12, 55, 58, 62, 64, 66
growth, viii, 10, 18, 29, 34, 37, 39–40, 43, 131, 136–38, 142–46, 153–54, 160–61, 165–67, 207, 214, 217–18, 220–21, 226, 228
gypsum, 188, 201

Haber, Fritz, 51–54, 58, 64, 68, 218
Haber-Bosch process, 53–54, 71, 111
habitats, 46, 66
Hall, Charles, 85–86, 152, 169
Hannon, Bruce, 169
Hansen, James, 149, 179, 222
Harvard, 158, 176, 201
heat, 13, 36, 81, 96, 107, 110–22, 127, 148, 160
heat capacity, 115
heat engines, 104, 111–13, 119
heat flows, 111–13, 120, 148
heating source, 82–83
Henry, Joseph, 77–78
herbivores, 153
Herendeen, Robert, 85, 169
holistic science, 17, 40, 153–54, 223
horses, 11, 82, 84
Hubbert, M. King, 85–86, 121
human development, 79, 85, 98, 229–30
hunger, 24, 30, 41, 133
hunters, 42–43, 45
hunting, 42–46
hydroelectric, 101
hydrogen, 75
hydropower, 101
hypothesis, 88–89, 127, 150, 154

ice sheet, 8
IIASA (International Institute for Applied Systems Analysis), 18, 154, 197
Imperial College London, 95, 133

296

Index

India, 49, 57–58, 67
Indigenous knowledge, 70–71, 218, 220
Indigenous Peoples, 27, 44, 183
industrial ecologists, 17, 20, 191–92, 201–2, 204–5, 207–8, 210–11, 214, 220, 223, 225
industrial ecology, viii–ix, 17, 20, 34, 40, 63, 130, 153, 155, 185, 187–214, 220, 223–24
industrial metabolism, 20, 191, 195, 197–201, 203, 214
Industrial Revolution, 11, 13, 25, 38, 73, 75, 78–81, 83–84, 87, 112
industrial symbiosis, 187–89, 213, 220
industrial systems, 17, 20, 135, 141, 185, 187, 191, 193–94, 199, 220
industry, 18, 20, 52–53, 103, 122, 130, 135, 142, 159, 162, 176–77, 179, 181, 187–89, 191, 193–96, 199, 201, 210, 226
innovation, 10, 38, 58, 71–73, 79, 88, 110, 118, 191, 206, 213, 216, 220
insecticides, 4
insects, 32, 184
institutions, 45, 166
integrated assessment models, 139, 155, 207, 223
Intergovernmental Panel on Climate Change. *See* IPCC
Intergovernmental Science-Policy Platform on Biodiversity and Ecosystem Services, 173, 203
internal combustion engines, 104, 121
International Energy Agency, 12
International Institute for Applied Systems Analysis. *See* IIASA
International Resource Panel, 20, 192, 203–7, 223

International Society for Ecological Economics, 168
International Society for Industrial Ecology. *See* ISIE
International Union for Conservation of Nature. *See* IUCN
IPAT Equation, 39
IPCC (Intergovernmental Panel on Climate Change), 7, 36, 100, 149, 154, 179, 192, 203, 224, 228
iron, 53, 83, 193, 208
ISIE (International Society for Industrial Ecology), 187, 190, 204
Italy, 28, 44, 131–32, 141
IUCN (International Union for Conservation of Nature), 46, 62–63, 99, 204

Jansson, AnnMari, 170
Jantsch, Erich, 134–35
Japan, 122, 179, 189, 193–94, 204–5, 211
Jevons, William Stanley, 15, 85, 87–89, 91, 217, 228
Jevons' Paradox, 15, 89–91
Joule, James, 110, 114, 117, 119–20, 129

K
Kahn, Herman, 33–34
Kalundborg, 187–88, 196, 220
Karlsruhe, 52–53
Keeling, Charles, 14, 96–98, 104, 222
Kenya, 70
kinetic theory, 110, 114
King, Alexander, 131, 133–35, 143–44, 154, 228
Kleidon, Axel, x, 126–28
Kleijn, René, 187
Kneese, Allen, 197–98
Kyoto Protocol, 19, 100, 179, 182

297

lakes, 6, 108, 141
land, viii, 12, 45, 58, 61, 66–67, 69, 72, 84, 86, 88, 96, 184, 203, 218
land use, 31, 48
Latin America, 6, 132–33, 202
laws, 13, 27, 117, 120, 125, 214, 226
  physical, 18–19, 27, 153, 155, 207
  scientific, 125, 136
laws of physics, 19, 40, 110, 125, 161, 191, 214
laws of thermodynamics, 13, 16, 18–19, 21, 80, 105, 110–11, 122–23, 125, 129, 163, 220–21, 223
Lenzen, Manfred, 63
Leontief, Wassily, 158, 201–2
Leopold, Aldo, 18, 29, 59
licences, 44
life cycle, 103, 197, 204–5
life expectancy, human, 48–49
Lifset, Reid, x, 189–90
limits to growth, 10, 18, 29, 34, 37, 40, 131, 136–38, 142–46, 153–54, 160, 207, 217, 220–21, 226
livestock, 62, 65, 70
Livingstone, David, 41–42, 69
living systems, 13, 40, 107, 111, 123, 125, 129, 199, 223
London, 42–45, 48, 82, 89, 96, 116, 179
Lord Kelvin, 13, 110–11, 115, 118–20, 125, 129, 223
Los Angeles, 7, 57, 174, 176, 178
Lovelock, James, 3, 125–26, 128
Lovins, Amory, 100–101, 219

MacArthur, Ellen, 208–10
malnutrition, 24
mammals, 7–8, 60–62, 64
manufacture, 13, 51, 54, 90, 102, 104, 132

manufacturing, 26, 43, 53, 80, 88, 137, 205
markets, 16, 19–20, 30, 144, 164, 166–68, 171–72, 174–75, 177–85, 214, 217, 227–28
  free, 37, 167, 227–28
  global, 164, 184
  pollution trading, 175, 178
markets for air pollution, 16, 19
markets for biodiversity, 16, 20, 164, 181–82, 185, 227
Martínez-Alier, Joan, 170
Massachusetts Institute of Technology, 29, 140, 187
mass extinction, 8, 63–64, 78, 171, 182
material flows, 20, 63, 153, 168, 187–88, 190–91, 193, 197–98, 205, 223
material resources, 10, 160, 207
materials, 17, 160, 166, 168, 188, 192, 194, 198, 200, 203, 205, 210–14, 218
mathematical models, 18, 39–40, 154, 207
mathematicians, 107–8
mathematics, 23, 109, 118, 158, 207
Meadows
  Dennis, 133, 135–39, 141–42, 144–45, 151, 153–54, 157, 207, 218
  Donella, 136
metals, viii, 2, 31, 192, 200, 205, 207–8, 212
meteorology, 147–48
Mexico, 56–58
Millennium Ecosystem Assessment, 8, 14, 58, 60–62, 72
mining, 87, 103, 198–99
Minnesota, 55–56, 65
models, 15, 39–40, 81, 136, 139, 141, 143–48, 150, 154–55, 196, 202, 207, 223
  conceptual, 38, 154

Index

economic, 48, 169, 198, 202
economy, 207, 223
output, 202–3, 207
physics-based, 18
scenario, 207
thermodynamic, 129
Montreal, 68, 78, 94, 99
Moon, 1–5, 21, 49, 128, 138, 146, 221
Moriguchi, Yuichi, 204

NASA, 125–26, 191, 195
natural capital, 20, 172–73, 184, 211, 228
natural gas company, 15, 99
natural resources, viii, 29–30, 32, 34, 45, 59, 149, 159–60, 188, 192, 203, 205
Nebraska, 140
net primary production (NPP), 36–38, 127–28, 222
Neumann, John von, 14, 147
Newcomen engines, 82–83
New Hampshire, 116, 151, 172
New York, 7, 45, 76–77, 190, 204
New York City, 149, 195
nickel, 31, 208
nitrates, 53, 200
nitrogen, 51, 54, 66, 68, 126, 129, 188, 199, 218, 223
nitrogen cycle, 14, 35, 50, 200
nitrogen fertilizers, 54, 66
nitrogen fixation, 51–54, 111
nitrogen oxides, 175–77
Nordhaus, Ted, 215, 219–20, 229
Nordhaus, William, 10, 142–45, 219
North America, 61, 116, 133, 140, 170, 180, 194
North Carolina, 151–52
Norwegian University of Science and Technology (NTNU), 196-7, 203
NPP. *See* net primary production
nuclear fallout, 5

nuclear holocaust, 133
nuclear power, 217, 220
nuclear reactors, 104
nuclear weapons, 4
nutrients, viii, 12, 27–28, 39, 129

ocean acidification, 35, 129, 223
oceans, 200, 222
Odum,
  Eugene, ix, 3, 17–18, 150–53, 155, 194
  Howard, ix, 17–19, 29, 139, 150–53, 155, 157, 163, 168–70, 185
OECD (Organisation for Economic Co-operation and Development), ix, 7, 133–34, 143, 205, 225
oil, 5, 65, 75, 85–86, 88, 99, 101, 103, 122, 188, 228
oil crises, 12, 85, 89, 121–22, 194
oil industry, 99
oil spill, public, 5
organic agriculture, 67–68, 220
Organisation for Economic Co-operation and Development. *See* OECD
organisms, 18, 60, 124, 199, 226
Osborn, Henry, 28, 39
Ostwald, Friedrich Wilhelm, 53, 92–93
oxygen, 75, 126

Pakistan, 57–58
Papua New Guinea, 63
Paris, ix, 5, 7, 100–101, 108–9, 116, 118, 132, 134, 158
Pearson, Karl, 158
Peccei, Aurelio, 18, 131–35, 137–38, 141, 143, 154, 229
Pestel, Eduard, 134–35, 144
pesticides, 4, 6, 67, 200
petroleum, 85, 96
Petroski, Henry, 38
Phillips, Norman, 147–48, 222

299

phosphorus, 51, 83, 129, 188, 201
photosynthesis, 36, 62, 123, 128
photovoltaic panels, 102, 122–23
photovoltaics, 102–4
physicists, 63, 122, 147, 217
physics, 19, 40, 78, 93, 105, 110, 113, 123, 125, 149, 161, 191–92, 195, 197, 214, 224
planet, x, 3, 6–10, 12–15, 21, 24, 30, 34–35, 37–38, 43, 45–47, 50, 58–60, 64–65, 72–73, 79, 102, 104, 111, 125–26, 128, 136, 164–65, 174, 210, 213–16, 218, 220
planetary boundaries, 9, 13, 35–40, 50, 104–5, 127, 167, 216, 222
plants, 28, 30, 32, 36, 38, 46, 56, 68, 123, 127, 148, 226
plant species, 17, 32, 43, 62, 69
Plass, Gilbert, 94, 222
plastics, 46, 193–94, 200
Poland, 52, 193
policies, 11, 15, 17, 20–21, 25–27, 89, 105, 130, 139, 163, 181, 211, 213–14, 226
policy layer, 27, 163, 227
political economy, 23, 26–27, 109, 165
pollination, 62, 172
pollution, 5–6, 29, 131, 135–37, 144, 146, 157, 159, 165–68, 171, 174–76, 178, 187, 191, 210–11, 218–19, 227
 chemical, 35
 particulate, 146
pollution markets, 183, 185
pollution trading, 178
Poor Laws, 24–26, 84, 226
population, viii, 6, 10, 13, 19, 23–27, 29–30, 32, 34, 39, 46, 48–51, 58, 65, 68–69, 72, 88, 111, 135–37, 159–60, 163, 202, 211, 216–20, 224–26
 increasing, 6, 71, 182
 stabilizing, 166

 world's, 32, 47, 49, 54
Population Bomb, 6, 39, 58, 142, 221
population growth, 6, 23–25, 27, 30, 46–49, 72, 88, 163, 205, 226, 229
population problem, 9, 23–24, 26–27, 29–30, 47, 80, 118, 159, 218, 221, 227
population trap, 11, 25, 71, 118
potassium, 51
Potsdam Institute for Climate Impact Research, ix
Preservation, 15, 45–46, 224
Prigogine, Ilya, 124, 135, 223
primary production, 36–37, 127
Princeton, 28, 147–48
Principle of Population, 23, 217, 226
productivity, 11–12, 51, 57, 70
 agricultural, 25, 52, 55, 58, 70, 136
Promethean III, 104
property rights, 166, 173
Puerto Rico, 151

radiation, 2, 36, 122–23, 127
rate, birth, 48–49, 137, 145
rebound effects, 15–16, 80, 84, 89–91, 212–13
RECLAIM, 177–78
recycling, 51, 136, 188, 194, 199–200, 211–12
Red List, 14, 46, 62–63
regenerative, 211
renewable energy technologies, 73, 101, 103–4
renewables, 86, 101–3, 181, 219
renewable sources, 85, 100–103
resilience, 35, 40, 169, 220
resource constraints, 11, 34, 47, 71, 105, 218
Resourceful Earth, 32–34, 37, 39–40, 221
resource markets, 43, 227

Index

resources, ix, 7, 9–10, 15, 24, 32, 34, 90, 99, 135–36, 144, 160, 166–67, 174, 176, 187, 197, 201–6, 210, 213, 216
   non-renewable, 50, 86, 136, 168, 204
   scarce, 43, 166
resource use, 146, 168, 192, 206
restorative, 211
Rio, 15, 99
risks, 1, 14
rivers, 41, 61, 82
Rockström, Johan, 13, 35–36, 38, 40, 128, 222
Rome, 131, 134 (also see Club of Rome)
Roosevelt, 45, 98
Rossby, Carl-Gustaf, 147–48
Running, Steven, 36, 127

SAGE (Semi-Automatic Ground Environment), 140
Sand County Almanac, 18, 59
sanitation, 48, 229
Santa Barbara, ix, 5, 172–73
Saxony, 102
scarcity, 30–31, 39, 171, 201
scenarios, 7, 66, 136, 154, 202, 206–7, 218
Schrödinger, Erwin, 123–25, 199, 223
science, 10–11, 13–15, 17, 30, 38, 72–73, 76–77, 85, 104–5, 107, 109–11, 119–21, 125–27, 129, 134, 148–50, 154–55, 169–70, 191, 195–96, 202–4, 207, 213–14, 220–23
   environmental, ix, 33
   holistic Earth, 139
   interdisciplinary, 63, 213–14
   natural, 214, 224
   physical, 19, 163
scientific discovery, 76, 78–79, 104, 110, 118

scientific struggle, 79, 91, 113–14, 118, 129
scientists, 2, 7, 13–14, 35–36, 38–39, 47, 57, 60, 63, 65, 72, 80, 92–93, 104, 107, 110, 114, 116–17, 132, 150, 154, 190, 193, 204, 216, 221–22
second law efficiency, 122
second law of thermodynamics, 19, 107, 110, 120–22, 124–25, 127, 160, 222
Semi-Automatic Ground Environment (SAGE), 140
Seufert, Verena, 12, 68
Shellenberger, Michael, 215, 219–20, 229
Sierra Club, 6
Silent Spring, 4, 28
Simon, Julian, 10, 29–31, 33–35, 37–39, 60, 64, 71, 192, 207, 219, 221–22
Smith, Adam, 26, 226–27
smog, 6, 176
social injustice, 177
socio-economic systems, 155, 223
sodium chloride, 92
soil, 27–28, 68, 216
solar, 12–13, 100–101, 122, 126, 204, 219
solar cells, 100, 102
solar radiation, 36, 94, 127–28, 153
Solow, Robert, 161
Sonoran Desert, 56–57
South Africa, 42–43, 58, 69
South America, 132
South Coast Air Quality Management District, 176–77
Southern Ocean, 208–10
South Pole, 97
Soviet Union, 54, 132–33, 135
Spaceship Earth, 9, 13, 98, 216, 230
Spash, Clive, 164, 184–85

301

species, 8, 14, 17, 40, 43, 46–47, 59–60, 62–64, 69, 78, 121, 124, 126, 153, 159, 164, 185, 187, 210, 216, 226–27
animal, 28, 46, 226
Stakman, Elvin, 41, 56
starvation, 7, 46, 56–57, 215
statistics, 87–88
Stavins, Robert, 176, 178, 183
Steady-State Economics, 19, 163, 165, 167–68, 174, 206, 227
steam, 27, 84, 88, 95, 114, 119, 121, 188, 221
steam engines, 8, 13, 15, 79, 81–84, 88, 91, 104, 111–13, 115, 121, 125, 217, 221
steel, 194, 208
steelmaking, 11
Stiglitz, Joseph, 10, 135
Stockholm, 5, 15, 92–93, 98–99, 147, 149
Strong, Maurice, 15, 98, 105, 149, 195
subsistence, 25, 27
succession, 18, 134
sulphur, 83, 188, 199
sulphur dioxide, 175–76, 180
sulphuric acid, 188
Surrey, 23, 167
Sussex, 143
sustainability, 6, 13, 29, 87, 107–29, 204, 210, 223
sustainable development, 13, 15, 23, 99, 174, 189
Sweden, 92–93, 170
Switzerland, viii, 23, 54, 108
systems approaches, 17, 29, 130, 139, 151, 191
systems dynamics, 140–41, 144
systems ecology, 4, 150–52, 169, 194
systems modelling, 131, 135–36, 153, 221
systems models, 18, 39–40, 63, 135, 139, 207

systems perspective, 39, 102, 163
systems thinking, 17, 20–21, 24, 39, 134, 155, 163

Tansley, Arthur, 18, 150
technological change, 9, 11, 46, 51, 87–88, 144, 155, 197, 217
technological development, 69, 105, 110–11, 189, 213, 215
technological progress, 12, 25, 47, 133, 160, 221
technological solutions, 12, 15, 21, 65, 219
technological systems, 16, 72, 111
technology, 10–13, 17, 20, 32, 47, 50–52, 71–73, 79, 83–84, 89–91, 101, 103–5, 121–23, 134, 187, 194–97, 203, 205–6, 213, 216, 219–21, 225
temperature change, 117
temperature difference, 113, 119
temperatures, viii, 13, 35–36, 53, 68, 75, 91–92, 95, 113–15, 120–21, 126
Thatcher, Margaret, 183, 228
thermodynamics, 8, 13, 18–19, 21, 39–40, 105, 107, 109–11, 114–30, 139, 147, 151–55, 159–63, 165, 171, 185, 191, 207, 220–23
non-equilibrium, 124, 169
Thermodynamics of Life, 123
thermodynamics of living systems, 13
Thiemann, Hugo, 134–35
Thompson, Benjamin (Count Rumford), 110, 116–17
Tibbs, Hardin, 196
Tilman, David, 12, 65–67, 69, 72, 202
tin, 31, 208
Torino, 132
trade, 89, 180, 191, 202, 214, 217
transport, 48, 148
transportation, 37, 84, 210
tropical, 61

Index

Tyndall, John, 78–79, 91–92

UNEP (United Nations Environment Programme), 5, 60, 99, 173, 195, 203–5
UNESCO (United Nations Educational, Scientific and Cultural Organization), 5
United Kingdom, 188
United Nations Educational, Scientific and Cultural Organization (UNESCO), 5
United Nations Environment Programme. *See* UNEP
United Nations Framework Convention on Climate Change, 35, 100
United States, 4–5, 27, 66, 86–87, 101–2, 121, 143, 151, 175–76, 179, 194–95, 199, 201, 203
University College London, 87, 95, 158
Uppsala University, 92

valuation of ecosystem services, 171, 173, 184
value, human, 21, 163–64
value of ecosystem services, 173, 228
Vienna, 18, 123, 154, 211
Vienna University of Economics and Business, 164
Vogt, William, 28, 39

war, 2, 25, 29, 48, 54, 57, 96, 107–8, 116, 132, 146, 174, 215, 226
Washington, 28, 97, 138, 195
wastes, 9, 12, 20, 51, 54, 160, 187–89, 191, 193, 198–201, 203, 210, 219
water, 6, 28, 36, 41–42, 59, 61–62, 66, 68, 82, 91–92, 107–8, 112, 114–15, 117, 120, 123, 129, 153, 188, 199, 202–3, 209–10

Watt, James, 13, 36, 81–83, 102, 110–11, 115, 217, 221
Watt's steam engine, 81–83, 103, 111–12
wealth, x, 19, 51, 56, 80, 89, 133, 164, 167, 225
wetlands, 151, 173
wheat, 11, 25, 55–58
wildlife, 15, 46
willingness-to-pay estimates, 172
Wilson, Carroll, 134–35, 141
wind, 12–13, 54, 80, 100–101, 103, 132, 204, 219
wind power, 100–101, 104, 122
wind turbines, 13, 100, 103, 140
wood, 27–28, 82–83, 193
World3 model, 136, 139, 141, 143–46, 207, 218, 226
World Bank, 50, 60, 99, 162, 204
World Resources Institute, 99, 189, 205
World War I, 53, 140, 158
World War II, 2, 28–29, 31, 56, 94–96, 98, 132, 146–47, 158, 222
World Wildlife Fund, 99, 132
Wuppertal Institute, 204

Yale University, 151, 189–90, 207

zinc, 201, 208
Zoological Society of London, 42, 45

303

Printed in Great Britain
by Amazon

636b67e7-95ab-4135-bc2c-21aae912c88dR01